Reeds Astro Navigation Tables

Lt Cdr Harry J Baker
RD RNR MRIN

2012 EDITION

ADLARD COLES NAUTICAL
London

ACKNOWLEDGEMENT

This product is derived in part from material obtained from the UK Hydrographic Office with the permission of the UK Hydrographic Office and Her Majestys Stationery Office.

Astronomical data generated by HM Nautical Almanac Office

2012 edition published by Adlard Coles Nautical
an imprint of Bloomsbury Publishing Plc
50 Bedford Square, London WC1B 3DP
www.adlardcoles.com

First edition published 2011
Reprinted 2012

ISBN 978-1-4081-4227-1

A CIP catalogue record for this book is available from the British Library.

Bloomsbury Publishing uses paper produced with elemental chlorine-free pulp, harvested from managed sustainable forests.

Printed and bound in Great Britain by the MPG Books Group Ltd

CONTENTS

Editor's Notes for the Cruising Yachtsman

GENERAL

You might call this the common sense almanac. It has been devised and produced by practical ocean navigators, and has these main advantages:

- All the information which you need in order to navigate by the heavenly bodies – Sun, Moon, Planets, Stars – is contained herein.
- The slim volume is designed to open flat on your chart table.

We have not attempted to make this a text book on astro-navigation, and we have excluded data which are not relevant to the task in hand, i.e. finding out where you are.

Nowadays nearly all ocean-going yachts are fitted with GPS. If you also carry a sextant (and know how to use it) plus our little volume you will be able to navigate confidently by the heavenly bodies; this will not only add to your safety but will give you great personal satisfaction. Perhaps one day you will achieve the ultimate observed position with simultaneous sights of Sun, Moon and Venus!

Of course there are other almanacs, and there are also dedicated navigational computers which carry ephemerides for some or all of the heavenly bodies. The latter are expensive and, by their nature, vulnerable. We believe that the cruising yachtsman should carry his almanac in indestructible book form and be able to reduce his sights without the need to refer to any data outside his almanac. We also believe that the method of sight reduction must be in simple, easily-explained, steps and must be the same for all bodies.

People who are learning astro-navigation have asked us how they can practise ashore with their sextants and with the tables; so we have included a section (page 6) which tells them how to go about it.

All navigational working is done using Greenwich Mean Time (GMT), which is the same as Universal Time (UT). GMT is therefore used in all examples and no reference is made to other definitions of time.

These descriptions, in accordance with traditional marine practice, are used for sextant values:

Sextant altitude – as read from the sextant

Observed altitude – SA corrected for index error

Apparent altitude – OA corrected for dip

True altitude – AA corrected for all other influences.

LAYOUT

The almanac is in five different parts, although it is thought better to adhere to simple page numbering, as shown on the contents page.

PART ONE is this section, the editor's notes, including a full description of the volume.

PART TWO starts with a comprehensive set of examples for all the normal use of the almanac by the practical navigator. The intention has been to group the examples so that they follow each other naturally. It is recommended that you should work through them in comfort ashore so that you can find your way around the almanac without difficulty at sea. The section concludes with supplementary information which may be useful.

PART THREE is the ephemerides. It consists of thirty-seven monthly pages – three for each month plus one for the Pole Star. The first of the monthly pages is devoted to Greenwich Hour Angle and Declination of the Sun, and Greenwich Hour Angle of the first point of Aries. Be careful not to confuse them – it is a fairly common error! On the next monthly page GHA and DEC of the Moon and the four navigational Planets are tabulated. The third and final monthly page carries all the necessary Star information – Magnitude, Transit, Declination and Sidereal Hour Angle for 60 selected Stars. Alongside the Stars are astronomical and complementary data for the Sun and Moon, thus completing all the necessary ephemeral details. This section ends on page 44 with the Pole Star table; this looks like a perennial correction table but in fact it changes annually.

PART FOUR is the necessary collection of tables to produce increments and corrections for the main tabulations in the ephemerides, and also to correct observed altitudes. Also slipped in are some notes to help you to identify and use the Planets, and some observations about the Moon's phases. There is a refraction table which is not actually used in any of the worked examples; it is there because when you practise ashore you will need this little table.

PART FIVE includes all the tables – Versines, Log Cosines and ABC, which will enable you to reduce your sights, finding calculated altitude and azimuth. These may look rather daunting at first; but after working through a few examples you will find that they come quite easily. We have resisted producing a sight form for workings because most people have their own ideas in this respect. Right at the back, on page 68, is a simple Star chart.

ALMANAC DETAILS

TO FIND THE GEOGRAPHICAL POSITION OF HEAVENLY BODIES

- Greenwich hour angle (GHA) and declination (DEC) of the SUN are tabulated at intervals of two hours.
- GHA and DEC of the MOON are tabulated at intervals of six hours.
- GHA and DEC of the four navigational PLANETS are tabulated at intervals of twenty-four hours.
- GHA of the first point of ARIES is tabulated at intervals of two hours.
- Sidereal hour angle (SHA) and DEC of the principal sixty navigational STARS are tabulated each month.

Incremental tables enable the navigator to obtain values of GHA and DEC for any given time. Note that, by their nature, these tables may not produce a precisely accurate value; but differences are trivial and of no importance for the practical yacht navigator.

TO FIND CALCULATED ALTITUDE AND AZIMUTH

The versine method is used to determine the calculated altitude (CA) of the body from the navigator's estimated position (EP).

An important advantage of the versine method is that your sight is plotted from your EP, rather than from an assumed position, thus eliminating large intercepts and the consequent risk of plotting errors.

ABC tables are used to determine azimuth (AZ).

For those navigators who prefer to use simple scientific calculators formulae are provided for obtaining CA and AZ.

TO FIND TRUE ALTITUDE

Tables are provided to correct observed altitude (OA) into true altitude (TA) for all bodies.

OTHER USEFUL DATA

- Daily pages for sunrise, sunset, twilight, moonrise, moonset.
- Alphabetical index of principal Stars.
- Eclipses in 2012.
- Meridian passage time for all bodies.
- Arc to time table.
- Limited use of the almanac in 2013.

EXAMPLES

Examples are included for all workings, using the 2012 almanac data.

OTHER RECOMMENDED BOOKS

Reeds Sextant Simplified by Dag Pike, published by Adlard Coles Nautical. ISBN 0 7136 6705 2.

This is a comprehensive description of the sextant and how to use it.

Celestial Navigation For Yachtsmen by Mary Blewitt, published by Adlard Coles Nautical. ISBN 0 7136 6422 3.

This is the classic primer, including a simple and elegant description of astro-navigation theory.

ERRORS AND OMISSIONS

We have carefully checked the text and tables of this volume. However, neither the publisher nor the editor can accept any responsibility for the consequences of any errors or omissions which might have occurred.

FINALLY

2012 is the sixteenth year of publication of these tables.

Our thanks are due to Mr George Whitchurch for his help in checking the calculations.

We shall be happy to receive comments and suggestions from teachers, students and users of astro-navigation.

Don t forget your 2013 almanac. It will be published in September, 2012. If you cannot obtain a copy from your normal source, contact:

Adlard Coles Nautical
36 Soho Square
London W1D 3QY

www.adlardcoles.com

Harry Baker, *Editor*

Examples

In some tables interpolation by eye is necessary.

The first four examples guide you through the process of finding GHA and DEC for all the heavenly bodies; these values are required for the sight reduction process. Then there are examples for applying corrections to observed altitude, followed by establishing meridian passage times for all bodies, finding times of sunrise, moonrise etc. and use of Polaris to find latitude. Finally, there are examples on the use of the tables for establishing CA and AZ in the sight reduction process.

To find GHA and DEC of the SUN on Tuesday, 10th January at 11h 23m 13s

Tabulated values are read off from the first monthly page. Aries is included in the same table so be careful not to confuse. GHA increments are found in the table on page 47. DEC can be interpolated by eye.

Tabulated value at 10h *monthly page*	328° 10.9
Increment for 1h 23m *page 47*	20° 45.0
Increment for 13s *page 47*	03.3
Required GHA	348° 59.2
DEC *monthly page, by inspection*	S 22° 00.2

When interpolating by eye for DEC take care to observe if it is increasing or decreasing.

To find GHA and DEC of the MOON on Saturday, 14th January at 9h 18m 41s

Tabulated values are read off from the second monthly page. At the same time note mean variation per hour for both GHA and DEC. The mean variation per hour for GHA is always 14 degrees plus a number of minutes of arc; but only the minutes are shown in the column. For GHA enter table on page 51 with arguments incremental whole hours and mean variation per hour; then go to the minutes table on page 52 to complete the increments. For DEC simply multiply whole hours by mean variation per hour and then obtain the increment for minutes from the table on page 53.

GHA mean variation per hour *monthly page*	14° 30.0
GHA tabulated value at 6h *monthly page*	29° 08.1
Increment for 3h *page 51*	43° 30.0
Increment for 18m *page 52*	4° 21.0
Increment for 41s *page 52*	09.9
Required GHA	77° 09.0
DEC mean variation per hour *monthly page*	+13.1
DEC tabulated value at 6h *monthly page*	S2° 54.5
Increment for 3h *mental arithmetic*	+39.3
Increment for 19m *page 53*	+04.1
Required DEC	S30° 37.9

When extracting mean variation per hour for DEC take care to check if DEC is increasing or decreasing.

To find GHA and DEC of VENUS on Sunday, 11th March at 14h 28m 19s

Tabulated values are read off from the second monthly page. At the same time note mean variation per hour for both GHA and DEC. For GHA enter table on page 48 with arguments incremental whole hours and mean variation per hour; then go to the minutes table on page 49 to complete the increments.

For DEC enter the single table on page 50 with arguments total incremental time and mean variation per hour.

GHA mean variation per hour *monthly page*	14° 59.9
Tabulated value at 0h *monthly page*	135° 55.2
Increment for 14h *page 48*	209° 58.6
Increment for 28m *page 49*	7° 00.0
Increment for 19s *page 49*	04.8
Required GHA	352° 58.6
DEC mean variation per hour *monthly page*	+1.1
Tabulated value at 0h *monthly page*	N15° 16.2
Increment for 14h 28m *page 50*	+15.8
Required DEC	N15° 32.0

When extracting mean variation per hour for DEC take care to check if DEC is increasing or decreasing.

To find GHA and DEC of SIRIUS on Sunday, 15th April at 17h 26m 14s

To find GHA of the Star it is necessary first to establish GHA Aries and then add sidereal hour angle (SHA) of the Star. Tabulated value for GHA Aries is read off from the first monthly page; the tabulation is in the same table as the Sun so be careful not to confuse. The table on page 47, right-hand side of page, is entered to establish the increments. SHA of the Star is then read off from the third monthly page, and DEC is read off at the same time. Neither SHA nor DEC change appreciably during the month so they are applied as read.

Tabulated value GHA ARIES at 16h *monthly page*	84°	12.9
Increment for 1h 26m *page 47*	21°	33.6
Increment for 14s *page 47*		03.5
GHA ARIES	105°	50.0
SHA SIRIUS *monthly page*	258°	34.5
Required GHA	364°	24.5
	–360°	
	4°	24.5
Required DEC *monthly page*	S 16°	44.3

To find TRUE ALTITUDE of the SUN

On Wednesday, 5th December the observed altitude (OA) of the Sun's lower limb was 19° 24'.8. Height of eye (HE) was 8 feet.

Enter table on page 45 (left hand side) with arguments OA and HE. Interpolate by eye. Remember to apply the monthly correction, found at bottom of the table. The total correction is always additive.

OA	19°	24.8
Main correction *page 45*		+10.7
Correction for month *at foot of page*		+00.3
TA	19°	35.8

Note that if the upper limb is observed (this is unusual) you must go to the monthly page to find the semi-diameter. Deduct twice this value from the OA and then use the table as described above.

To find TRUE ALTITUDE of PLANET or STAR

OA of body 37° 28.4'. HE 10 feet.

Enter the table on page 45, right hand side, with arguments OA and HE. Interpolate by eye. The correction is always subtractive.

OA	37°	28.4
Total correction *page 45*		–04.3
TA	37°	24.1

To find TRUE ALTITUDE of MOON

On Thursday, 16th August the OA of the Moon s upper limb was 51° 14.6'. HE 10 feet.

Table on page 46 is entered with arguments OA and horizontal parallax (HP). HP is read from the Moon data on the monthly page; the value is for 12h and interpolation may be necessary. Note that when using the left hand main table, i.e. upper limb, the sign changes from plus to minus when the OA is greater than 62 . Finally a HE correction, found at the foot of the table, must be made.

HP *monthly page*	56.7
OA	51° 14.6
Main correction *page 46, interpolation by eye*	+10.2
HE correction *at foot of table*	+06.7
TA	51° 31.5

To find the time of MERIDIAN PASSAGE of the SUN on Monday, 19th November in longitude 54° 15' W

For all practical purposes the GMT of transit of Sun and Planets can be found simply by converting vessel's longitude to time, using the arc to time table. One degree of longitude equals four minutes of time. In West longitudes transit occurs after Greenwich transit, in East longitudes before.

Read off from the monthly page the GMT of transit at Greenwich; then convert arc (your longitude) to time from the table on page 53.

GMT of transit at Greenwich *monthly page*	11h 45m
Arc to time 54° *page 53*	+3 36
Arc to time 15' *page 53*	+01
GMT of transit in 54° 15' W	15h 22m

To find the time of MERIDIAN PASSAGE of JUPITER on Monday, 8th October in longitude 33° 19'W

The method is similar to that for the Sun. Read off from the second monthly page the GMT of transit at Greenwich then do the arc to time conversion from the table on page 53.

GMT of transit at Greenwich *monthly page*	03h 52m
Arc to time 33° *page 53*	+2 12
Arc to time 19' *page 53, nearest minute*	+01
GMT of transit in 33° 19'W	06h 05m

To find the time of MERIDIAN PASSAGE of Star ARCTURUS on Saturday, 8th December in longitude 84° 55' E

On the monthly Star pages the time of the meridian passage of the star at Greenwich is given once only, and this is for the first day of the month. The star transits approximately four minutes earlier each day. Thus you must first calculate the Greenwich transit time for the day in question by deducting four minutes for each completed day, and then apply the arc to time adjustment.

Transit at Greenwich on 1st December (monthly page) is at 09h 34m. Deduct 28m for seven completed days; so the time of transit at Greenwich on 8th December is 09h 06m.

GMT of transit at Greenwich *monthly page less correction*	9h 06m
Arc to time 84° *page 53*	–5 36
Arc to time 55' *page 53, nearest minute*	–04
GMT of transit in 84° 55'E	3h 26m

To find the time of MERIDIAN PASSAGE of the MOON on Wednesday, 25th April in longitude 64° W

The monthly page shows the time of meridian passage at Greenwich. The basic calculation of the GMT of meridian passage in your longitude is the same as for Sun and Planets; but a further correction must be applied because of the Moon's continuous movement around the earth. This correction is derived from a value called Diff. in minutes of time, and is used, together with longitude, to enter the table on page 54. Diff. changes daily and is found on the third monthly page.

GMT of transit at Greenwich *monthly page, also note Diff. 49m*	15h 13m
Arc to time 64 *page 53*	+4 16
Diff./longitude correction *page 54*	+08
GMT of transit	19h 37m

To find times of SUNRISE, SUNSET and CIVIL TWILIGHT

These times are tabulated on the third monthly page for the Greenwich meridian and latitude 52° N. To obtain the times for other latitudes you must enter the correction table which is alongside. Note that the correction table is accurate for the 15th of each month; thus interpolation is sometimes necessary.

To find GMT of SUNRISE on Tuesday, 15th May in 30° N 42° E

GMT of sunrise at Greenwich *monthly page*	4h 06m
Latitude correction 30° N *table alongside*	+1 00
Arc to time 42° E *page 53*	–2 48
GMT of sunrise	2h 18m

Supposing you wish to find sunrise on 31st May in the same latitude. You must note the correction for the middle of May, i.e. +1h 00m. Then look up the correction for mid June, which is +1h 19m. Since end May lies midway between these two values the accurate correction, by interpolation, is +1h 10m.

Exactly the same working is used for sunset, morning civil twilight and evening civil twilight.

To find times of MOONRISE and MOONSET

The yachtsman is unlikely to need the precise times of moonrise and moonset. The phase of the Moon will give him a rough idea (see notes on page 51). The monthly pages give the exact times for the Greenwich meridian in latitude 52° N. By applying longitude (arc to time) a rough idea can be obtained in moderate latitudes, although this can be an hour or more out depending on the position of the observer and the declination of the Moon.

To find approximate GMT of MOONRISE on Wednesday, 23rd May in position 40° N 20° E.

GMT of moonrise at Greenwich *monthly page*	05h 43m
Arc to time 20° E *Page 53*	–1 20
Approx. GMT of moonrise	04h 23m

To find LATITUDE from sextant observation of POLARIS

Obtain LHA ARIES and enter table, page 44, to obtain value Q. Apply this value to your TA to establish your latitude.

To find latitude by POLARIS on Monday, 22nd October at 18h 49m. Estimated position 24° 30'N 22° 15"W. OA 24° 44.0". HE 7 feet.

GHA ARIES at 18h *monthly page*	301° 34.1
Increment for 49m *page 47*	12° 17.0
Longitude W	–22° 15.0
LHA ARIES at 18h 49m	291° 36.1
OA	24° 44.0
Altitude total correction *page 45*	–04.5
TA	24° 39.5
Q correction *page 44*	+15.0
LAT	24° 54.5

Note that the Pole Star table changes annually. It is therefore part of the ephemerides rather than the correction tables.

Using the VERSINE formula to find CALCULATED ALTITUDE

The formula is used for all bodies.

Vers CZD = Vers LHA x Cos LAT x Cos DEC + Vers (LAT ± DEC)

In the brackets, LAT and DEC are added if different names, i.e. one is N and the other is S, and subtracted if same names.

CZD is calculated zenith distance. Calculated altitude (CA) is 90° – CZD.

Solution requires tables of natural versines and log versines. These are combined in the same table; so take care to differentiate. A table of log cosines is also provided; so the arithmetic required is only addition.

The following example shows the process. You will probably wish to produce your own blank pro formas in order to facilitate the work.

LAT 25° 30.0'N. DEC of body S 15° 25.0'. LHA of body 335° 24.0'.

1. Log vers LHA 8.9579
   When degrees are printed at the bottom minutes are read (upwards) on the right. Take care to enter the units, i.e. 8.9579 not 9579. Units are not printed in every column (for space considerations) so be careful when units are changing, e.g. from 8 to 9.
2. Log cos LAT 9.9555
   Again, units are not printed in every column. Ignore the integrated table of log sines, with degrees on the bottom and minutes on the right.
3. Log cos DEC 9.9841
4. Add 1 + 2 + 3 28.8975
5. Discard ten's digit 8.8975
6. Natural versine of 5 0.0789
   By inspection of versine table. Interpolation may be necessary. Again, units are not printed in every column.
7. LAT + DEC 40° 55.0'
   Adding LAT and DEC because different names.
8. Natural versine of 7 0.2443
9. Add 6 + 8 0.3232
10. CZD 47° 24.5
    By inspection of versine table.
11. CA is 90° – CZD = 42° 35.5

Using the ABC tables to find AZIMUTH

The tables are arranged so that A + B = C. C is the azimuth, and is in quadrantal notation. Table A is entered using arguments LHA and LAT. Table B is entered using arguments LHA and DEC. Table C is entered using arguments A + B and LAT.

This example shows the process, and is for the same sight as the previous (versine) example.

Precision in interpolation is not necessary because plus or minus one degree of bearing in plotting is not critical.

LAT 25° 30.0'N. DEC of body S 15° 25.0'. LHA of body 335° 24.0'.

1. From table A +1.040
   LHA in column, LAT in row. Labelled + because of note at top of table. Interpolation necessary.
2. From table B +0.665
   LHA in column, DEC in row. Labelled + because of note at top of table. Interpolation necessary.
3. Add A+ B +1.705
   In cases where this is a minus quantity ignore sign; but the sign is used in step 5.
4. From table C, AZ is 33°
   A + B in row, LAT in column.

The AZ is in quadrantal notation, which is expressed from North to East or West or from South to East or West. The method of applying the AZ in order to obtain the true bearing is explained at the bottom of table C.

5. The value of A + B was positive; so the AZ is named South.
6. LHA is greater than 180° so AZ is named South towards East.
7. AZ is 33° so true bearing of body is 180° – 33° = 147°.

In practice you know that the observed body was in the South-East so you can apply the AZ without resorting to this working. It is only really necessary if the body bears almost East or almost West.

PRACTISING ASHORE

It can be very useful to practise ashore. If you are new to astro-navigation it gives you the opportunity to become familiar with your sextant and with the routine of recording altitudes and times. It will also help you to become adept at using the tables and perhaps to develop your own pro forma for the sight reduction process.

What you have to do is create an artificial horizon. This is easy. Choose a calm day or night and put a large bucket or bowl of water on the ground. The surface of the water provides the artificial horizon because it is at a tangent to the curvature of the earth. Windy weather is no good because the water has to be completely smooth so that it can act as a mirror.

To make a start, the Moon when full or nearly full is the best object. There is usually less wind at night, and the Moon can be picked up quite easily without dazzle. The Sun is also an easy object on a calm day; but you must remember to use your sextant shades, as you would at sea, in order to avoid risk of eye damage. It is a good deal more difficult, but still feasible, to isolate the reflections of bright Planets and Stars.

To take the altitude, stand so that you can see the reflection of the body on the water. With your sextant bring the body down (as you would at sea) and superimpose it on the water reflection. You are using the sextant in the normal way except that instead of bringing the body down until it kisses the horizon you must bring it down so that it *covers* the water reflection of the body.

Just as you would at sea, take several observations and then calculate average times and sextant altitudes. To obtain observed altitude we first apply your sextant's index error and then halve the result. Here is an example:

Sextant altitude	104° 23.8
Index error	–6.4
	104° 17.4
Divide by 2 =	52° 8.7

which is observed altitude.

You will not be able to observe altitudes much above 55° because your sextant arc covers only about 110°. It is difficult to handle low altitudes so the best range is about 15° to 55°.

The sight reduction process is standard; but in finding true altitudes we have to vary the method because the only corrections required are for refraction (all bodies) and, for the Moon, horizontal parallax.

For Sun, Planet and Star go to the Mean Refraction table on page 46 and enter it with apparent altitude which, for our purposes here, is observed altitude. The correction is always subtractive, and this is the only correction required.

The Moon is a little more complicated. From the daily page make a note of horizontal parallax; the value is for 1200 GMT so you will probably need to interpolate. Then go to the Moon Altitude Total Correction table on page 46. There is a section for the upper limb and another for the lower limb. Enter each table with arguments observed altitude and horizontal parallax. Take the mean of these two values and add 9.8 minutes. The result is the total and only correction. Ignore the height of eye correction at the bottom of the table.

Here's wishing you very small intercepts!

Compass Checking by AMPLITUDE

The integrated table of LOG COSINES and LOG SINES makes it easy to calculate amplitudes by using the formula: SIN AMPLITUDE = SIN DEC (of the observed body) ÷ COS LAT (of the observer).

Note that SINES are read at the bottom of the table, with minutes read upwards from the right.

This example shows the method: LAT 46° 20'N, Sunset, Sun's DEC N 9° 30'.

1. Log sin 9° 30'	9.2176
2. Add 10	19.2176
3. Log cosine 46° 20'	9.8391
4. 2–3	9.3785

This is the log sin of the amplitude. Inspect the table and read off a value of 13° 50. The amplitude is converted to a true bearing like this: the bearing is named E for a rising body and W for a setting body; the bearing is named N or S according to the body's DEC. The true bearing of the body is therefore W 13° 50'N or, more familiarly, 283° 50'.

This is the bearing your compass would show if there were no compass error. You can now apply known variation and therefore identify the deviation for your present heading.

Formulae for SCIENTIFIC CALCULATORS

Simple formulae exist for finding CA and AZ using a scientific calculator.

Sin CA = Cos LAT x Cos DEC x Cos LHA ± (Sin LAT x Sin DEC)

The final bracketed value is added if LAT and DEC are same names, subtracted if LAT and DEC are different names.

Sin AZ = (Sin LHA x Cos DEC) ÷ Cos CA

Azimuth is in quadrantal notation.

GMT	SUN GHA ° '	SUN Dec ° '	ARIES GHA ° '
Sunday, 1st January			
00	179 13.7	**S23 03.8**	100 03.9
02	209 13.1	**23 03.4**	130 08.8
04	239 12.5	**23 03.0**	160 13.7
06	269 11.9	**23 02.6**	190 18.7
08	299 11.3	**23 02.2**	220 23.6
10	329 10.8	**23 01.8**	250 28.5
12	359 10.2	**23 01.4**	280 33.4
14	29 09.6	**23 01.0**	310 38.4
16	59 09.0	**23 00.6**	340 43.3
18	89 08.4	**23 00.2**	10 48.2
20	119 07.8	**22 59.8**	40 53.2
22	149 07.2	**S22 59.4**	70 58.1
Monday, 2nd January			
00	179 06.6	**S22 59.0**	101 03.0
02	209 06.0	**22 58.6**	131 07.9
04	239 05.4	**22 58.1**	161 12.9
06	269 04.9	**22 57.7**	191 17.8
08	299 04.3	**22 57.3**	221 22.7
10	329 03.7	**22 56.8**	251 27.7
12	359 03.1	**22 56.4**	281 32.6
14	29 02.5	**22 56.0**	311 37.5
16	59 01.9	**22 55.5**	341 42.4
18	89 01.4	**22 55.1**	11 47.4
20	119 00.8	**22 54.6**	41 52.3
22	149 00.2	**S22 54.2**	71 57.2
Tuesday, 3rd January			
00	178 59.6	**S22 53.7**	102 02.1
02	208 59.0	**22 53.3**	132 07.1
04	238 58.4	**22 52.8**	162 12.0
06	268 57.9	**22 52.3**	192 16.9
08	298 57.3	**22 51.9**	222 21.9
10	328 56.7	**22 51.4**	252 26.8
12	358 56.1	**22 50.9**	282 31.7
14	28 55.6	**22 50.4**	312 36.6
16	58 55.0	**22 50.0**	342 41.6
18	88 54.4	**22 49.5**	12 46.5
20	118 53.8	**22 49.0**	42 51.4
22	148 53.3	**S22 48.5**	72 56.4
Wednesday, 4th January			
00	178 52.7	**S22 48.0**	103 01.3
02	208 52.1	**22 47.5**	133 06.2
04	238 51.5	**22 47.0**	163 11.1
06	268 51.0	**22 46.5**	193 16.1
08	298 50.4	**22 46.0**	223 21.0
10	328 49.8	**22 45.5**	253 25.9
12	358 49.3	**22 45.0**	283 30.9
14	28 48.7	**22 44.5**	313 35.8
16	58 48.1	**22 43.9**	343 40.7
18	88 47.6	**22 43.4**	13 45.6
20	118 47.0	**22 42.9**	43 50.6
22	148 46.4	**S22 42.4**	73 55.5
Thursday, 5th January			
00	178 45.9	**S22 41.8**	104 00.4
02	208 45.3	**22 41.3**	134 05.4
04	238 44.7	**22 40.8**	164 10.3
06	268 44.2	**22 40.2**	194 15.2
08	298 43.6	**22 39.7**	224 20.1
10	328 43.1	**22 39.1**	254 25.1
12	358 42.5	**22 38.6**	284 30.0
14	28 41.9	**22 38.0**	314 34.9
16	58 41.4	**22 37.5**	344 39.9
18	88 40.8	**22 36.9**	14 44.8
20	118 40.3	**22 36.4**	44 49.7
22	148 39.7	**S22 35.8**	74 54.6

GMT	SUN GHA ° '	SUN Dec ° '	ARIES GHA ° '
Friday, 6th January			
00	178 39.1	**S22 35.2**	104 59.6
02	208 38.6	**22 34.6**	135 04.5
04	238 38.0	**22 34.1**	165 09.4
06	268 37.5	**22 33.5**	195 14.4
08	298 36.9	**22 32.9**	225 19.3
10	328 36.4	**22 32.3**	255 24.2
12	358 35.8	**22 31.7**	285 29.1
14	28 35.3	**22 31.2**	315 34.1
16	58 34.7	**22 30.6**	345 39.0
18	88 34.2	**22 30.0**	15 43.9
20	118 33.6	**22 29.4**	45 48.9
22	148 33.1	**S22 28.8**	75 53.8
Saturday, 7th January			
00	178 32.5	**S22 28.2**	105 58.7
02	208 32.0	**22 27.6**	136 03.6
04	238 31.5	**22 26.9**	166 08.6
06	268 30.9	**22 26.3**	196 13.5
08	298 30.4	**22 25.7**	226 18.4
10	328 29.8	**22 25.1**	256 23.3
12	358 29.3	**22 24.5**	286 28.3
14	28 28.8	**22 23.8**	316 33.2
16	58 28.2	**22 23.2**	346 38.1
18	88 27.7	**22 22.6**	16 43.1
20	118 27.1	**22 21.9**	46 48.0
22	148 26.6	**S22 21.3**	76 52.9
Sunday, 8th January			
00	178 26.1	**S22 20.7**	106 57.8
02	208 25.5	**22 20.0**	137 02.8
04	238 25.0	**22 19.4**	167 07.7
06	268 24.5	**22 18.7**	197 12.6
08	298 23.9	**22 18.1**	227 17.6
10	328 23.4	**22 17.4**	257 22.5
12	358 22.9	**22 16.7**	287 27.4
14	28 22.3	**22 16.1**	317 32.3
16	58 21.8	**22 15.4**	347 37.3
18	88 21.3	**22 14.7**	17 42.2
20	118 20.8	**22 14.1**	47 47.1
22	148 20.2	**S22 13.4**	77 52.1
Monday, 9th January			
00	178 19.7	**S22 12.7**	107 57.0
02	208 19.2	**22 12.0**	138 01.9
04	238 18.7	**22 11.3**	168 06.8
06	268 18.1	**22 10.7**	198 11.8
08	298 17.6	**22 10.0**	228 16.7
10	328 17.1	**22 09.3**	258 21.6
12	358 16.6	**22 08.6**	288 26.6
14	28 16.1	**22 07.9**	318 31.5
16	58 15.5	**22 07.2**	348 36.4
18	88 15.0	**22 06.5**	18 41.3
20	118 14.5	**22 05.8**	48 46.3
22	148 14.0	**S22 05.1**	78 51.2
Tuesday, 10th January			
00	178 13.5	**S22 04.3**	108 56.1
02	208 13.0	**22 03.6**	139 01.1
04	238 12.5	**22 02.9**	169 06.0
06	268 11.9	**22 02.2**	199 10.9
08	298 11.4	**22 01.4**	229 15.8
10	328 10.9	**22 00.7**	259 20.8
12	358 10.4	**22 00.0**	289 25.7
14	28 09.9	**21 59.2**	319 30.6
16	58 09.4	**21 58.5**	349 35.6
18	88 08.9	**21 57.8**	19 40.5
20	118 08.4	**21 57.0**	49 45.4
22	148 07.9	**S21 56.3**	79 50.3

GMT	SUN GHA ° '	SUN Dec ° '	ARIES GHA ° '
Wednesday, 11th January			
00	178 07.4	**S21 55.5**	109 55.3
02	208 06.9	**21 54.8**	140 00.2
04	238 06.4	**21 54.0**	170 05.1
06	268 05.9	**21 53.3**	200 10.1
08	298 05.4	**21 52.5**	230 15.0
10	328 04.9	**21 51.7**	260 19.9
12	358 04.4	**21 51.0**	290 24.8
14	28 03.9	**21 50.2**	320 29.8
16	58 03.4	**21 49.4**	350 34.7
18	88 02.9	**21 48.6**	20 39.6
20	118 02.4	**21 47.9**	50 44.6
22	148 01.9	**S21 47.1**	80 49.5
Thursday, 12th January			
00	178 01.4	**S21 46.3**	110 54.4
02	208 01.0	**21 45.5**	140 59.3
04	238 00.5	**21 44.7**	171 04.3
06	268 00.0	**21 43.9**	201 09.2
08	297 59.5	**21 43.1**	231 14.1
10	327 59.0	**21 42.3**	261 19.0
12	357 58.5	**21 41.5**	291 24.0
14	27 58.0	**21 40.7**	321 28.9
16	57 57.6	**21 39.9**	351 33.8
18	87 57.1	**21 39.1**	21 38.8
20	117 56.6	**21 38.3**	51 43.7
22	147 56.1	**S21 37.4**	81 48.6
Friday, 13th January			
00	177 55.6	**S21 36.6**	111 53.5
02	207 55.2	**21 35.8**	141 58.5
04	237 54.7	**21 35.0**	172 03.4
06	267 54.2	**21 34.1**	202 08.3
08	297 53.7	**21 33.3**	232 13.3
10	327 53.3	**21 32.5**	262 18.2
12	357 52.8	**21 31.6**	292 23.1
14	27 52.3	**21 30.8**	322 28.0
16	57 51.9	**21 30.0**	352 33.0
18	87 51.4	**21 29.1**	22 37.9
20	117 50.9	**21 28.3**	52 42.8
22	147 50.5	**S21 27.4**	82 47.8
Saturday, 14th January			
00	177 50.0	**S21 26.5**	112 52.7
02	207 49.5	**21 25.7**	142 57.6
04	237 49.1	**21 24.8**	173 02.5
06	267 48.6	**21 24.0**	203 07.5
08	297 48.1	**21 23.1**	233 12.4
10	327 47.7	**21 22.2**	263 17.3
12	357 47.2	**21 21.4**	293 22.3
14	27 46.8	**21 20.5**	323 27.2
16	57 46.3	**21 19.6**	353 32.1
18	87 45.8	**21 18.7**	23 37.0
20	117 45.4	**21 17.8**	53 42.0
22	147 44.9	**S21 16.9**	83 46.9
Sunday, 15th January			
00	177 44.5	**S21 16.1**	113 51.8
02	207 44.0	**21 15.2**	143 56.8
04	237 43.6	**21 14.3**	174 01.7
06	267 43.1	**21 13.4**	204 06.6
08	297 42.7	**21 12.5**	234 11.5
10	327 42.2	**21 11.6**	264 16.5
12	357 41.8	**21 10.6**	294 21.4
14	27 41.4	**21 09.7**	324 26.3
16	57 40.9	**21 08.8**	354 31.2
18	87 40.5	**21 07.9**	24 36.2
20	117 40.0	**21 07.0**	54 41.1
22	147 39.6	**S21 06.1**	84 46.0

GMT	SUN GHA ° '	SUN Dec ° '	ARIES GHA ° '
Monday, 16th January			
00	177 39.1	**S21 05.1**	114 51.0
02	207 38.7	**21 04.2**	144 55.9
04	237 38.3	**21 03.3**	175 00.8
06	267 37.8	**21 02.4**	205 05.7
08	297 37.4	**21 01.4**	235 10.7
10	327 37.0	**21 00.5**	265 15.6
12	357 36.5	**20 59.5**	295 20.5
14	27 36.1	**20 58.6**	325 25.5
16	57 35.7	**20 57.6**	355 30.4
18	87 35.2	**20 56.7**	25 35.3
20	117 34.8	**20 55.7**	55 40.2
22	147 34.4	**S20 54.8**	85 45.2
Tuesday, 17th January			
00	177 34.0	**S20 53.8**	115 50.1
02	207 33.5	**20 52.9**	145 55.0
04	237 33.1	**20 51.9**	176 00.0
06	267 32.7	**20 50.9**	206 04.9
08	297 32.3	**20 50.0**	236 09.8
10	327 31.9	**20 49.0**	266 14.7
12	357 31.4	**20 48.0**	296 19.7
14	27 31.0	**20 47.0**	326 24.6
16	57 30.6	**20 46.1**	356 29.5
18	87 30.2	**20 45.1**	26 34.5
20	117 29.8	**20 44.1**	56 39.4
22	147 29.4	**S20 43.1**	86 44.3
Wednesday, 18th January			
00	177 29.0	**S20 42.1**	116 49.2
02	207 28.5	**20 41.1**	146 54.2
04	237 28.1	**20 40.1**	176 59.1
06	267 27.7	**20 39.1**	207 04.0
08	297 27.3	**20 38.1**	237 09.0
10	327 26.9	**20 37.1**	267 13.9
12	357 26.5	**20 36.1**	297 18.8
14	27 26.1	**20 35.1**	327 23.7
16	57 25.7	**20 34.1**	357 28.7
18	87 25.3	**20 33.1**	27 33.6
20	117 24.9	**20 32.1**	57 38.5
22	147 24.5	**S20 31.0**	87 43.5
Thursday, 19th January			
00	177 24.1	**S20 30.0**	117 48.4
02	207 23.7	**20 29.0**	147 53.3
04	237 23.3	**20 28.0**	177 58.2
06	267 22.9	**20 26.9**	208 03.2
08	297 22.5	**20 25.9**	238 08.1
10	327 22.2	**20 24.9**	268 13.0
12	357 21.8	**20 23.8**	298 17.9
14	27 21.4	**20 22.8**	328 22.9
16	57 21.0	**20 21.7**	358 27.8
18	87 20.6	**20 20.7**	28 32.7
20	117 20.2	**20 19.6**	58 37.7
22	147 19.8	**S20 18.6**	88 42.6
Friday, 20th January			
00	177 19.5	**S20 17.5**	118 47.5
02	207 19.1	**20 16.5**	148 52.4
04	237 18.7	**20 15.4**	178 57.4
06	267 18.3	**20 14.3**	209 02.3
08	297 17.9	**20 13.3**	239 07.2
10	327 17.6	**20 12.2**	269 12.2
12	357 17.2	**20 11.1**	299 17.1
14	27 16.8	**20 10.1**	329 22.0
16	57 16.4	**20 09.0**	359 26.9
18	87 16.1	**20 07.9**	29 31.9
20	117 15.7	**20 06.8**	59 36.8
22	147 15.3	**S20 05.7**	89 41.7
Tuesday, 31st January			
00	176 40.9	**S17 36.2**	129 38.0
02	206 40.7	**17 34.8**	159 43.0
04	236 40.5	**17 33.5**	189 47.9
06	266 40.3	**S17 32.1**	219 52.8

GMT	SUN GHA ° '	SUN Dec ° '	ARIES GHA ° '
Saturday, 21st January			
00	177 15.0	**S20 04.6**	119 46.7
02	207 14.6	**20 03.6**	149 51.6
04	237 14.2	**20 02.5**	179 56.5
06	267 13.9	**20 01.4**	210 01.4
08	297 13.5	**20 00.3**	240 06.4
10	327 13.2	**19 59.2**	270 11.3
12	357 12.8	**19 58.1**	300 16.2
14	27 12.4	**19 57.0**	330 21.2
16	57 12.1	**19 55.8**	0 26.1
18	87 11.7	**19 54.7**	30 31.0
20	117 11.4	**19 53.6**	60 35.9
22	147 11.0	**S19 52.5**	90 40.9
Sunday, 22nd January			
00	177 10.7	**S19 51.4**	120 45.8
02	207 10.3	**19 50.3**	150 50.7
04	237 10.0	**19 49.1**	180 55.7
06	267 09.6	**19 48.0**	211 00.6
08	297 09.3	**19 46.9**	241 05.5
10	327 08.9	**19 45.8**	271 10.4
12	357 08.6	**19 44.6**	301 15.4
14	27 08.3	**19 43.5**	331 20.3
16	57 07.9	**19 42.3**	1 25.2
18	87 07.6	**19 41.2**	31 30.2
20	117 07.2	**19 40.1**	61 35.1
22	147 06.9	**S19 38.9**	91 40.0
Monday, 23rd January			
00	177 06.6	**S19 37.8**	121 44.9
02	207 06.2	**19 36.6**	151 49.9
04	237 05.9	**19 35.5**	181 54.8
06	267 05.6	**19 34.3**	211 59.7
08	297 05.2	**19 33.1**	242 04.7
10	327 04.9	**19 32.0**	272 09.6
12	357 04.6	**19 30.8**	302 14.5
14	27 04.3	**19 29.7**	332 19.4
16	57 03.9	**19 28.5**	2 24.4
18	87 03.6	**19 27.3**	32 29.3
20	117 03.3	**19 26.1**	62 34.2
22	147 03.0	**S19 25.0**	92 39.2
Tuesday, 24th January			
00	177 02.6	**S19 23.8**	122 44.1
02	207 02.3	**19 22.6**	152 49.0
04	237 02.0	**19 21.4**	182 53.9
06	267 01.7	**19 20.2**	212 58.9
08	297 01.4	**19 19.0**	243 03.8
10	327 01.1	**19 17.8**	273 08.7
12	357 00.8	**19 16.7**	303 13.7
14	27 00.5	**19 15.5**	333 18.6
16	57 00.1	**19 14.3**	3 23.5
18	86 59.8	**19 13.1**	33 28.4
20	116 59.5	**19 11.8**	63 33.4
22	146 59.2	**S19 10.6**	93 38.3
Wednesday, 25th January			
00	176 58.9	**S19 09.4**	123 43.2
02	206 58.6	**19 08.2**	153 48.1
04	236 58.3	**19 07.0**	183 53.1
06	266 58.0	**19 05.8**	213 58.0
08	296 57.7	**19 04.6**	244 02.9
10	326 57.4	**19 03.4**	274 07.9
12	356 57.1	**19 02.1**	304 12.8
14	26 56.8	**19 00.9**	334 17.7
16	56 56.6	**18 59.7**	4 22.6
18	86 56.3	**18 58.4**	34 27.6
20	116 56.0	**18 57.2**	64 32.5
22	146 55.7	**S18 56.0**	94 37.4
Tuesday, 31st January			
08	296 40.1	**S17 30.7**	249 57.8
10	326 39.9	**17 29.3**	280 02.7
12	356 39.7	**17 27.9**	310 07.6
14	26 39.5	**S17 26.5**	340 12.5

GMT	SUN GHA ° '	SUN Dec ° '	ARIES GHA ° '
Thursday, 26th January			
00	176 55.4	**S18 54.7**	124 42.4
02	206 55.1	**18 53.5**	154 47.3
04	236 54.8	**18 52.3**	184 52.2
06	266 54.6	**18 51.0**	214 57.1
08	296 54.3	**18 49.8**	245 02.1
10	326 54.0	**18 48.5**	275 07.0
12	356 53.7	**18 47.3**	305 11.9
14	26 53.4	**18 46.0**	335 16.9
16	56 53.2	**18 44.7**	5 21.8
18	86 52.9	**18 43.5**	35 26.7
20	116 52.6	**18 42.2**	65 31.6
22	146 52.4	**S18 41.0**	95 36.6
Friday, 27th January			
00	176 52.1	**S18 39.7**	125 41.5
02	206 51.8	**18 38.4**	155 46.4
04	236 51.6	**18 37.2**	185 51.4
06	266 51.3	**18 35.9**	215 56.3
08	296 51.0	**18 34.6**	246 01.2
10	326 50.8	**18 33.3**	276 06.1
12	356 50.5	**18 32.0**	306 11.1
14	26 50.2	**18 30.8**	336 16.0
16	56 50.0	**18 29.5**	6 20.9
18	86 49.7	**18 28.2**	36 25.8
20	116 49.5	**18 26.9**	66 30.8
22	146 49.2	**S18 25.6**	96 35.7
Saturday, 28th January			
00	176 49.0	**S18 24.3**	126 40.6
02	206 48.7	**18 23.0**	156 45.6
04	236 48.5	**18 21.7**	186 50.5
06	266 48.2	**18 20.4**	216 55.4
08	296 48.0	**18 19.1**	247 00.3
10	326 47.7	**18 17.8**	277 05.3
12	356 47.5	**18 16.5**	307 10.2
14	26 47.2	**18 15.2**	337 15.1
16	56 47.0	**18 13.9**	7 20.1
18	86 46.8	**18 12.6**	37 25.0
20	116 46.5	**18 11.2**	67 29.9
22	146 46.3	**S18 09.9**	97 34.8
Sunday, 29th January			
00	176 46.1	**S18 08.6**	127 39.8
02	206 45.8	**18 07.3**	157 44.7
04	236 45.6	**18 06.0**	187 49.6
06	266 45.4	**18 04.6**	217 54.6
08	296 45.1	**18 03.3**	247 59.5
10	326 44.9	**18 02.0**	278 04.4
12	356 44.7	**18 00.6**	308 09.3
14	26 44.5	**17 59.3**	338 14.3
16	56 44.2	**17 58.0**	8 19.2
18	86 44.0	**17 56.6**	38 24.1
20	116 43.8	**17 55.3**	68 29.1
22	146 43.6	**S17 53.9**	98 34.0
Monday, 30th January			
00	176 43.4	**S17 52.6**	128 38.9
02	206 43.1	**17 51.2**	158 43.8
04	236 42.9	**17 49.9**	188 48.8
06	266 42.7	**17 48.5**	218 53.7
08	296 42.5	**17 47.2**	248 58.6
10	326 42.3	**17 45.8**	279 03.5
12	356 42.1	**17 44.4**	309 08.5
14	26 41.9	**17 43.1**	339 13.4
16	56 41.7	**17 41.7**	9 18.3
18	86 41.5	**17 40.3**	39 23.3
20	116 41.3	**17 39.0**	69 28.2
22	146 41.1	**S17 37.6**	99 33.1
Tuesday, 31st January			
16	56 39.3	**S17 25.1**	10 17.5
18	86 39.1	**17 23.7**	40 22.4
20	116 38.9	**17 22.3**	70 27.3
22	146 38.8	**S17 20.9**	100 32.3

MOON

Day	GMT hr	GHA ° ′	Mean Var/hr 14°+	Dec ° ′	Mean Var/hr ′
1 Sun	0	95 27.2	34.4	N 7 15.2	10.8
	6	182 53.7	34.4	N 8 19.9	10.6
	12	270 19.7	34.2	N 9 23.2	10.3
	18	357 45.0	34.0	N10 25.1	10.0
2 Mon	0	85 09.5	33.9	N11 25.3	9.8
	6	172 33.0	33.7	N12 23.9	9.4
	12	259 55.4	33.5	N13 20.6	9.1
	18	347 16.6	33.3	N14 15.3	8.7
3 Tu	0	74 36.3	33.0	N15 07.9	8.4
	6	161 54.4	32.8	N15 58.3	7.9
	12	249 11.0	32.5	N16 46.4	7.5
	18	336 25.8	32.1	N17 31.9	7.1
4 Wed	0	63 38.8	31.8	N18 14.8	6.6
	6	150 50.0	31.5	N18 54.9	6.2
	12	237 59.3	31.2	N19 32.0	5.7
	18	325 06.7	30.8	N20 06.1	5.1
5 Th	0	52 12.1	30.5	N20 36.9	4.5
	6	139 15.7	30.3	N21 04.4	4.0
	12	226 17.5	29.9	N21 28.4	3.4
	18	313 17.4	29.7	N21 48.8	2.8
6 Fri	0	40 15.7	29.4	N22 05.4	2.1
	6	127 12.4	29.1	N22 18.1	1.4
	12	214 07.7	28.9	N22 26.9	0.7
	18	301 01.7	28.8	N22 31.5	0.1
7 Sat	0	27 54.6	28.7	N22 32.1	0.7
	6	114 46.5	28.5	N22 28.4	1.3
	12	201 37.6	28.4	N22 20.4	2.1
	18	288 28.2	28.4	N22 08.2	2.9
8 Sun	0	15 18.4	28.4	N21 51.6	3.5
	6	102 08.4	28.4	N21 30.8	4.3
	12	188 58.5	28.3	N21 05.7	4.9
	18	275 48.7	28.4	N20 36.3	5.7
9 Mon	0	2 39.3	28.5	N20 02.8	6.3
	6	89 30.4	28.6	N19 25.3	7.0
	12	176 22.2	28.8	N18 43.7	7.6
	18	263 14.7	28.9	N17 58.4	8.2
10 Tu	0	350 08.1	29.1	N17 09.4	8.8
	6	77 02.4	29.2	N16 16.8	9.4
	12	163 57.7	29.4	N15 20.9	9.9
	18	250 53.9	29.6	N14 21.8	10.3
11 Wed	0	337 51.2	29.8	N13 19.8	10.9
	6	64 49.4	29.9	N12 14.9	11.2
	12	151 48.5	30.0	N11 07.5	11.7
	18	238 48.5	30.2	N 9 57.8	12.0
12 Th	0	325 49.2	30.3	N 8 46.0	12.3
	6	52 50.6	30.3	N 7 32.4	12.6
	12	139 52.5	30.4	N 6 17.1	12.8
	18	226 54.8	30.4	N 5 00.4	13.0
13 Fri	0	313 57.4	30.5	N 3 42.6	13.1
	6	41 00.1	30.4	N 2 23.9	13.2
	12	128 02.6	30.4	N 1 04.6	13.3
	18	215 05.0	30.3	S 0 15.1	13.3
14 Sat	0	302 06.8	30.2	S 1 34.8	13.3
	6	29 08.1	30.0	S 2 54.5	13.2
	12	116 08.6	29.9	S 4 13.7	13.0
	18	203 08.0	29.7	S 5 32.2	12.9
15 Sun	0	290 06.3	29.4	S 6 49.8	12.7
	6	17 03.3	29.2	S 8 06.2	12.5
	12	103 58.7	28.9	S 9 21.0	12.2
	18	190 52.4	28.6	S10 34.1	11.8
16 Mon	0	277 44.3	28.3	S11 45.2	11.4
	6	4 34.3	27.9	S12 53.9	11.0
	12	91 22.1	27.6	S13 59.9	10.5
	18	178 07.9	27.2	S15 03.1	9.9
17 Tu	0	264 51.4	26.9	S16 03.2	9.4
	6	351 32.7	26.5	S16 59.8	8.8
	12	78 11.8	26.1	S17 52.7	8.1
	18	164 48.7	25.8	S18 41.6	7.4
18 Wed	0	251 23.5	25.5	S19 26.3	6.6
	6	337 56.3	25.2	S20 06.6	5.8
	12	64 27.3	24.9	S20 42.2	5.0
	18	150 56.7	24.6	S21 12.9	4.3
19 Th	0	237 24.7	24.5	S21 38.7	3.4
	6	323 51.6	24.3	S21 59.3	2.5
	12	50 17.7	24.3	S22 14.6	1.5
	18	136 43.4	24.3	S22 24.6	0.7
20 Fri	0	223 08.9	24.3	S22 29.2	0.2
	6	309 34.6	24.4	S22 28.4	1.1
	12	36 00.9	24.6	S22 22.2	2.0
	18	122 28.2	24.7	S22 10.7	2.9
21 Sat	0	208 56.7	25.1	S21 54.1	3.7
	6	295 26.8	25.4	S21 32.4	4.5
	12	21 58.8	25.7	S21 05.8	5.2
	18	108 32.9	26.2	S20 34.5	6.1
22 Sun	0	195 09.4	26.5	S19 58.7	6.7
	6	281 48.4	27.0	S19 18.7	7.4
	12	8 30.1	27.4	S18 34.7	8.0
	18	95 14.7	28.0	S17 47.0	8.6
23 Mon	0	182 02.1	28.4	S16 55.8	9.1
	6	268 52.4	28.9	S16 01.4	9.6
	12	355 45.7	29.4	S15 04.2	10.0
	18	82 41.8	29.9	S14 04.4	10.4
24 Tu	0	169 40.8	30.3	S13 02.2	10.7
	6	256 42.5	30.7	S11 58.0	11.0
	12	343 46.9	31.2	S10 52.0	11.2
	18	70 53.8	31.6	S 9 44.5	11.5
25 Wed	0	158 03.0	32.0	S 8 35.8	11.7
	6	245 14.6	32.3	S 7 26.0	11.7
	12	332 28.2	32.6	S 6 15.4	11.8
	18	59 43.7	32.9	S 5 04.2	12.0
26 Th	0	147 01.0	33.2	S 3 52.7	11.9
	6	234 19.8	33.4	S 2 41.0	11.9
	12	321 40.1	33.6	S 1 29.3	11.9
	18	49 01.5	33.7	S 0 17.8	11.9
27 Fri	0	136 23.9	33.9	N 0 53.3	11.7
	6	223 47.2	34.0	N 2 03.9	11.7
	12	311 11.1	34.0	N 3 13.8	11.5
	18	38 35.5	34.2	N 4 22.8	11.3
28 Sat	0	126 00.2	34.1	N 5 30.9	11.2
	6	213 25.0	34.1	N 6 37.9	10.9
	12	300 49.8	34.1	N 7 43.6	10.7
	18	28 14.3	34.1	N 8 47.9	10.5
29 Sun	0	115 38.4	34.0	N 9 50.8	10.1
	6	203 02.1	33.8	N10 52.0	9.9
	12	290 25.0	33.7	N11 51.4	9.6
	18	17 47.0	33.5	N12 49.0	9.3
30 Mon	0	105 08.1	33.3	N13 44.6	8.8
	6	192 28.1	33.1	N14 38.0	8.5
	12	279 46.9	32.9	N15 29.2	8.1
	18	7 04.4	32.7	N16 18.1	7.7
31 Tu	0	94 20.4	32.4	N17 04.5	7.2
	6	181 34.9	32.2	N17 48.2	6.8
	12	268 47.8	31.9	N18 29.2	6.3
	18	355 59.0	31.6	N19 07.4	5.9

PLANETS

VENUS						JUPITER				
Mer Pass h m	GHA ° ′	Mean Var/hr 14°+	Dec ° ′	Mean Var/hr ′	Day	GHA ° ′	Mean Var/hr 15°+	Dec ° ′	Mean Var/hr ′	Mer Pass h m
14 28	143 13.3	59.4	S18 25.9	0.9	1 SUN	71 18.3	2.4	N10 28.2	0.0	19 12
14 29	142 58.5	59.4	S18 03.8	0.9	2 Mon	72 16.3	2.4	N10 28.9	0.0	19 08
14 30	142 44.1	59.4	S17 41.3	1.0	3 Tu	73 14.0	2.4	N10 29.8	0.0	19 04
14 31	142 30.0	59.4	S17 18.3	1.0	4 Wed	74 11.6	2.4	N10 30.7	0.0	19 00
14 31	142 16.3	59.4	S16 54.9	1.0	5 Th	75 08.9	2.4	N10 31.7	0.0	18 56
14 32	142 03.0	59.5	S16 31.0	1.0	6 Fri	76 06.1	2.4	N10 32.7	0.0	18 53
14 33	141 50.0	59.5	S16 06.7	1.0	7 Sat	77 03.0	2.4	N10 33.8	0.1	18 49
14 34	141 37.5	59.5	S15 41.9	1.0	8 SUN	77 59.8	2.4	N10 35.0	0.1	18 45
14 35	141 25.2	59.5	S15 16.8	1.1	9 Mon	78 56.4	2.4	N10 36.3	0.1	18 41
14 36	141 13.3	59.5	S14 51.2	1.1	10 Tu	79 52.8	2.3	N10 37.6	0.1	18 38
14 36	141 01.8	59.5	S14 25.3	1.1	11 Wed	80 49.0	2.3	N10 39.0	0.1	18 34
14 37	140 50.6	59.5	S13 59.0	1.1	12 Th	81 45.0	2.3	N10 40.5	0.1	18 30
14 38	140 39.8	59.6	S13 32.3	1.1	13 Fri	82 40.9	2.3	N10 42.0	0.1	18 26
14 38	140 29.2	59.6	S13 05.3	1.1	14 Sat	83 36.5	2.3	N10 43.6	0.1	18 23
14 39	140 19.0	59.6	S12 38.0	1.2	15 SUN	84 32.0	2.3	N10 45.2	0.1	18 19
14 40	140 09.2	59.6	S12 10.3	1.2	16 Mon	85 27.2	2.3	N10 47.0	0.1	18 15
14 40	139 59.6	59.6	S11 42.4	1.2	17 Tu	86 22.3	2.3	N10 48.7	0.1	18 12
14 41	139 50.4	59.6	S11 14.1	1.2	18 Wed	87 17.3	2.3	N10 50.6	0.1	18 08
14 42	139 41.4	59.6	S10 45.6	1.2	19 Th	88 12.0	2.3	N10 52.5	0.1	18 04
14 42	139 32.8	59.7	S10 16.7	1.2	20 Fri	89 06.6	2.3	N10 54.5	0.1	18 01
14 43	139 24.4	59.7	S 9 47.7	1.2	21 Sat	90 01.0	2.3	N10 56.5	0.1	17 57
14 43	139 16.3	59.7	S 9 18.4	1.2	22 SUN	90 55.2	2.3	N10 58.6	0.1	17 54
14 44	139 08.5	59.7	S 8 48.9	1.2	23 Mon	91 49.2	2.2	N11 00.7	0.1	17 50
14 44	139 01.0	59.7	S 8 19.1	1.2	24 Tu	92 43.1	2.2	N11 02.9	0.1	17 46
14 45	138 53.7	59.7	S 7 49.2	1.3	25 Wed	93 36.8	2.2	N11 05.2	0.1	17 43
14 45	138 46.7	59.7	S 7 19.0	1.3	26 Th	94 30.3	2.2	N11 07.5	0.1	17 39
14 46	138 39.9	59.7	S 6 48.7	1.3	27 Fri	95 23.7	2.2	N11 09.9	0.1	17 36
14 46	138 33.3	59.7	S 6 18.3	1.3	28 Sat	96 16.9	2.2	N11 12.3	0.1	17 32
14 46	138 27.0	59.7	S 5 47.7	1.3	29 SUN	97 09.9	2.2	N11 14.8	0.1	17 29
14 47	138 20.9	59.8	S 5 16.9	1.3	30 Mon	98 02.8	2.2	N11 17.3	0.1	17 25
14 47	138 15.0	59.8	S 4 46.0	1.3	31 Tu	98 55.5	2.2	N11 19.9	0.1	17 22

VENUS, Av. Mag. –4.0
SHA January 5 38; 10 32; 15 26; 20 21; 25 15; 30 10

JUPITER, Av. Mag. –2.5
SHA January 5 331; 10 331; 15 331; 20 330; 25 330; 30 329

MARS						SATURN				
Mer Pass h m	GHA ° ′	Mean Var/hr 15°+	Dec ° ′	Mean Var/hr ′	Day	GHA ° ′	Mean Var/hr 15°+	Dec ° ′	Mean Var/hr ′	Mer Pass h m
04 47	287 59.2	1.9	N 6 37.2	0.2	1 SUN	252 54.7	2.3	S 8 36.2	0.0	07 07
04 44	288 44.6	1.9	N 6 33.1	0.2	2 Mon	253 50.2	2.3	S 8 37.3	0.0	07 04
04 41	289 30.3	1.9	N 6 29.2	0.2	3 Tu	254 45.8	2.3	S 8 38.3	0.0	07 00
04 38	290 16.6	2.0	N 6 25.5	0.1	4 Wed	255 41.5	2.3	S 8 39.4	0.0	06 56
04 35	291 03.4	2.0	N 6 22.1	0.1	5 Th	256 37.3	2.3	S 8 40.4	0.0	06 52
04 32	291 50.7	2.0	N 6 18.8	0.1	6 Fri	257 33.2	2.3	S 8 41.3	0.0	06 49
04 29	292 38.4	2.0	N 6 15.8	0.1	7 Sat	258 29.1	2.3	S 8 42.2	0.0	06 45
04 26	293 26.7	2.0	N 6 13.0	0.1	8 SUN	259 25.2	2.3	S 8 43.1	0.0	06 41
04 22	294 15.5	2.1	N 6 10.5	0.1	9 Mon	260 21.3	2.3	S 8 44.0	0.0	06 38
04 19	295 04.9	2.1	N 6 08.1	0.1	10 Tu	261 17.5	2.3	S 8 44.8	0.0	06 34
04 16	295 54.8	2.1	N 6 06.1	0.1	11 Wed	262 13.8	2.4	S 8 45.6	0.0	06 30
04 12	296 45.3	2.1	N 6 04.2	0.1	12 Th	263 10.2	2.4	S 8 46.3	0.0	06 26
04 09	297 36.3	2.2	N 6 02.7	0.1	13 Fri	264 06.7	2.4	S 8 47.1	0.0	06 23
04 06	298 27.9	2.2	N 6 01.4	0.0	14 Sat	265 03.3	2.4	S 8 47.7	0.0	06 19
04 02	299 20.1	2.2	N 6 00.3	0.0	15 SUN	266 00.0	2.4	S 8 48.4	0.0	06 15
03 59	300 12.9	2.2	N 5 59.5	0.0	16 Mon	266 56.7	2.4	S 8 49.0	0.0	06 11
03 55	301 06.4	2.3	N 5 59.0	0.0	17 Tu	267 53.6	2.4	S 8 49.5	0.0	06 07
03 51	302 00.4	2.3	N 5 58.7	0.0	18 Wed	268 50.6	2.4	S 8 50.1	0.0	06 04
03 48	302 55.2	2.3	N 5 58.7	0.0	19 Th	269 47.6	2.4	S 8 50.6	0.0	06 00
03 44	303 50.5	2.3	N 5 59.0	0.0	20 Fri	270 44.7	2.4	S 8 51.0	0.0	05 56
03 40	304 46.5	2.4	N 5 59.6	0.0	21 Sat	271 42.0	2.4	S 8 51.5	0.0	05 52
03 37	305 43.2	2.4	N 6 00.5	0.0	22 SUN	272 39.3	2.4	S 8 51.8	0.0	05 48
03 33	306 40.6	2.4	N 6 01.6	0.1	23 Mon	273 36.8	2.4	S 8 52.2	0.0	05 45
03 29	307 38.7	2.4	N 6 03.1	0.1	24 Tu	274 34.3	2.4	S 8 52.5	0.0	05 41
03 25	308 37.5	2.5	N 6 04.8	0.1	25 Wed	275 31.9	2.4	S 8 52.8	0.0	05 37
03 21	309 36.9	2.5	N 6 06.8	0.1	26 Th	276 29.7	2.4	S 8 53.0	0.0	05 33
03 17	310 37.1	2.5	N 6 09.1	0.1	27 Fri	277 27.5	2.4	S 8 53.2	0.0	05 29
03 13	311 38.0	2.6	N 6 11.7	0.1	28 Sat	278 25.4	2.4	S 8 53.4	0.0	05 25
03 09	312 39.6	2.6	N 6 14.6	0.1	29 SUN	279 23.5	2.4	S 8 53.5	0.0	05 22
03 05	313 42.0	2.6	N 6 17.8	0.1	30 Mon	280 21.6	2.4	S 8 53.6	0.0	05 18
03 00	314 45.1	2.7	N 6 21.3	0.2	31 Tu	281 19.8	2.4	S 8 53.6	0.0	05 14

MARS, Av. Mag. –0.1
SHA January 5 187; 10 186; 15 185; 20 185; 25 185; 30 185

SATURN, Av. Mag. +0.7
SHA January 5 153; 10 152; 15 152; 20 152; 25 152; 30 152

JANUARY 2012

STARS

No.	Name	Mag	Transit h m	Dec ° '	SHA ° '
	Oh GMT January 1				
Ψ	ARIES	–	17 17	–	–
1	Alpheratz	2.1	17 26	N29 09.6	357 44.7
2	Ankaa	2.4	17 44	S42 14.6	353 16.8
3	Schedar	2.2	17 58	N56 36.5	349 41.8
4	Diphda	2.0	18 01	S17 55.3	348 56.9
5	Achernar	0.5	18 55	S57 10.8	335 27.4
6	POLARIS	2.0	20 04	N89 19.3	318 06.8
7	Hamal	2.0	19 24	N23 31.3	328 01.8
8	Acamar	3.2	20 15	S40 15.6	315 18.9
9	Menkar	2.5	20 19	N 4 08.2	314 15.9
10	Mirfak	1.8	20 42	N49 54.4	308 41.5
11	Aldebaran	0.9	21 53	N16 32.0	290 50.2
12	Rigel	0.1	22 31	S 8 11.4	281 12.7
13	Capella	0.1	22 34	N46 00.6	280 35.5
14	Bellatrix	1.6	22 42	N 6 21.5	278 32.8
15	Elnath	1.7	22 43	N28 37.0	278 13.5
16	Alnilam	1.7	22 53	S 1 11.8	275 47.1
17	Betelgeuse	0.1–1.2	23 12	N 7 24.4	271 02.0
18	Canopus	–0.7	23 40	S52 42.3	263 56.0
19	Sirius	–1.5	0 05	S16 44.2	258 34.2
20	Adhara	1.5	0 19	S28 59.5	255 12.9
21	Castor	1.6	0 55	N31 51.5	246 08.8
22	Procyon	0.4	1 00	N 5 11.4	245 00.4
23	Pollux	1.1	1 06	N27 59.6	243 28.5
24	Avior	1.9	1 42	S59 33.0	234 17.8
25	Suhail	2.2	2 28	S43 29.0	222 52.8
26	Miaplacidus	1.7	2 33	S69 46.0	221 39.0
27	Alphard	2.0	2 47	S 8 42.8	217 56.8
28	Regulus	1.4	3 28	N11 54.3	207 44.3
29	Dubhe	1.8	4 24	N61 40.8	193 52.4
30	Denebola	2.1	5 09	N14 30.0	182 34.5
31	Gienah	2.6	5 35	S17 36.6	175 53.2
32	Acrux	1.3	5 46	S63 09.8	173 10.2
33	Gacrux	1.6	5 51	S57 10.7	172 01.9
34	Mimosa	1.3	6 07	S59 45.1	167 53.0
35	Alioth	1.8	6 13	N55 53.3	166 21.4
36	Spica	1.0	6 44	S11 13.5	158 32.3
37	Alkaid	1.9	7 07	N49 14.8	152 59.7
38	Hadar	0.6	7 23	S60 25.6	148 49.4
39	Menkent	2.1	7 26	S36 25.6	148 08.8
40	Arcturus	0.0	7 35	N19 07.0	145 56.7
41	Rigil Kent	–0.3	7 59	S60 52.8	139 53.3
42	Zuben'ubi	2.8	8 10	S16 05.5	137 06.6
43	Kochab	2.1	8 09	N74 06.0	137 20.3
44	Alphecca	2.2	8 53	N26 40.3	126 12.0
45	Antares	1.0	9 48	S26 27.4	112 27.7
46	Atria	1.9	10 08	S69 02.7	107 30.9
47	Sabik	2.4	10 29	S15 44.3	102 14.0
48	Shaula	1.6	10 52	S37 06.6	96 23.6
49	Rasalhague	2.1	10 53	N12 33.1	96 07.7
50	Eltanin	2.2	11 15	N51 29.2	90 47.1
51	Kaus Aust.	1.9	11 43	S34 22.6	83 45.5
52	Vega	0.0	11 55	N38 47.7	80 40.0
53	Nunki	2.0	12 14	S26 16.8	75 59.9
54	Altair	0.8	13 09	N 8 54.1	62 09.5
55	Peacock	1.9	13 44	S56 41.7	53 21.4
56	Deneb	1.3	13 59	N45 19.6	49 32.6
57	Enif	2.4	15 02	N 9 55.9	33 48.4
58	Al Na'ir	1.7	15 26	S46 54.2	27 45.3
59	Fomalhaut	1.2	16 15	S29 33.6	15 25.3
60	Markab	2.5	16 22	N15 16.4	13 39.5

SUN AND MOON

SUN

Yr	Day of Mth	Week	Transit h m	Semi-Diam	Lat 52°N Twilight h m	Sunrise h m	Sunset h m	Twilight h m
1	1	Sun	12 03	16.3	07 28	08 08	15 59	16 39
2	2	Mon	12 04	16.3	07 28	08 08	16 00	16 40
3	3	Tu	12 04	16.3	07 27	08 08	16 01	16 41
4	4	Wed	12 05	16.3	07 27	08 08	16 02	16 42
5	5	Th	12 05	16.3	07 27	08 07	16 03	16 43
6	6	Fri	12 06	16.3	07 27	08 07	16 04	16 45
7	7	Sat	12 06	16.3	07 27	08 07	16 06	16 46
8	8	Sun	12 06	16.3	07 26	08 06	16 07	16 47
9	9	Mon	12 07	16.3	07 26	08 06	16 08	16 48
10	10	Tu	12 07	16.3	07 25	08 05	16 10	16 50
11	11	Wed	12 08	16.3	07 25	08 05	16 11	16 51
12	12	Th	12 08	16.3	07 24	08 04	16 13	16 52
13	13	Fri	12 08	16.3	07 24	08 03	16 14	16 54
14	14	Sat	12 09	16.3	07 23	08 02	16 16	16 55
15	15	Sun	12 09	16.3	07 22	08 02	16 17	16 56
16	16	Mon	12 10	16.3	07 22	08 01	16 19	16 58
17	17	Tu	12 10	16.3	07 21	08 00	16 20	16 59
18	18	Wed	12 10	16.3	07 20	07 59	16 22	17 01
19	19	Th	12 11	16.3	07 19	07 58	16 24	17 02
20	20	Fri	12 11	16.3	07 18	07 57	16 25	17 04
21	21	Sat	12 11	16.3	07 17	07 56	16 27	17 05
22	22	Sun	12 11	16.3	07 16	07 55	16 29	17 07
23	23	Mon	12 12	16.3	07 15	07 53	16 31	17 09
24	24	Tu	12 12	16.3	07 14	07 52	16 32	17 10
25	25	Wed	12 12	16.3	07 13	07 51	16 34	17 12
26	26	Th	12 12	16.3	07 12	07 50	16 36	17 13
27	27	Fri	12 13	16.3	07 11	07 48	16 38	17 15
28	28	Sat	12 13	16.3	07 10	07 47	16 39	17 17
29	29	Sun	12 13	16.3	07 08	07 45	16 41	17 18
30	30	Mon	12 13	16.3	07 07	07 44	16 43	17 20
31	31	Tu	12 13	16.3	07 06	07 42	16 45	17 22

Lat Corr to Sunrise, Sunset etc.

Lat °	Twilight h m	Sunrise h m	Sunset h m	Twilight h m
N70	+1 57	SBH	SBH	–1 56
68	+1 32	+2 34	–2 33	–1 32
66	+1 13	+1 53	–1 53	–1 13
64	+0 57	+1 25	–1 25	–0 57
62	+0 44	+1 04	–1 04	–0 44
N60	+0 32	+0 47	–0 47	–0 33
58	+0 22	+0 33	–0 32	–0 23
56	+0 14	+0 21	–0.20	–0 14
54	+0 07	+0 10	–0 09	–0 07
50	–0 06	–0 08	+0 09	+0 06
N45	–0 20	–0 26	+0 27	+0 20
40	–0 33	–0 41	+0 42	+0 32
35	–0 43	–0 54	+0 54	+0 42
30	–0 52	–1 05	+1 05	+0 52
20	–1 08	–1 24	+1 24	+1 08
N10	–1 24	–1 41	+1 41	+1 24
0	–1 39	–1 56	+1 56	+1 39
S10	–1 55	–2 13	+2 11	+1 55
20	–2 14	–2 30	+2 28	+2 14
30	–2 36	–2 49	+2 48	+2 36
S35	–2 49	–3 01	+3 00	+2 49
40	–3 06	–3 15	+3 13	+3 06
45	–3 26	–3 31	+3 29	+3 26
S50	–3 51	–3 51	+3 49	+3 52

NOTES
The corrections to sunrise etc. are for middle of January.
SBH means Sun below Horizon.

MOON

Yr	Day of Mth	Week	Age days	Transit (Upper) h m	Diff m	Semi-diam '	Hor Par '	Lat 52°N Moonrise h m	Moonset h m
1	1	Sun	08	18 09	43	14.8	54.4	11 10	00 16
2	2	Mon	09	18 52	45	14.8	54.2	11 30	01 22
3	3	Tu	10	19 37	47	14.8	54.2	11 53	02 28
4	4	Wed	11	20 24	49	14.8	54.4	12 21	03 34
5	5	Th	12	21 13	51	14.9	54.7	12 56	04 37
6	6	Fri	13	22 04	53	15.0	55.2	13 40	05 36
7	7	Sat	14	22 57	52	15.2	55.7	14 33	06 29
8	8	Sun	15	23 49	52	15.3	56.2	15 36	07 14
9	9	Mon	16	24 41	–	15.5	56.8	16 46	07 51
10	10	Tu	17	00 41	51	15.6	57.3	18 01	08 21
11	11	Wed	18	01 32	49	15.8	57.8	19 18	08 47
12	12	Th	19	02 21	49	15.9	58.2	20 36	09 09
13	13	Fri	20	03 10	49	16.0	58.6	21 54	09 30
14	14	Sat	21	03 59	50	16.0	58.8	23 13	09 51
15	15	Sun	22	04 49	52	16.1	59.1	– –	10 13
16	16	Mon	23	05 41	54	16.1	59.2	00 34	10 37
17	17	Tu	24	06 35	57	16.2	59.3	01 54	11 07
18	18	Wed	25	07 32	59	16.1	59.3	03 13	11 45
19	19	Th	26	08 31	59	16.1	59.2	04 26	12 32
20	20	Fri	27	09 30	59	16.1	58.9	05 30	13 31
21	21	Sat	28	10 29	56	16.0	58.6	06 21	14 40
22	22	Sun	29	11 25	53	15.8	58.1	07 02	15 55
23	23	Mon	00	12 18	49	15.7	57.5	07 34	17 11
24	24	Tu	01	13 07	47	15.5	56.9	07 59	18 26
25	25	Wed	02	13 54	44	15.3	56.3	08 20	19 39
26	26	Th	03	14 38	43	15.2	55.6	08 39	20 50
27	27	Fri	04	15 21	43	15.0	55.1	08 57	21 58
28	28	Sat	05	16 04	43	14.9	54.7	09 15	23 06
29	29	Sun	06	16 47	44	14.8	54.4	09 35	– –
30	30	Mon	07	17 31	46	14.8	54.2	09 56	00 12
31	31	Tu	08	18 17	47	14.8	54.3	10 22	01 18

Phases of the Moon

	d	h	m
☾ First Quarter	1	06	15
○ Full Moon	9	07	30
☽ Last Quarter	16	09	08
● New Moon	23	07	39
☾ First Quarter	31	04	10

	d	h
Apogee	2	20
Perigee	17	21
Apogee	30	18

FEBRUARY 2012

SUN AND ARIES

Wednesday, 1st February

GMT	SUN GHA ° '	SUN Dec ° '	ARIES GHA ° '
00	176 38.6	**S17 19.5**	130 37.2
02	206 38.4	**17 18.1**	160 42.1
04	236 38.2	**17 16.7**	190 47.0
06	266 38.0	**17 15.3**	220 52.0
08	296 37.9	**17 13.9**	250 56.9
10	326 37.7	**17 12.5**	281 01.8
12	356 37.5	**17 11.1**	311 06.8
14	26 37.3	**17 09.7**	341 11.7
16	56 37.2	**17 08.3**	11 16.6
18	86 37.0	**17 06.8**	41 21.5
20	116 36.8	**17 05.4**	71 26.5
22	146 36.7	**S17 04.0**	101 31.4

Thursday, 2nd February

GMT	SUN GHA ° '	SUN Dec ° '	ARIES GHA ° '
00	176 36.5	**S17 02.6**	131 36.3
02	206 36.3	**17 01.1**	161 41.3
04	236 36.2	**16 59.7**	191 46.2
06	266 36.0	**16 58.3**	221 51.1
08	296 35.9	**16 56.9**	251 56.0
10	326 35.7	**16 55.4**	282 01.0
12	356 35.5	**16 54.0**	312 05.9
14	26 35.4	**16 52.5**	342 10.8
16	56 35.2	**16 51.1**	12 15.8
18	86 35.1	**16 49.7**	42 20.7
20	116 34.9	**16 48.2**	72 25.6
22	146 34.8	**S16 46.8**	102 30.5

Friday, 3rd February

GMT	SUN GHA ° '	SUN Dec ° '	ARIES GHA ° '
00	176 34.6	**S16 45.3**	132 35.5
02	206 34.5	**16 43.9**	162 40.4
04	236 34.3	**16 42.4**	192 45.3
06	266 34.2	**16 40.9**	222 50.2
08	296 34.1	**16 39.5**	252 55.2
10	326 33.9	**16 38.0**	283 00.1
12	356 33.8	**16 36.6**	313 05.0
14	26 33.6	**16 35.1**	343 10.0
16	56 33.5	**16 33.6**	13 14.9
18	86 33.4	**16 32.2**	43 19.8
20	116 33.2	**16 30.7**	73 24.7
22	146 33.1	**S16 29.2**	103 29.7

Saturday, 4th February

GMT	SUN GHA ° '	SUN Dec ° '	ARIES GHA ° '
00	176 33.0	**S16 27.7**	133 34.6
02	206 32.8	**16 26.3**	163 39.5
04	236 32.7	**16 24.8**	193 44.5
06	266 32.6	**16 23.3**	223 49.4
08	296 32.5	**16 21.8**	253 54.3
10	326 32.3	**16 20.3**	283 59.2
12	356 32.2	**16 18.9**	314 04.2
14	26 32.1	**16 17.4**	344 09.1
16	56 32.0	**16 15.9**	14 14.0
18	86 31.9	**16 14.4**	44 19.0
20	116 31.7	**16 12.9**	74 23.9
22	146 31.6	**S16 11.4**	104 28.8

Sunday, 5th February

GMT	SUN GHA ° '	SUN Dec ° '	ARIES GHA ° '
00	176 31.5	**S16 09.9**	134 33.7
02	206 31.4	**16 08.4**	164 38.7
04	236 31.3	**16 06.9**	194 43.6
06	266 31.2	**16 05.4**	224 48.5
08	296 31.1	**16 03.9**	254 53.5
10	326 31.0	**16 02.4**	284 58.4
12	356 30.9	**16 00.9**	315 03.3
14	26 30.8	**15 59.4**	345 08.2
16	56 30.7	**15 57.8**	15 13.2
18	86 30.6	**15 56.3**	45 18.1
20	116 30.5	**15 54.8**	75 23.0
22	146 30.4	**S15 53.3**	105 28.0

Monday, 6th February

GMT	SUN GHA ° '	SUN Dec ° '	ARIES GHA ° '
00	176 30.3	**S15 51.8**	135 32.9
02	206 30.2	**15 50.3**	165 37.8
04	236 30.1	**15 48.7**	195 42.7
06	266 30.0	**15 47.2**	225 47.7
08	296 29.9	**15 45.7**	255 52.6
10	326 29.8	**15 44.1**	285 57.5
12	356 29.7	**15 42.6**	316 02.5
14	26 29.6	**15 41.1**	346 07.4
16	56 29.5	**15 39.5**	16 12.3
18	86 29.5	**15 38.0**	46 17.2
20	116 29.4	**15 36.5**	76 22.2
22	146 29.3	**S15 34.9**	106 27.1

Tuesday, 7th February

GMT	SUN GHA ° '	SUN Dec ° '	ARIES GHA ° '
00	176 29.2	**S15 33.4**	136 32.0
02	206 29.1	**15 31.8**	166 37.0
04	236 29.1	**15 30.3**	196 41.9
06	266 29.0	**15 28.7**	226 46.8
08	296 28.9	**15 27.2**	256 51.7
10	326 28.8	**15 25.6**	286 56.7
12	356 28.8	**15 24.1**	317 01.6
14	26 28.7	**15 22.5**	347 06.5
16	56 28.6	**15 21.0**	17 11.5
18	86 28.6	**15 19.4**	47 16.4
20	116 28.5	**15 17.8**	77 21.3
22	146 28.4	**S15 16.3**	107 26.2

Wednesday, 8th February

GMT	SUN GHA ° '	SUN Dec ° '	ARIES GHA ° '
00	176 28.4	**S15 14.7**	137 31.2
02	206 28.3	**15 13.1**	167 36.1
04	236 28.3	**15 11.6**	197 41.0
06	266 28.2	**15 10.0**	227 45.9
08	296 28.1	**15 08.4**	257 50.9
10	326 28.1	**15 06.9**	287 55.8
12	356 28.0	**15 05.3**	318 00.7
14	26 28.0	**15 03.7**	348 05.7
16	56 27.9	**15 02.1**	18 10.6
18	86 27.9	**15 00.5**	48 15.5
20	116 27.8	**14 59.0**	78 20.4
22	146 27.8	**S14 57.4**	108 25.4

Thursday, 9th February

GMT	SUN GHA ° '	SUN Dec ° '	ARIES GHA ° '
00	176 27.7	**S14 55.8**	138 30.3
02	206 27.7	**14 54.2**	168 35.2
04	236 27.6	**14 52.6**	198 40.2
06	266 27.6	**14 51.0**	228 45.1
08	296 27.6	**14 49.4**	258 50.0
10	326 27.5	**14 47.8**	288 54.9
12	356 27.5	**14 46.2**	318 59.9
14	26 27.4	**14 44.6**	349 04.8
16	56 27.4	**14 43.0**	19 09.7
18	86 27.4	**14 41.4**	49 14.7
20	116 27.3	**14 39.8**	79 19.6
22	146 27.3	**S14 38.2**	109 24.5

Friday, 10th February

GMT	SUN GHA ° '	SUN Dec ° '	ARIES GHA ° '
00	176 27.3	**S14 36.6**	139 29.4
02	206 27.2	**14 35.0**	169 34.4
04	236 27.2	**14 33.4**	199 39.3
06	266 27.2	**14 31.8**	229 44.2
08	296 27.2	**14 30.2**	259 49.2
10	326 27.1	**14 28.6**	289 54.1
12	356 27.1	**14 26.9**	319 59.0
14	26 27.1	**14 25.3**	350 03.9
16	56 27.1	**14 23.7**	20 08.9
18	86 27.1	**14 22.1**	50 13.8
20	116 27.1	**14 20.5**	80 18.7
22	146 27.0	**S14 18.8**	110 23.6

Saturday, 11th February

GMT	SUN GHA ° '	SUN Dec ° '	ARIES GHA ° '
00	176 27.0	**S14 17.2**	140 28.6
02	206 27.0	**14 15.6**	170 33.5
04	236 27.0	**14 13.9**	200 38.4
06	266 27.0	**14 12.3**	230 43.4
08	296 27.0	**14 10.7**	260 48.3
10	326 27.0	**14 09.0**	290 53.2
12	356 27.0	**14 07.4**	320 58.1
14	26 27.0	**14 05.8**	351 03.1
16	56 27.0	**14 04.1**	21 08.0
18	86 27.0	**14 02.5**	51 12.9
20	116 27.0	**14 00.8**	81 17.9
22	146 27.0	**S13 59.2**	111 22.8

Sunday, 12th February

GMT	SUN GHA ° '	SUN Dec ° '	ARIES GHA ° '
00	176 27.0	**S13 57.5**	141 27.7
02	206 27.0	**13 55.9**	171 32.6
04	236 27.0	**13 54.2**	201 37.6
06	266 27.0	**13 52.6**	231 42.5
08	296 27.0	**13 50.9**	261 47.4
10	326 27.0	**13 49.3**	291 52.4
12	356 27.0	**13 47.6**	321 57.3
14	26 27.0	**13 46.0**	352 02.2
16	56 27.0	**13 44.3**	22 07.1
18	86 27.0	**13 42.7**	52 12.1
20	116 27.0	**13 41.0**	82 17.0
22	146 27.1	**S13 39.3**	112 21.9

Monday, 13th February

GMT	SUN GHA ° '	SUN Dec ° '	ARIES GHA ° '
00	176 27.1	**S13 37.7**	142 26.9
02	206 27.1	**13 36.0**	172 31.8
04	236 27.1	**13 34.3**	202 36.7
06	266 27.1	**13 32.6**	232 41.6
08	296 27.2	**13 31.0**	262 46.6
10	326 27.2	**13 29.3**	292 51.5
12	356 27.2	**13 27.6**	322 56.4
14	26 27.2	**13 25.9**	353 01.3
16	56 27.3	**13 24.3**	23 06.3
18	86 27.3	**13 22.6**	53 11.2
20	116 27.3	**13 20.9**	83 16.1
22	146 27.3	**S13 19.2**	113 21.1

Tuesday, 14th February

GMT	SUN GHA ° '	SUN Dec ° '	ARIES GHA ° '
00	176 27.4	**S13 17.5**	143 26.0
02	206 27.4	**13 15.9**	173 30.9
04	236 27.4	**13 14.2**	203 35.8
06	266 27.5	**13 12.5**	233 40.8
08	296 27.5	**13 10.8**	263 45.7
10	326 27.6	**13 09.1**	293 50.6
12	356 27.6	**13 07.4**	323 55.6
14	26 27.6	**13 05.7**	354 00.5
16	56 27.7	**13 04.0**	24 05.4
18	86 27.7	**13 02.3**	54 10.3
20	116 27.8	**13 00.6**	84 15.3
22	146 27.8	**S12 58.9**	114 20.2

Wednesday, 15th February

GMT	SUN GHA ° '	SUN Dec ° '	ARIES GHA ° '
00	176 27.9	**S12 57.2**	144 25.1
02	206 27.9	**12 55.5**	174 30.1
04	236 28.0	**12 53.8**	204 35.0
06	266 28.0	**12 52.1**	234 39.9
08	296 28.1	**12 50.4**	264 44.8
10	326 28.1	**12 48.7**	294 49.8
12	356 28.2	**12 47.0**	324 54.7
14	26 28.2	**12 45.2**	354 59.6
16	56 28.3	**12 43.5**	25 04.6
18	86 28.3	**12 41.8**	55 09.5
20	116 28.4	**12 40.1**	85 14.4
22	146 28.5	**S12 38.4**	115 19.3

Thursday, 16th February

GMT	SUN GHA ° '	SUN Dec ° '	ARIES GHA ° '
00	176 28.5	**S12 36.7**	145 24.3
02	206 28.6	**12 34.9**	175 29.2
04	236 28.6	**12 33.2**	205 34.1
06	266 28.7	**12 31.5**	235 39.1
08	296 28.8	**12 29.8**	265 44.0
10	326 28.8	**12 28.0**	295 48.9
12	356 28.9	**12 26.3**	325 53.8
14	26 29.0	**12 24.6**	355 58.8
16	56 29.1	**12 22.8**	26 03.7
18	86 29.1	**12 21.1**	56 08.6
20	116 29.2	**12 19.4**	86 13.6
22	146 29.3	**S12 17.6**	116 18.5

Friday, 17th February

GMT	SUN GHA ° '	SUN Dec ° '	ARIES GHA ° '
00	176 29.4	**S12 15.9**	146 23.4
02	206 29.4	**12 14.2**	176 28.3
04	236 29.5	**12 12.4**	206 33.3
06	266 29.6	**12 10.7**	236 38.2
08	296 29.7	**12 08.9**	266 43.1
10	326 29.8	**12 07.2**	296 48.1
12	356 29.8	**12 05.5**	326 53.0
14	26 29.9	**12 03.7**	356 57.9
16	56 30.0	**12 02.0**	27 02.8
18	86 30.1	**12 00.2**	57 07.8
20	116 30.2	**11 58.5**	87 12.7
22	146 30.3	**S11 56.7**	117 17.6

Saturday, 18th February

GMT	SUN GHA ° '	SUN Dec ° '	ARIES GHA ° '
00	176 30.4	**S11 55.0**	147 22.5
02	206 30.5	**11 53.2**	177 27.5
04	236 30.6	**11 51.4**	207 32.4
06	266 30.6	**11 49.7**	237 37.3
08	296 30.7	**11 47.9**	267 42.3
10	326 30.8	**11 46.2**	297 47.2
12	356 30.9	**11 44.4**	327 52.1
14	26 31.0	**11 42.6**	357 57.0
16	56 31.1	**11 40.9**	28 02.0
18	86 31.2	**11 39.1**	58 06.9
20	116 31.3	**11 37.4**	88 11.8
22	146 31.4	**S11 35.6**	118 16.8

Sunday, 19th February

GMT	SUN GHA ° '	SUN Dec ° '	ARIES GHA ° '
00	176 31.5	**S11 33.8**	148 21.7
02	206 31.7	**11 32.0**	178 26.6
04	236 31.8	**11 30.3**	208 31.5
06	266 31.9	**11 28.5**	238 36.5
08	296 32.0	**11 26.7**	268 41.4
10	326 32.1	**11 25.0**	298 46.3
12	356 32.2	**11 23.2**	328 51.3
14	26 32.3	**11 21.4**	358 56.2
16	56 32.4	**11 19.6**	29 01.1
18	86 32.5	**11 17.8**	59 06.0
20	116 32.7	**11 16.1**	89 11.0
22	146 32.8	**S11 14.3**	119 15.9

Monday, 20th February

GMT	SUN GHA ° '	SUN Dec ° '	ARIES GHA ° '
00	176 32.9	**S11 12.5**	149 20.8
02	206 33.0	**11 10.7**	179 25.8
04	236 33.1	**11 08.9**	209 30.7
06	266 33.3	**11 07.1**	239 35.6
08	296 33.4	**11 05.3**	269 40.5
10	326 33.5	**11 03.6**	299 45.5
12	356 33.6	**11 01.8**	329 50.4
14	26 33.8	**11 00.0**	359 55.3
16	56 33.9	**10 58.2**	30 00.3
18	86 34.0	**10 56.4**	60 05.2
20	116 34.1	**10 54.6**	90 10.1
22	146 34.3	**S10 52.8**	120 15.0

Tuesday, 21st February

GMT	SUN GHA ° '	SUN Dec ° '	ARIES GHA ° '
00	176 34.4	**S10 51.0**	150 20.0
02	206 34.5	**10 49.2**	180 24.9
04	236 34.7	**10 47.4**	210 29.8
06	266 34.8	**10 45.6**	240 34.8
08	296 34.9	**10 43.8**	270 39.7
10	326 35.1	**10 42.0**	300 44.6
12	356 35.2	**10 40.2**	330 49.5
14	26 35.4	**10 38.4**	0 54.5
16	56 35.5	**10 36.6**	30 59.4
18	86 35.7	**10 34.8**	61 04.3
20	116 35.8	**10 33.0**	91 09.2
22	146 35.9	**S10 31.1**	121 14.2

Wednesday, 22nd February

GMT	SUN GHA ° '	SUN Dec ° '	ARIES GHA ° '
00	176 36.1	**S10 29.3**	151 19.1
02	206 36.2	**10 27.5**	181 24.0
04	236 36.4	**10 25.7**	211 29.0
06	266 36.5	**10 23.9**	241 33.9
08	296 36.7	**10 22.1**	271 38.8
10	326 36.8	**10 20.3**	301 43.7
12	356 37.0	**10 18.4**	331 48.7
14	26 37.1	**10 16.6**	1 53.6
16	56 37.3	**10 14.8**	31 58.5
18	86 37.4	**10 13.0**	62 03.5
20	116 37.6	**10 11.2**	92 08.4
22	146 37.8	**S10 09.3**	122 13.3

Thursday, 23rd February

GMT	SUN GHA ° '	SUN Dec ° '	ARIES GHA ° '
00	176 37.9	**S10 07.5**	152 18.2
02	206 38.1	**10 05.7**	182 23.2
04	236 38.2	**10 03.8**	212 28.1
06	266 38.4	**10 02.0**	242 33.0
08	296 38.6	**10 00.2**	272 38.0
10	326 38.7	**9 58.4**	302 42.9
12	356 38.9	**9 56.5**	332 47.8
14	26 39.1	**9 54.7**	2 52.7
16	56 39.2	**9 52.9**	32 57.7
18	86 39.4	**9 51.0**	63 02.6
20	116 39.6	**9 49.2**	93 07.5
22	146 39.7	**S 9 47.4**	123 12.4

Friday, 24th February

GMT	SUN GHA ° '	SUN Dec ° '	ARIES GHA ° '
00	176 39.9	**S 9 45.5**	153 17.4
02	206 40.1	**9 43.7**	183 22.3
04	236 40.3	**9 41.8**	213 27.2
06	266 40.4	**9 40.0**	243 32.2
08	296 40.6	**9 38.2**	273 37.1
10	326 40.8	**9 36.3**	303 42.0
12	356 41.0	**9 34.5**	333 46.9
14	26 41.1	**9 32.6**	3 51.9
16	56 41.3	**9 30.8**	33 56.8
18	86 41.5	**9 28.9**	64 01.7
20	116 41.7	**9 27.1**	94 06.7
22	146 41.9	**S 9 25.2**	124 11.6

Saturday, 25th February

GMT	SUN GHA ° '	SUN Dec ° '	ARIES GHA ° '
00	176 42.1	**S 9 23.4**	154 16.5
02	206 42.2	**9 21.5**	184 21.4
04	236 42.4	**9 19.7**	214 26.4
06	266 42.6	**9 17.8**	244 31.3
08	296 42.8	**9 16.0**	274 36.2
10	326 43.0	**9 14.1**	304 41.2
12	356 43.2	**9 12.3**	334 46.1
14	26 43.4	**9 10.4**	4 51.0
16	56 43.6	**9 08.6**	34 55.9
18	86 43.8	**9 06.7**	65 00.9
20	116 44.0	**9 04.8**	95 05.8
22	146 44.2	**S 9 03.0**	125 10.7

Sunday, 26th February

GMT	SUN GHA ° '	SUN Dec ° '	ARIES GHA ° '
00	176 44.4	**S 9 01.1**	155 15.7
02	206 44.6	**8 59.3**	185 20.6
04	236 44.8	**8 57.4**	215 25.5
06	266 45.0	**8 55.5**	245 30.4
08	296 45.2	**8 53.7**	275 35.4
10	326 45.4	**8 51.8**	305 40.3
12	356 45.6	**8 49.9**	335 45.2
14	26 45.8	**8 48.1**	5 50.1
16	56 46.0	**8 46.2**	35 55.1
18	86 46.2	**8 44.3**	66 00.0
20	116 46.4	**8 42.5**	96 04.9
22	146 46.6	**S 8 40.6**	126 09.9

Monday, 27th February

GMT	SUN GHA ° '	SUN Dec ° '	ARIES GHA ° '
00	176 46.8	**S 8 38.7**	156 14.8
02	206 47.0	**8 36.8**	186 19.7
04	236 47.2	**8 35.0**	216 24.6
06	266 47.4	**8 33.1**	246 29.6
08	296 47.6	**8 31.2**	276 34.5
10	326 47.9	**8 29.3**	306 39.4
12	356 48.1	**8 27.5**	336 44.4
14	26 48.3	**8 25.6**	6 49.3
16	56 48.5	**8 23.7**	36 54.2
18	86 48.7	**8 21.8**	66 59.1
20	116 48.9	**8 20.0**	97 04.1
22	146 49.2	**S 8 18.1**	127 09.0

Tuesday, 28th February

GMT	SUN GHA ° '	SUN Dec ° '	ARIES GHA ° '
00	176 49.4	**S 8 16.2**	157 13.9
02	206 49.6	**8 14.3**	187 18.9
04	236 49.8	**8 12.4**	217 23.8
06	266 50.0	**8 10.5**	247 28.7
08	296 50.3	**8 08.7**	277 33.6
10	326 50.5	**8 06.8**	307 38.6
12	356 50.7	**8 04.9**	337 43.5
14	26 51.0	**8 03.0**	7 48.4
16	56 51.2	**8 01.1**	37 53.4
18	86 51.4	**7 59.2**	67 58.3
20	116 51.6	**7 57.3**	98 03.2
22	146 51.9	**S 7 55.4**	128 08.1

Wednesday, 29th February

GMT	SUN GHA ° '	SUN Dec ° '	ARIES GHA ° '
00	176 52.1	**S 7 53.5**	158 13.1
02	206 52.3	**7 51.6**	188 18.0
04	236 52.6	**7 49.8**	218 22.9
06	266 52.8	**7 47.9**	248 27.8
08	296 53.0	**7 46.0**	278 32.8
10	326 53.3	**7 44.1**	308 37.7
12	356 53.5	**7 42.2**	338 42.6
14	26 53.8	**7 40.3**	8 47.6
16	56 54.0	**7 38.4**	38 52.5
18	86 54.2	**7 36.5**	68 57.4
20	116 54.5	**7 34.6**	99 02.3
22	146 54.7	**S 7 32.7**	129 07.3

MOON

Day	GMT hr	GHA ° '	Mean Var/hr 14°+	Dec ° '	Mean Var/hr '
1 Wed	0	83 08.6	31.2	N19 42.5	5.3
	6	170 16.5	31.0	N20 14.6	4.7
	12	257 22.6	30.7	N20 43.3	4.2
	18	344 27.0	30.4	N21 08.8	3.6
2 Th	0	71 29.8	30.2	N21 30.7	3.0
	6	158 30.8	29.9	N21 49.0	2.4
	12	245 30.4	29.7	N22 03.6	1.7
	18	332 28.4	29.5	N22 14.4	1.1
3 Fri	0	59 25.0	29.2	N22 21.2	0.4
	6	146 20.3	29.1	N22 24.1	0.3
	12	233 14.5	28.8	N22 22.8	1.0
	18	320 07.6	28.7	N22 17.4	1.7
4 Sat	0	46 59.9	28.6	N22 07.8	2.3
	6	133 51.5	28.5	N21 54.0	3.1
	12	220 42.5	28.4	N21 35.9	3.8
	18	307 33.1	28.3	N21 13.5	4.5
5 Sun	0	34 23.5	28.4	N20 46.9	5.2
	6	121 13.8	28.4	N20 16.1	5.9
	12	208 04.2	28.4	N19 41.1	6.5
	18	294 54.7	28.5	N19 02.0	7.2
6 Mon	0	21 45.5	28.6	N18 18.9	7.9
	6	108 36.7	28.6	N17 31.9	8.5
	12	195 28.4	28.7	N16 41.2	9.1
	18	282 20.7	28.8	N15 46.9	9.7
7 Tu	0	9 13.5	28.9	N14 49.1	10.2
	6	96 07.0	29.1	N13 48.0	10.7
	12	183 01.0	29.1	N12 43.9	11.2
	18	269 55.7	29.2	N11 36.9	11.7
8 Wed	0	356 51.0	29.3	N10 27.2	12.0
	6	83 46.8	29.3	N 9 15.2	12.4
	12	170 43.0	29.4	N 8 01.1	12.7
	18	257 39.6	29.5	N 6 45.0	12.9
9 Th	0	344 36.5	29.5	N 5 27.3	13.2
	6	71 33.5	29.5	N 4 08.3	13.4
	12	158 30.6	29.5	N 2 48.2	13.5
	18	245 27.6	29.4	N 1 27.4	13.6
10 Fri	0	332 24.3	29.4	N 0 06.1	13.6
	6	59 20.6	29.3	S 1 15.4	13.5
	12	146 16.4	29.2	S 2 36.7	13.5
	18	233 11.5	29.0	S 3 57.6	13.3
11 Sat	0	320 05.8	28.9	S 5 17.7	13.2
	6	46 59.2	28.7	S 6 36.9	13.0
	12	133 51.4	28.5	S 7 54.6	12.6
	18	220 42.3	28.2	S 9 10.7	12.4
12 Sun	0	307 31.9	28.0	S10 24.8	11.9
	6	34 20.1	27.8	S11 36.7	11.6
	12	121 06.6	27.5	S12 46.1	11.1
	18	207 51.6	27.2	S13 52.6	10.5
13 Mon	0	294 34.9	26.9	S14 56.0	9.9
	6	21 16.5	26.7	S15 56.1	9.4
	12	107 56.4	26.3	S16 52.5	8.7
	18	194 34.6	26.1	S17 45.0	8.0
14 Tu	0	281 11.3	25.9	S18 33.5	7.3
	6	7 46.4	25.6	S19 17.6	6.6
	12	94 20.2	25.4	S19 57.3	5.8
	18	180 52.8	25.3	S20 32.2	5.0
15 Wed	0	267 24.4	25.1	S21 02.4	4.1
	6	353 55.2	25.0	S21 27.6	3.3
	12	80 25.4	25.0	S21 47.8	2.5
	18	166 55.2	25.0	S22 02.8	1.6
16 Th	0	253 25.1	25.0	S22 12.7	0.7
	6	339 55.1	25.1	S22 17.5	0.1
	12	66 25.7	25.3	S22 17.1	1.0
	18	152 57.1	25.5	S22 11.6	1.8

Day	GMT hr	GHA ° '	Mean Var/hr 14°+	Dec ° '	Mean Var/hr '
17 Fri	0	239 29.6	25.6	S22 01.0	2.6
	6	326 03.5	26.0	S21 45.6	3.5
	12	52 38.9	26.3	S21 25.4	4.2
	18	139 16.1	26.6	S21 00.5	5.0
18 Sat	0	225 55.3	26.9	S20 31.2	5.7
	6	312 36.6	27.3	S19 57.6	6.3
	12	39 20.3	27.7	S19 19.9	7.0
	18	126 06.4	28.2	S18 38.3	7.6
19 Sun	0	212 54.9	28.5	S17 53.1	8.2
	6	299 46.0	29.0	S17 04.6	8.7
	12	26 39.7	29.4	S16 12.9	9.2
	18	113 35.8	29.8	S15 18.2	9.6
20 Mon	0	200 34.5	30.2	S14 21.0	10.0
	6	287 35.6	30.6	S13 21.3	10.3
	12	14 39.2	31.0	S12 19.4	10.7
	18	101 45.0	31.4	S11 15.7	10.9
21 Tu	0	188 53.0	31.7	S10 10.2	11.2
	6	276 03.1	32.1	S 9 03.3	11.3
	12	3 15.1	32.3	S 7 55.2	11.5
	18	90 28.9	32.6	S 6 46.1	11.7
22 Wed	0	177 44.4	32.9	S 5 36.3	11.7
	6	265 01.4	33.1	S 4 25.8	11.8
	12	352 19.7	33.3	S 3 15.0	11.8
	18	79 39.3	33.4	S 2 04.0	11.8
23 Th	0	166 59.9	33.6	S 0 53.1	11.8
	6	254 21.4	33.8	N 0 17.7	11.7
	12	341 43.6	33.8	N 1 28.1	11.6
	18	69 06.5	33.9	N 2 37.9	11.5
24 Fri	0	156 29.7	33.9	N 3 47.1	11.4
	6	243 53.2	33.9	N 4 55.3	11.2
	12	331 16.8	33.9	N 6 02.6	11.0
	18	58 40.4	33.9	N 7 08.7	10.8
25 Sat	0	146 03.8	33.8	N 8 13.4	10.5
	6	233 26.9	33.8	N 9 16.7	10.3
	12	320 49.6	33.7	N10 18.4	9.9
	18	48 11.7	33.5	N11 18.4	9.7
26 Sun	0	135 33.1	33.4	N12 16.5	9.3
	6	222 53.7	33.3	N13 12.7	8.9
	12	310 13.4	33.1	N14 06.7	8.6
	18	37 32.1	32.9	N14 58.5	8.2
27 Mon	0	124 49.7	32.7	N15 47.9	7.8
	6	212 06.2	32.5	N16 34.9	7.3
	12	299 21.3	32.3	N17 19.2	6.9
	18	26 35.2	32.1	N18 00.8	6.4
28 Tu	0	113 47.7	31.8	N18 39.6	6.0
	6	200 58.9	31.6	N19 15.5	5.4
	12	288 08.6	31.3	N19 48.3	4.9
	18	15 16.9	31.1	N20 17.9	4.3
29 Wed	0	102 23.8	30.9	N20 44.2	3.8
	6	189 29.4	30.7	N21 07.2	3.2
	12	276 33.5	30.4	N21 26.7	2.6
	18	3 36.3	30.2	N21 42.6	2.0

PLANETS

VENUS					JUPITER					
Mer Pass h m	GHA ° '	Mean Var/hr 14°+	Dec ° '	Mean Var/hr '	Day	GHA ° '	Mean Var/hr 15°+	Dec ° '	Mean Var/hr '	Mer Pass h m
14 48	138 09.4	59.8	S 4 15.0	1.3	1 Wed	99 48.0	2.2	N11 22.5	0.1	17 18
14 48	138 03.9	59.8	S 3 43.9	1.3	2 Th	100 40.4	2.2	N11 25.2	0.1	17 15
14 48	137 58.6	59.8	S 3 12.8	1.3	3 Fri	101 32.7	2.2	N11 28.0	0.1	17 11
14 49	137 53.6	59.8	S 2 41.5	1.3	4 Sat	102 24.8	2.2	N11 30.7	0.1	17 08
14 49	137 48.7	59.8	S 2 10.2	1.3	5 SUN	103 16.7	2.2	N11 33.6	0.1	17 04
14 49	137 43.9	59.8	S 1 38.8	1.3	6 Mon	104 08.5	2.2	N11 36.4	0.1	17 01
14 50	137 39.3	59.8	S 1 07.4	1.3	7 Tu	105 00.1	2.1	N11 39.4	0.1	16 58
14 50	137 34.9	59.8	S 0 36.0	1.3	8 Wed	105 51.7	2.1	N11 42.3	0.1	16 54
14 50	137 30.7	59.8	S 0 04.5	1.3	9 Th	106 43.0	2.1	N11 45.3	0.1	16 51
14 50	137 26.5	59.8	N 0 26.9	1.3	10 Fri	107 34.2	2.1	N11 48.4	0.1	16 47
14 51	137 22.5	59.8	N 0 58.4	1.3	11 Sat	108 25.3	2.1	N11 51.5	0.1	16 44
14 51	137 18.6	59.8	N 1 29.8	1.3	12 SUN	109 16.3	2.1	N11 54.6	0.1	16 41
14 51	137 14.9	59.8	N 2 01.2	1.3	13 Mon	110 07.1	2.1	N11 57.8	0.1	16 37
14 51	137 11.2	59.8	N 2 32.5	1.3	14 Tu	110 57.7	2.1	N12 01.0	0.1	16 34
14 52	137 07.6	59.9	N 3 03.8	1.3	15 Wed	111 48.3	2.1	N12 04.2	0.1	16 30
14 52	137 04.2	59.9	N 3 35.1	1.3	16 Th	112 38.7	2.1	N12 07.5	0.1	16 27
14 52	137 00.8	59.9	N 4 06.3	1.3	17 Fri	113 29.0	2.1	N12 10.8	0.1	16 24
14 52	136 57.4	59.9	N 4 37.3	1.3	18 Sat	114 19.1	2.1	N12 14.2	0.1	16 20
14 53	136 54.2	59.9	N 5 08.3	1.3	19 SUN	115 09.1	2.1	N12 17.6	0.1	16 17
14 53	136 51.0	59.9	N 5 39.2	1.3	20 Mon	115 59.0	2.1	N12 21.0	0.1	16 14
14 53	136 47.9	59.9	N 6 09.9	1.3	21 Tu	116 48.8	2.1	N12 24.5	0.1	16 11
14 53	136 44.8	59.9	N 6 40.5	1.3	22 Wed	117 38.5	2.1	N12 28.0	0.1	16 07
14 53	136 41.8	59.9	N 7 11.0	1.3	23 Th	118 28.0	2.1	N12 31.5	0.2	16 04
14 54	136 38.8	59.9	N 7 41.3	1.3	24 Fri	119 17.4	2.1	N12 35.1	0.1	16 01
14 54	136 35.8	59.9	N 8 11.5	1.2	25 Sat	120 06.7	2.0	N12 38.6	0.2	15 57
14 54	136 32.9	59.9	N 8 41.4	1.2	26 SUN	120 55.9	2.0	N12 42.3	0.2	15 54
14 54	136 30.0	59.9	N 9 11.2	1.2	27 Mon	121 44.9	2.0	N12 45.9	0.2	15 51
14 54	136 27.2	59.9	N 9 40.8	1.2	28 Tu	122 33.9	2.0	N12 49.6	0.2	15 48
14 54	136 24.4	59.9	N10 10.2	1.2	29 Wed	123 22.7	2.0	N12 53.2	0.2	15 44

VENUS, Av. Mag. –4.2
SHA February 5 3; 10 358; 15 353; 20 348; 25 342; 29 338

JUPITER, Av. Mag. –2.3
SHA February 5 329; 10 328; 15 327; 20 327; 25 326; 29 325

MARS					SATURN					
Mer Pass h m	GHA ° '	Mean Var/hr 15°+	Dec ° '	Mean Var/hr '	Day	GHA ° '	Mean Var/hr 15°+	Dec ° '	Mean Var/hr '	Mer Pass h m
02 56	315 48.9	2.7	N 6 25.1	0.2	1 Wed	282 18.2	2.4	S 8 53.6	0.0	05 10
02 52	316 53.5	2.7	N 6 29.1	0.2	2 Th	283 16.6	2.4	S 8 53.6	0.0	05 06
02 48	317 58.8	2.8	N 6 33.4	0.2	3 Fri	284 15.1	2.4	S 8 53.5	0.0	05 02
02 43	319 04.8	2.8	N 6 38.0	0.2	4 Sat	285 13.8	2.4	S 8 53.4	0.0	04 58
02 39	320 11.6	2.8	N 6 42.9	0.2	5 SUN	286 12.5	2.4	S 8 53.3	0.0	04 54
02 34	321 19.1	2.8	N 6 48.1	0.2	6 Mon	287 11.3	2.5	S 8 53.1	0.0	04 50
02 30	322 27.4	2.9	N 6 53.5	0.2	7 Tu	288 10.3	2.5	S 8 52.9	0.0	04 47
02 25	323 36.3	2.9	N 6 59.2	0.3	8 Wed	289 09.3	2.5	S 8 52.7	0.0	04 43
02 20	324 46.1	2.9	N 7 05.2	0.3	9 Th	290 08.5	2.5	S 8 52.4	0.0	04 39
02 16	325 56.5	3.0	N 7 11.4	0.3	10 Fri	291 07.7	2.5	S 8 52.1	0.0	04 35
02 11	327 07.7	3.0	N 7 17.9	0.3	11 Sat	292 07.0	2.5	S 8 51.7	0.0	04 31
02 06	328 19.6	3.0	N 7 24.6	0.3	12 SUN	293 06.5	2.5	S 8 51.3	0.0	04 27
02 01	329 32.1	3.1	N 7 31.5	0.3	13 Mon	294 06.0	2.5	S 8 50.9	0.0	04 23
01 57	330 45.4	3.1	N 7 38.6	0.3	14 Tu	295 05.6	2.5	S 8 50.4	0.0	04 19
01 52	331 59.4	3.1	N 7 46.0	0.3	15 Wed	296 05.4	2.5	S 8 49.9	0.0	04 15
01 47	333 14.0	3.1	N 7 53.6	0.3	16 Th	297 05.2	2.5	S 8 49.4	0.0	04 11
01 42	334 29.3	3.2	N 8 01.3	0.3	17 Fri	298 05.1	2.5	S 8 48.9	0.0	04 07
01 37	335 45.2	3.2	N 8 09.2	0.3	18 Sat	299 05.2	2.5	S 8 48.3	0.0	04 03
01 32	337 01.7	3.2	N 8 17.3	0.3	19 SUN	300 05.3	2.5	S 8 47.6	0.0	03 59
01 26	338 18.7	3.2	N 8 25.6	0.3	20 Mon	301 05.5	2.5	S 8 46.9	0.0	03 55
01 21	339 36.4	3.3	N 8 33.9	0.4	21 Tu	302 05.8	2.5	S 8 46.2	0.0	03 51
01 16	340 54.5	3.3	N 8 42.4	0.4	22 Wed	303 06.3	2.5	S 8 45.5	0.0	03 47
01 11	342 13.2	3.3	N 8 51.0	0.4	23 Th	304 06.8	2.5	S 8 44.7	0.0	03 43
01 06	343 32.3	3.3	N 8 59.6	0.4	24 Fri	305 07.4	2.5	S 8 43.9	0.0	03 39
01 00	344 51.8	3.3	N 9 08.4	0.4	25 Sat	306 08.1	2.5	S 8 43.1	0.0	03 35
00 55	346 11.8	3.3	N 9 17.1	0.4	26 SUN	307 08.9	2.5	S 8 42.2	0.0	03 31
00 50	347 32.1	3.4	N 9 25.9	0.4	27 Mon	308 09.8	2.5	S 8 41.3	0.0	03 27
00 44	348 52.7	3.4	N 9 34.7	0.4	28 Tu	309 10.8	2.5	S 8 40.4	0.0	03 23
00 39	350 13.5	3.4	N 9 43.5	0.4	29 Wed	310 11.8	2.6	S 8 39.4	0.0	03 19

MARS, Av. Mag. –0.9
SHA February 5 186; 10 186; 15 188; 20 189; 25 191; 29 192

SATURN, Av. Mag. +0.5
SHA February 5 152; 10 152; 15 152; 20 152; 25 152; 29 152

STARS

No.	Name	Mag	Transit h m	Dec ° '	SHA ° '
	Oh GMT February 1				
Ψ	**ARIES**	–	15 15	–	–
1	**Alpheratz**	2.1	15 24	N29 09.6	357 44.7
2	**Ankaa**	2.4	15 42	S42 14.5	353 16.9
3	**Schedar**	2.2	15 56	N56 36.5	349 42.0
4	**Diphda**	2.0	15 59	S17 55.3	348 57.0
5	**Achernar**	0.5	16 53	S57 10.7	335 27.7
6	**POLARIS**	2.0	18 01	N89 19.3	318 19.2
7	**Hamal**	2.0	17 23	N23 31.2	328 01.9
8	**Acamar**	3.2	18 13	S40 15.6	315 19.1
9	**Menkar**	2.5	18 17	N 4 08.2	314 16.0
10	**Mirfak**	1.8	18 40	N49 54.4	308 41.7
11	**Aldebaran**	0.9	19 51	N16 31.9	290 50.3
12	**Rigel**	0.1	20 29	S 8 11.5	281 12.8
13	**Capella**	0.1	20 32	N46 00.7	280 35.6
14	**Bellatrix**	1.6	20 40	N 6 21.5	278 32.8
15	**Elnath**	1.7	20 41	N28 37.0	278 13.6
16	**Alnilam**	1.7	20 51	S 1 11.9	275 47.1
17	**Betelgeuse**	0.1–1.2	21 10	N 7 24.4	271 02.1
18	**Canopus**	–0.7	21 38	S52 42.5	263 56.2
19	**Sirius**	–1.5	22 00	S16 44.3	258 34.3
20	**Adhara**	1.5	22 13	S28 59.6	255 13.0
21	**Castor**	1.6	22 49	N31 51.5	246 08.8
22	**Procyon**	0.4	22 54	N 5 11.4	245 00.4
23	**Pollux**	1.1	23 00	N27 59.6	243 28.5
24	**Avior**	1.9	23 36	S59 33.2	234 17.8
25	**Suhail**	2.2	0 26	S43 29.1	222 52.7
26	**Miaplacidus**	1.7	0 31	S69 46.2	221 39.0
27	**Alphard**	2.0	0 46	S 8 42.9	217 56.7
28	**Regulus**	1.4	1 26	N11 54.2	207 44.2
29	**Dubhe**	1.8	2 22	N61 40.8	193 52.1
30	**Denebola**	2.1	3 07	N14 30.0	182 34.3
31	**Gienah**	2.6	3 33	S17 36.7	175 53.0
32	**Acrux**	1.3	3 44	S63 09.9	173 09.8
33	**Gacrux**	1.6	3 49	S57 10.8	172 01.5
34	**Mimosa**	1.3	4 05	S59 45.2	167 52.6
35	**Alioth**	1.8	4 11	N55 53.3	166 21.1
36	**Spica**	1.0	4 43	S11 13.6	158 32.1
37	**Alkaid**	1.9	5 05	N49 14.8	152 59.4
38	**Hadar**	0.6	5 21	S60 25.7	148 49.0
39	**Menkent**	2.1	5 24	S36 25.7	148 08.5
40	**Arcturus**	0.0	5 33	N19 06.9	145 56.4
41	**Rigil Kent**	–0.3	5 57	S60 52.9	139 52.8
42	**Zuben'ubi**	2.8	6 08	S16 05.6	137 06.4
43	**Kochab**	2.1	6 07	N74 06.0	137 19.7
44	**Alphecca**	2.2	6 52	N26 40.2	126 11.8
45	**Antares**	1.0	7 46	S26 27.4	112 27.5
46	**Atria**	1.9	8 06	S69 02.6	107 30.4
47	**Sabik**	2.4	8 27	S15 44.3	102 13.7
48	**Shaula**	1.6	8 51	S37 06.6	96 23.4
49	**Rasalhague**	2.1	8 52	N12 33.0	96 07.5
50	**Eltanin**	2.2	9 13	N51 29.0	90 46.9
51	**Kaus Aust.**	1.9	9 41	S34 22.6	83 45.3
52	**Vega**	0.0	9 53	N38 47.6	80 39.9
53	**Nunki**	2.0	10 12	S26 16.8	75 59.7
54	**Altair**	0.8	11 07	N 8 54.0	62 09.4
55	**Peacock**	1.9	11 42	S56 41.6	53 21.2
56	**Deneb**	1.3	11 57	N45 19.4	49 32.6
57	**Enif**	2.4	13 00	N 9 55.9	33 48.3
58	**Al Na'ir**	1.7	13 24	S46 54.1	27 45.3
59	**Fomalhaut**	1.2	14 13	S29 33.5	15 25.3
60	**Markab**	2.5	14 21	N15 16.3	13 39.6

SUN

Yr	Day of Mth	Week	Transit h m	Semi-Diam	Twilight h m (Lat 52°N)	Sunrise h m (Lat 52°N)	Sunset h m (Lat 52°N)	Twilight h m (Lat 52°N)
32	1	Wed	12 13	16.3	07 04	07 41	16 47	17 23
33	2	Th	12 14	16.3	07 03	07 39	16 49	17 25
34	3	Fri	12 14	16.3	07 01	07 38	16 50	17 27
35	4	Sat	12 14	16.3	07 00	07 36	16 52	17 29
36	5	Sun	12 14	16.2	06 58	07 34	16 54	17 30
37	6	Mon	12 14	16.2	06 57	07 33	16 56	17 32
38	7	Tu	12 14	16.2	06 55	07 31	16 58	17 34
39	8	Wed	12 14	16.2	06 53	07 29	17 00	17 36
40	9	Th	12 14	16.2	06 52	07 27	17 02	17 37
41	10	Fri	12 14	16.2	06 50	07 26	17 03	17 39
42	11	Sat	12 14	16.2	06 48	07 24	17 05	17 41
43	12	Sun	12 14	16.2	06 47	07 22	17 07	17 43
44	13	Mon	12 14	16.2	06 45	07 20	17 09	17 44
45	14	Tu	12 14	16.2	06 43	07 18	17 11	17 46
46	15	Wed	12 14	16.2	06 41	07 16	17 13	17 48
47	16	Th	12 14	16.2	06 39	07 14	17 15	17 50
48	17	Fri	12 14	16.2	06 38	07 12	17 17	17 51
49	18	Sat	12 14	16.2	06 36	07 10	17 18	17 53
50	19	Sun	12 14	16.2	06 34	07 08	17 20	17 55
51	20	Mon	12 14	16.2	06 32	07 06	17 22	17 57
52	21	Tu	12 14	16.2	06 30	07 04	17 24	17 58
53	22	Wed	12 14	16.2	06 28	07 02	17 26	18 00
54	23	Th	12 13	16.2	06 26	07 00	17 28	18 02
55	24	Fri	12 13	16.2	06 24	06 58	17 29	18 04
56	25	Sat	12 13	16.2	06 22	06 56	17 31	18 05
57	26	Sun	12 13	16.2	06 20	06 54	17 33	18 07
58	27	Mon	12 13	16.2	06 18	06 52	17 35	18 09
59	28	Tu	12 13	16.2	06 15	06 49	17 37	18 11
60	29	Wed	12 12	16.2	06 13	06 47	17 39	18 12

Lat Corr to Sunrise, Sunset etc.

Lat °	Twilight h m	Sunrise h m	Sunset h m	Twilight h m
N70	+0 48	+1 23	−1 22	−0 49
68	+0 39	+1 06	−1 06	−0 40
66	+0 32	+0 53	−0 53	−0 32
64	+0 26	+0 42	−0 41	−0 26
62	+0 20	+0 32	−0 32	−0 20
N60	+0 15	+0 25	−0 25	−0 16
58	+0 10	+0 17	−0 17	−0 11
56	+0 06	+0 11	−0 11	−0 07
54	+0 03	+0 05	−0 05	−0 04
50	−0 03	−0 05	+0 04	+0 03
N45	−0 10	−0 15	+0 15	+0 10
40	−0 16	−0 23	+0 23	+0 16
35	−0 22	−0 30	+0 31	+0 21
30	−0 27	−0 37	+0 37	+0 26
20	−0 36	−0 48	+0 48	+0 35
N10	−0 45	−0 59	+0 58	+0 44
0	−0 55	−1 08	+1 07	+0 54
S10	−1 05	−1 08	+1 16	+1 04
20	−1 16	−1 28	+1 26	+1 15
30	−1 31	−1 40	+1 38	+1 30
S35	−1 40	−1 48	+1 45	+1 49
40	−1 50	−1 55	+1 53	+1 49
45	−2 02	−2 05	+2 02	+2 00
S50	−2 17	−2 16	+2 13	+2 16

NOTES
The corrections to sunrise etc. are for middle of February.

MOON

Yr	Day of Mth	Week	Age days	Transit (Upper) h m	Diff m	Semi-diam '	Hor Par '	Moonrise h m (Lat 52°N)	Moonset h m (Lat 52°N)
32	1	Wed	09	19 04	50	14.9	54.5	10 54	02 22
33	2	Th	10	19 54	51	15.0	54.9	11 33	03 22
34	3	Fri	11	20 45	52	15.1	55.4	12 21	04 18
35	4	Sat	12	21 37	53	15.3	56.1	13 20	05 06
36	5	Sun	13	22 30	52	15.5	56.8	14 26	05 47
37	6	Mon	14	23 22	51	15.7	57.5	15 40	06 20
38	7	Tu	15	24 13	–	15.8	58.2	16 57	06 49
39	8	Wed	16	00 13	51	16.0	58.8	18 16	07 13
40	9	Th	17	01 04	50	16.1	59.2	19 37	07 35
41	10	Fri	18	01 54	51	16.2	59.5	20 58	07 57
42	11	Sat	19	02 45	53	16.2	59.6	22 20	08 19
43	12	Sun	20	03 38	54	16.2	59.6	23 42	08 43
44	13	Mon	21	04 32	56	16.2	59.4	– –	09 12
45	14	Tu	22	05 28	57	16.1	59.2	01 01	09 47
46	15	Wed	23	06 25	59	16.0	58.9	02 16	10 31
47	16	Th	24	07 24	57	15.9	58.5	03 22	11 25
48	17	Fri	25	08 21	56	15.8	58.1	04 16	12 29
49	18	Sat	26	09 17	53	15.7	57.7	05 00	13 40
50	19	Sun	27	10 10	49	15.6	57.2	05 34	14 54
51	20	Mon	28	10 59	48	15.5	56.7	06 01	16 08
52	21	Tu	29	11 47	45	15.3	56.2	06 24	17 20
53	22	Wed	01	12 32	43	15.2	55.7	06 44	18 32
54	23	Th	02	13 15	43	15.1	55.2	07 03	19 41
55	24	Fri	03	13 58	43	14.9	54.8	07 21	20 49
56	25	Sat	04	14 41	44	14.8	54.5	07 40	21 56
57	26	Sun	05	15 25	45	14.8	54.2	08 01	23 02
58	27	Mon	06	16 10	47	14.8	54.2	08 25	– –
59	28	Tu	07	16 57	48	14.8	54.2	08 54	00 07
60	29	Wed	08	17 45	50	14.8	54.5	09 30	01 08

Phases of the Moon

	d	h	m
○ Full Moon	7	21	54
☽ Last Quarter	14	17	04
● New Moon	21	22	35

	d	h
Perigee	11	19
Apogee	27	14

GMT	SUN GHA ° ′	SUN Dec ° ′	ARIES GHA ° ′
Thursday, 1st March			
00	176 55.0	S 7 30.8	159 12.2
02	206 55.2	7 28.9	189 17.1
04	236 55.4	7 27.0	219 22.1
06	266 55.7	7 25.1	249 27.0
08	296 55.9	7 23.2	279 31.9
10	326 56.2	7 21.3	309 36.8
12	356 56.4	7 19.4	339 41.8
14	26 56.7	7 17.4	9 46.7
16	56 56.9	7 15.5	39 51.6
18	86 57.2	7 13.6	69 56.6
20	116 57.4	7 11.7	100 01.5
22	146 57.7	S 7 09.8	130 06.4
Friday, 2nd March			
00	176 57.9	S 7 07.9	160 11.3
02	206 58.2	7 06.0	190 16.3
04	236 58.5	7 04.1	220 21.2
06	266 58.7	7 02.2	250 26.1
08	296 59.0	7 00.3	280 31.1
10	326 59.2	6 58.3	310 36.0
12	356 59.5	6 56.4	340 40.9
14	26 59.7	6 54.5	10 45.8
16	57 00.0	6 52.6	40 50.8
18	87 00.3	6 50.7	70 55.7
20	117 00.5	6 48.8	101 00.6
22	147 00.8	S 6 46.9	131 05.6
Saturday, 3rd March			
00	177 01.1	S 6 44.9	161 10.5
02	207 01.3	6 43.0	191 15.4
04	237 01.6	6 41.1	221 20.3
06	267 01.9	6 39.2	251 25.3
08	297 02.1	6 37.3	281 30.2
10	327 02.4	6 35.3	311 35.1
12	357 02.7	6 33.4	341 40.1
14	27 02.9	6 31.5	11 45.0
16	57 03.2	6 29.6	41 49.9
18	87 03.5	6 27.7	71 54.8
20	117 03.7	6 25.7	101 59.8
22	147 04.0	S 6 23.8	132 04.7
Sunday, 4th March			
00	177 04.3	S 6 21.9	162 09.6
02	207 04.6	6 20.0	192 14.6
04	237 04.8	6 18.0	222 19.5
06	267 05.1	6 16.1	252 24.4
08	297 05.4	6 14.2	282 29.3
10	327 05.7	6 12.2	312 34.3
12	357 06.0	6 10.3	342 39.2
14	27 06.2	6 08.4	12 44.1
16	57 06.5	6 06.5	42 49.0
18	87 06.8	6 04.5	72 54.0
20	117 07.1	6 02.6	102 58.9
22	147 07.4	S 6 00.7	133 03.8
Monday, 5th March			
00	177 07.6	S 5 58.7	163 08.8
02	207 07.9	5 56.8	193 13.7
04	237 08.2	5 54.9	223 18.6
06	267 08.5	5 52.9	253 23.5
08	297 08.8	5 51.0	283 28.5
10	327 09.1	5 49.1	313 33.4
12	357 09.4	5 47.1	343 38.3
14	27 09.6	5 45.2	13 43.3
16	57 09.9	5 43.3	43 48.2
18	87 10.2	5 41.3	73 53.1
20	117 10.5	5 39.4	103 58.0
22	147 10.8	S 5 37.4	134 03.0

GMT	SUN GHA ° ′	SUN Dec ° ′	ARIES GHA ° ′
Tuesday, 6th March			
00	177 11.1	S 5 35.5	164 07.9
02	207 11.4	5 33.6	194 12.8
04	237 11.7	5 31.6	224 17.8
06	267 12.0	5 29.7	254 22.7
08	297 12.3	5 27.7	284 27.6
10	327 12.6	5 25.8	314 32.5
12	357 12.9	5 23.9	344 37.5
14	27 13.2	5 21.9	14 42.4
16	57 13.5	5 20.0	44 47.3
18	87 13.8	5 18.0	74 52.3
20	117 14.1	5 16.1	104 57.2
22	147 14.4	S 5 14.1	135 02.1
Wednesday, 7th March			
00	177 14.7	S 5 12.2	165 07.0
02	207 15.0	5 10.3	195 12.0
04	237 15.3	5 08.3	225 16.9
06	267 15.6	5 06.4	255 21.8
08	297 15.9	5 04.4	285 26.7
10	327 16.2	5 02.5	315 31.7
12	357 16.5	5 00.5	345 36.6
14	27 16.8	4 58.6	15 41.5
16	57 17.1	4 56.6	45 46.5
18	87 17.4	4 54.7	75 51.4
20	117 17.7	4 52.7	105 56.3
22	147 18.0	S 4 50.8	136 01.2
Thursday, 8th March			
00	177 18.3	S 4 48.8	166 06.2
02	207 18.6	4 46.9	196 11.1
04	237 19.0	4 44.9	226 16.0
06	267 19.3	4 43.0	256 21.0
08	297 19.6	4 41.0	286 25.9
10	327 19.9	4 39.1	316 30.8
12	357 20.2	4 37.1	346 35.7
14	27 20.5	4 35.2	16 40.7
16	57 20.8	4 33.2	46 45.6
18	87 21.1	4 31.3	76 50.5
20	117 21.5	4 29.3	106 55.5
22	147 21.8	S 4 27.3	137 00.4
Friday, 9th March			
00	177 22.1	S 4 25.4	167 05.3
02	207 22.4	4 23.4	197 10.2
04	237 22.7	4 21.5	227 15.2
06	267 23.0	4 19.5	257 20.1
08	297 23.4	4 17.6	287 25.0
10	327 23.7	4 15.6	317 30.0
12	357 24.0	4 13.6	347 34.9
14	27 24.3	4 11.7	17 39.8
16	57 24.6	4 09.7	47 44.7
18	87 25.0	4 07.8	77 49.7
20	117 25.3	4 05.8	107 54.6
22	147 25.6	S 4 03.9	137 59.5
Saturday, 10th March			
00	177 25.9	S 4 01.9	168 04.4
02	207 26.3	3 59.9	198 09.4
04	237 26.6	3 58.0	228 14.3
06	267 26.9	3 56.0	258 19.2
08	297 27.2	3 54.1	288 24.2
10	327 27.6	3 52.1	318 29.1
12	357 27.9	3 50.1	348 34.0
14	27 28.2	3 48.2	18 38.9
16	57 28.5	3 46.2	48 43.9
18	87 28.9	3 44.2	78 48.8
20	117 29.2	3 42.3	108 53.7
22	147 29.5	S 3 40.3	138 58.7

GMT	SUN GHA ° ′	SUN Dec ° ′	ARIES GHA ° ′
Sunday, 11th March			
00	177 29.9	S3 38.4	169 03.6
02	207 30.2	3 36.4	199 08.5
04	237 30.5	3 34.4	229 13.4
06	267 30.8	3 32.5	259 18.4
08	297 31.2	3 30.5	289 23.3
10	327 31.5	3 28.5	319 28.2
12	357 31.8	3 26.6	349 33.2
14	27 32.2	3 24.6	19 38.1
16	57 32.5	3 22.6	49 43.0
18	87 32.8	3 20.7	79 47.9
20	117 33.2	3 18.7	109 52.9
22	147 33.5	S 3 16.7	139 57.8
Monday, 12th March			
00	177 33.9	S 3 14.8	170 02.7
02	207 34.2	3 12.8	200 07.7
04	237 34.5	3 10.8	230 12.6
06	267 34.9	3 08.9	260 17.5
08	297 35.2	3 06.9	290 22.4
10	327 35.5	3 04.9	320 27.4
12	357 35.9	3 03.0	350 32.3
14	27 36.2	3 01.0	20 37.2
16	57 36.6	2 59.0	50 42.1
18	87 36.9	2 57.0	80 47.1
20	117 37.2	2 55.1	110 52.0
22	147 37.6	S 2 53.1	140 56.9
Tuesday, 13th March			
00	177 37.9	S 2 51.1	171 01.9
02	207 38.3	2 49.2	201 06.8
04	237 38.6	2 47.2	231 11.7
06	267 39.0	2 45.2	261 16.6
08	297 39.3	2 43.3	291 21.6
10	327 39.6	2 41.3	321 26.5
12	357 40.0	2 39.3	351 31.4
14	27 40.3	2 37.3	21 36.4
16	57 40.7	2 35.4	51 41.3
18	87 41.0	2 33.4	81 46.2
20	117 41.4	2 31.4	111 51.1
22	147 41.7	S 2 29.4	141 56.1
Wednesday, 14th March			
00	177 42.1	S 2 27.5	172 01.0
02	207 42.4	2 25.5	202 05.9
04	237 42.7	2 23.5	232 10.9
06	267 43.1	2 21.6	262 15.8
08	297 43.4	2 19.6	292 20.7
10	327 43.8	2 17.6	322 25.6
12	357 44.1	2 15.6	352 30.6
14	27 44.5	2 13.7	22 35.5
16	57 44.8	2 11.7	52 40.4
18	87 45.2	2 09.7	82 45.4
20	117 45.5	2 07.7	112 50.3
22	147 45.9	S 2 05.8	142 55.2
Thursday, 15th March			
00	177 46.2	S 2 03.8	173 00.1
02	207 46.6	2 01.8	203 05.1
04	237 46.9	1 59.8	233 10.0
06	267 47.3	1 57.9	263 14.9
08	297 47.7	1 55.9	293 19.9
10	327 48.0	1 53.9	323 24.8
12	357 48.4	1 51.9	353 29.7
14	27 48.7	1 50.0	23 34.6
16	57 49.1	1 48.0	53 39.6
18	87 49.4	1 46.0	83 44.5
20	117 49.8	1 44.0	113 49.4
22	147 50.1	S 1 42.1	143 54.4

GMT	SUN GHA ° ′	SUN Dec ° ′	ARIES GHA ° ′
Friday, 16th March			
00	177 50.5	S 1 40.1	173 59.3
02	207 50.8	1 38.1	204 04.2
04	237 51.2	1 36.1	234 09.1
06	267 51.6	1 34.1	264 14.1
08	297 51.9	1 32.2	294 19.0
10	327 52.3	1 30.2	324 23.9
12	357 52.6	1 28.2	354 28.9
14	27 53.0	1 26.2	24 33.8
16	57 53.3	1 24.3	54 38.7
18	87 53.7	1 22.3	84 43.6
20	117 54.1	1 20.3	114 48.6
22	147 54.4	S 1 18.3	144 53.5
Saturday, 17th March			
00	177 54.8	S 1 16.4	174 58.4
02	207 55.1	1 14.4	205 03.3
04	237 55.5	1 12.4	235 08.3
06	267 55.9	1 10.4	265 13.2
08	297 56.2	1 08.4	295 18.1
10	327 56.6	1 06.5	325 23.1
12	357 56.9	1 04.5	355 28.0
14	27 57.3	1 02.5	25 32.9
16	57 57.7	1 00.5	55 37.8
18	87 58.0	0 58.6	85 42.8
20	117 58.4	0 56.6	115 47.7
22	147 58.7	S 0 54.6	145 52.6
Sunday, 18th March			
00	177 59.1	S 0 52.6	175 57.6
02	207 59.5	0 50.6	206 02.5
04	237 59.8	0 48.7	236 07.4
06	268 00.2	0 46.7	266 12.3
08	298 00.6	0 44.7	296 17.3
10	328 00.9	0 42.7	326 22.2
12	358 01.3	0 40.8	356 27.1
14	28 01.6	0 38.8	26 32.1
16	58 02.0	0 36.8	56 37.0
18	88 02.4	0 34.8	86 41.9
20	118 02.7	0 32.8	116 46.8
22	148 03.1	S 0 30.9	146 51.8
Monday, 19th March			
00	178 03.5	S 0 28.9	176 56.7
02	208 03.8	0 26.9	207 01.6
04	238 04.2	0 24.9	237 06.6
06	268 04.6	0 23.0	267 11.5
08	298 04.9	0 21.0	297 16.4
10	328 05.3	0 19.0	327 21.3
12	358 05.7	0 17.0	357 26.3
14	28 06.0	0 15.1	27 31.2
16	58 06.4	0 13.1	57 36.1
18	88 06.8	0 11.1	87 41.0
20	118 07.1	0 09.1	117 46.0
22	148 07.5	S 0 07.1	147 50.9
Tuesday, 20th March			
00	178 07.9	S 0 05.2	177 55.8
02	208 08.2	0 03.2	208 00.8
04	238 08.6	0 01.2	238 05.7
06	268 09.0	N 0 00.8	268 10.6
08	298 09.4	0 02.7	298 15.5
10	328 09.7	0 04.7	328 20.5
12	358 10.1	0 06.7	358 25.4
14	28 10.5	0 08.7	28 30.3
16	58 10.8	0 10.6	58 35.3
18	88 11.2	0 12.6	88 40.2
20	118 11.6	0 14.6	118 45.1
22	148 11.9	N 0 16.6	148 50.0

GMT	SUN GHA ° ′	SUN Dec ° ′	ARIES GHA ° ′
Wednesday, 21st March			
00	178 12.3	N 0 18.5	178 55.0
02	208 12.7	0 20.5	208 59.9
04	238 13.1	0 22.5	239 04.8
06	268 13.4	0 24.5	269 09.8
08	298 13.8	0 26.4	299 14.7
10	328 14.2	0 28.4	329 19.6
12	358 14.5	0 30.4	359 24.5
14	28 14.9	0 32.4	29 29.5
16	58 15.3	0 34.3	59 34.4
18	88 15.7	0 36.3	89 39.3
20	118 16.0	0 38.3	119 44.3
22	148 16.4	N 0 40.3	149 49.2
Thursday, 22nd March			
00	178 16.8	N 0 42.2	179 54.1
02	208 17.1	0 44.2	209 59.0
04	238 17.5	0 46.2	240 04.0
06	268 17.9	0 48.2	270 08.9
08	298 18.3	0 50.1	300 13.8
10	328 18.6	0 52.1	330 18.7
12	358 19.0	0 54.1	0 23.7
14	28 19.4	0 56.1	30 28.6
16	58 19.8	0 58.0	60 33.5
18	88 20.1	1 00.0	90 38.5
20	118 20.5	1 02.0	120 43.4
22	148 20.9	N 1 03.9	150 48.3
Friday, 23rd March			
00	178 21.2	N 1 05.9	180 53.2
02	208 21.6	1 07.9	210 58.2
04	238 22.0	1 09.9	241 03.1
06	268 22.4	1 11.8	271 08.0
08	298 22.7	1 13.8	301 13.0
10	328 23.1	1 15.8	331 17.9
12	358 23.5	1 17.7	1 22.8
14	28 23.9	1 19.7	31 27.7
16	58 24.2	1 21.7	61 32.7
18	88 24.6	1 23.6	91 37.6
20	118 25.0	1 25.6	121 42.5
22	148 25.4	N 1 27.6	151 47.5
Saturday, 24th March			
00	178 25.7	N 1 29.6	181 52.4
02	208 26.1	1 31.5	211 57.3
04	238 26.5	1 33.5	242 02.2
06	268 26.9	1 35.5	272 07.2
08	298 27.2	1 37.4	302 12.1
10	328 27.6	1 39.4	332 17.0
12	358 28.0	1 41.4	2 21.9
14	28 28.4	1 43.3	32 26.9
16	58 28.7	1 45.3	62 31.8
18	88 29.1	1 47.3	92 36.7
20	118 29.5	1 49.2	122 41.7
22	148 29.9	N 1 51.2	152 46.6
Sunday, 25th March			
00	178 30.2	N 1 53.2	182 51.5
02	208 30.6	1 55.1	212 56.4
04	238 31.0	1 57.1	243 01.4
06	268 31.4	1 59.1	273 06.3
08	298 31.8	2 01.0	303 11.2
10	328 32.1	2 03.0	333 16.2
12	358 32.5	2 04.9	3 21.1
14	28 32.9	2 06.9	33 26.0
16	58 33.3	2 08.9	63 30.9
18	88 33.6	2 10.8	93 35.9
20	118 34.0	2 12.8	123 40.8
22	148 34.4	N 2 14.8	153 45.7

GMT	SUN GHA ° ′	SUN Dec ° ′	ARIES GHA ° ′
Monday, 26th March			
00	178 34.8	N 2 16.7	183 50.7
02	208 35.1	2 18.7	213 55.6
04	238 35.5	2 20.6	244 00.5
06	268 35.9	2 22.6	274 05.4
08	298 36.3	2 24.6	304 10.4
10	328 36.6	2 26.5	334 15.3
12	358 37.0	2 28.5	4 20.2
14	28 37.4	2 30.4	34 25.2
16	58 37.8	2 32.4	64 30.1
18	88 38.2	2 34.4	94 35.0
20	118 38.5	2 36.3	124 39.9
22	148 38.9	N 2 38.3	154 44.9
Tuesday, 27th March			
00	178 39.3	N 2 40.2	184 49.8
02	208 39.7	2 42.2	214 54.7
04	238 40.0	2 44.1	244 59.7
06	268 40.4	2 46.1	275 04.6
08	298 40.8	2 48.1	305 09.5
10	328 41.2	2 50.0	335 14.4
12	358 41.5	2 52.0	5 19.4
14	28 41.9	2 53.9	35 24.3
16	58 42.3	2 55.9	65 29.2
18	88 42.7	2 57.8	95 34.1
20	118 43.1	2 59.8	125 39.1
22	148 43.4	N 3 01.7	155 44.0
Wednesday, 28th March			
00	178 43.8	N 3 03.7	185 48.9
02	208 44.2	3 05.6	215 53.9
04	238 44.6	3 07.6	245 58.8
06	268 44.9	3 09.5	276 03.7
08	298 45.3	3 11.5	306 08.6
10	328 45.7	3 13.4	336 13.6
12	358 46.1	3 15.4	6 18.5
14	28 46.4	3 17.3	36 23.4
16	58 46.8	3 19.3	66 28.4
18	88 47.2	3 21.2	96 33.3
20	118 47.6	3 23.2	126 38.2
22	148 47.9	N 3 25.1	156 43.1
Thursday, 29th March			
00	178 48.3	N 3 27.1	186 48.1
02	208 48.7	3 29.0	216 53.0
04	238 49.1	3 31.0	246 57.9
06	268 49.4	3 32.9	277 02.9
08	298 49.8	3 34.9	307 07.8
10	328 50.2	3 36.8	337 12.7
12	358 50.6	3 38.8	7 17.6
14	28 50.9	3 40.7	37 22.6
16	58 51.3	3 42.6	67 27.5
18	88 51.7	3 44.6	97 32.4
20	118 52.1	3 46.5	127 37.4
22	148 52.4	N 3 48.5	157 42.3
Friday, 30th March			
00	178 52.8	N 3 50.4	187 47.2
02	208 53.2	3 52.4	217 52.1
04	238 53.6	3 54.3	247 57.1
06	268 53.9	3 56.2	278 02.0
08	298 54.3	3 58.2	308 06.9
10	328 54.7	4 00.1	338 11.9
12	358 55.1	4 02.1	8 16.8
14	28 55.4	4 04.0	38 21.7
16	58 55.8	4 05.9	68 26.6
18	88 56.2	4 07.9	98 31.6
20	118 56.6	4 09.8	128 36.5
22	148 56.9	N 4 11.7	158 41.4

GMT	SUN GHA ° ′	SUN Dec ° ′	ARIES GHA ° ′
Saturday, 31st March			
00	178 57.3	N 4 13.7	188 46.4
02	208 57.7	4 15.6	218 51.3
04	238 58.1	4 17.5	248 56.2
06	268 58.4	N 4 19.5	279 01.1
08	298 58.8	N 4 21.4	309 06.1
10	328 59.2	4 23.3	339 11.0
12	358 59.5	4 25.3	9 15.9
14	28 59.9	N 4 27.2	39 20.8
16	59 00.3	N 4 29.1	69 25.8
18	89 00.7	4 31.1	99 30.7
20	119 01.0	4 33.0	129 35.6
22	149 01.4	N 4 34.9	159 40.6

MARCH 2012

MOON

Day	GMT hr	GHA ° '	Mean Var/hr 14°+	Dec ° '	Mean Var/hr '	Day	GMT hr	GHA ° '	Mean Var/hr 14°+	Dec ° '	Mean Var/hr '
1 Th	0	90 37.9	30.1	N21 54.8	1.4	17 Sat	0	242 43.5	28.5	S18 22.3	7.4
	6	177 38.3	29.9	N22 03.3	0.7		6	329 34.7	29.0	S17 38.0	8.0
	12	264 37.5	29.7	N22 08.0	0.0		12	56 28.5	29.4	S16 50.4	8.5
	18	351 35.8	29.6	N22 08.8	0.6		18	143 25.1	29.9	S15 59.9	9.0
2 Fri	0	78 33.1	29.4	N22 05.6	1.2	18 Sun	0	230 24.2	30.3	S15 06.6	9.3
	6	165 29.5	29.3	N21 58.5	1.9		6	317 25.8	30.7	S14 10.8	9.7
	12	252 25.3	29.2	N21 47.3	2.6		12	44 29.9	31.1	S13 12.7	10.0
	18	339 20.4	29.1	N21 32.0	3.3		18	131 36.3	31.5	S12 12.6	10.3
3 Sat	0	66 15.1	29.1	N21 12.7	3.9	19 Mon	0	218 45.0	31.8	S11 10.6	10.6
	6	153 09.3	29.0	N20 49.3	4.6		6	305 55.7	32.1	S10 07.0	10.8
	12	240 03.3	28.9	N20 21.9	5.3		12	33 08.4	32.4	S 9 02.0	11.1
	18	326 57.1	28.9	N19 50.4	5.9		18	120 22.8	32.7	S 7 55.7	11.2
4 Sun	0	53 50.7	29.0	N19 14.9	6.7	20 Tu	0	207 39.0	32.9	S 6 48.5	11.3
	6	140 44.4	28.9	N18 35.4	7.3		6	294 56.6	33.2	S 5 40.4	11.4
	12	227 38.0	29.0	N17 52.1	7.9		12	22 15.5	33.4	S 4 31.7	11.5
	18	314 31.8	28.9	N17 05.1	8.5		18	109 35.7	33.5	S 3 22.6	11.6
5 Mon	0	41 25.7	29.0	N16 14.3	9.1	21 Wed	0	196 56.9	33.7	S 2 13.2	11.6
	6	128 19.8	29.1	N15 20.0	9.7		6	284 18.9	33.8	S 1 03.7	11.6
	12	215 14.1	29.1	N14 22.2	10.2		12	11 41.7	33.8	N 0 05.7	11.5
	18	302 08.5	29.1	N13 21.2	10.7		18	99 05.0	33.9	N 1 14.8	11.4
6 Tu	0	29 03.0	29.1	N12 17.1	11.2	22 Th	0	186 28.7	34.0	N 2 23.5	11.3
	6	115 57.7	29.1	N11 10.0	11.7		6	273 52.7	34.0	N 3 31.7	11.3
	12	202 52.4	29.1	N10 00.3	12.0		12	1 16.7	34.0	N 4 39.0	11.1
	18	289 47.2	29.1	N 8 48.0	12.4		18	88 40.8	34.0	N 5 45.5	10.9
7 Wed	0	16 41.8	29.1	N 7 33.5	12.8	23 Fri	0	176 04.7	33.9	N 6 50.9	10.7
	6	103 36.2	29.0	N 6 17.0	13.1		6	263 28.3	33.8	N 7 55.1	10.5
	12	190 30.4	28.9	N 4 58.7	13.3		12	350 51.5	33.8	N 8 58.0	10.2
	18	277 24.2	28.9	N 3 39.0	13.5		18	78 14.2	33.7	N 9 59.3	9.9
8 Th	0	4 17.5	28.7	N 2 18.1	13.6	24 Sat	0	165 36.2	33.5	N10 59.0	9.6
	6	91 10.1	28.6	N 0 56.3	13.7		6	252 57.5	33.4	N11 56.9	9.3
	12	178 02.0	28.5	S 0 26.0	13.8		12	340 18.0	33.2	N12 52.9	8.9
	18	264 53.0	28.3	S 1 48.5	13.7		18	67 37.6	33.0	N13 46.9	8.6
9 Fri	0	351 43.0	28.1	S 3 10.9	13.7	25 Sun	0	154 56.1	32.9	N14 38.6	8.2
	6	78 31.8	27.9	S 4 32.9	13.5		6	242 13.6	32.8	N15 28.0	7.8
	12	165 19.4	27.7	S 5 54.0	13.3		12	329 30.0	32.5	N16 15.0	7.4
	18	252 05.6	27.4	S 7 14.0	13.0		18	56 45.2	32.3	N16 59.4	6.9
10 Sat	0	338 50.4	27.1	S 8 32.5	12.7	26 Mon	0	143 59.3	32.1	N17 41.1	6.4
	6	65 33.6	27.0	S 9 49.1	12.4		6	231 12.1	32.0	N18 20.0	6.0
	12	152 15.1	26.6	S11 03.5	11.9		12	318 23.7	31.7	N18 55.9	5.4
	18	238 55.0	26.4	S12 15.3	11.4		18	45 34.1	31.5	N19 28.9	4.9
11 Sun	0	325 33.2	26.1	S13 24.3	10.9	27 Tu	0	132 43.2	31.3	N19 58.7	4.3
	6	52 09.7	25.8	S14 30.0	10.3		6	219 51.2	31.1	N20 25.2	3.8
	12	138 44.6	25.5	S15 32.2	9.7		12	306 58.1	31.0	N20 48.4	3.2
	18	225 17.9	25.3	S16 30.5	9.0		18	34 03.9	30.7	N21 08.2	2.7
12 Mon	0	311 49.6	25.0	S17 24.8	8.3	28 Wed	0	121 08.6	30.6	N21 24.6	2.0
	6	38 20.0	24.8	S18 14.7	7.5		6	208 12.3	30.5	N21 37.3	1.5
	12	124 49.1	24.7	S19 00.1	6.7		12	295 15.2	30.4	N21 46.4	0.8
	18	211 17.2	24.6	S19 40.7	5.8		18	22 17.3	30.2	N21 51.8	0.2
13 Tu	0	297 44.5	24.4	S20 16.4	5.1	29 Th	0	109 18.6	30.1	N21 53.5	0.4
	6	24 11.2	24.4	S20 47.0	4.1		6	196 19.3	30.1	N21 51.3	1.0
	12	110 37.5	24.4	S21 12.4	3.3		12	283 19.5	29.9	N21 45.4	1.7
	18	197 03.9	24.4	S21 32.6	2.4		18	10 19.2	29.9	N21 35.6	2.3
14 Wed	0	283 30.4	24.5	S21 47.6	1.5	30 Fri	0	97 18.5	29.8	N21 22.0	2.9
	6	9 57.5	24.7	S21 57.2	0.7		6	184 17.6	29.8	N21 04.5	3.6
	12	96 25.4	24.9	S22 01.6	0.2		12	271 16.5	29.8	N20 43.2	4.3
	18	182 54.5	25.1	S22 00.9	1.0		18	358 15.3	29.8	N20 18.0	4.8
15 Th	0	269 24.9	25.4	S21 55.0	1.9	31 Sat	0	85 14.0	29.8	N19 49.1	5.5
	6	355 57.0	25.7	S21 44.1	2.7		6	172 12.7	29.8	N19 16.4	6.1
	12	82 30.9	26.0	S21 28.4	3.4		12	259 11.5	29.8	N18 39.9	6.7
	18	169 07.0	26.4	S21 08.0	4.3		18	346 10.3	29.8	N17 59.8	7.3
16 Fri	0	255 45.3	26.8	S20 43.0	4.9						
	6	342 26.0	27.3	S20 13.8	5.6						
	12	69 09.2	27.6	S19 40.5	6.3						
	18	155 55.0	28.1	S19 03.2	6.9						

PLANETS

VENUS					JUPITER					
Mer Pass h m	GHA ° '	Mean Var/hr 14°+	Dec ° '	Mean Var/hr '	Day	GHA ° '	Mean Var/hr 15°+	Dec ° '	Mean Var/hr '	Mer Pass h m
14 55	136 21.6	59.9	N10 39.3	1.2	1 Th	124 11.5	2.0	N12 57.0	0.2	15 41
14 55	136 18.9	59.9	N11 08.2	1.2	2 Fri	125 00.1	2.0	N13 00.7	0.2	15 38
14 55	136 16.1	59.9	N11 36.9	1.2	3 Sat	125 48.6	2.0	N13 04.5	0.2	15 35
14 55	136 13.4	59.9	N12 05.3	1.2	4 SUN	126 37.1	2.0	N13 08.3	0.2	15 31
14 55	136 10.7	59.9	N12 33.5	1.2	5 Mon	127 25.4	2.0	N13 12.1	0.2	15 28
14 56	136 08.1	59.9	N13 01.3	1.2	6 Tu	128 13.6	2.0	N13 15.9	0.2	15 25
14 56	136 05.5	59.9	N13 28.9	1.1	7 Wed	129 01.8	2.0	N13 19.7	0.2	15 22
14 56	136 02.9	59.9	N13 56.2	1.1	8 Th	129 49.8	2.0	N13 23.6	0.2	15 19
14 56	136 00.3	59.9	N14 23.2	1.1	9 Fri	130 37.7	2.0	N13 27.5	0.2	15 15
14 56	135 57.8	59.9	N14 49.9	1.1	10 Sat	131 25.6	2.0	N13 31.4	0.2	15 12
14 56	135 55.2	59.9	N15 16.2	1.1	11 SUN	132 13.4	2.0	N13 35.3	0.2	15 09
14 57	135 52.8	59.9	N15 42.2	1.1	12 Mon	133 01.0	2.0	N13 39.2	0.2	15 06
14 57	135 50.3	59.9	N16 07.9	1.1	13 Tu	133 48.6	2.0	N13 43.2	0.2	15 03
14 57	135 47.9	59.9	N16 33.2	1.0	14 Wed	134 36.1	2.0	N13 47.1	0.2	15 00
14 57	135 45.6	59.9	N16 58.1	1.0	15 Th	135 23.5	2.0	N13 51.1	0.2	14 56
14 57	135 43.3	59.9	N17 22.7	1.0	16 Fri	136 10.8	2.0	N13 55.1	0.2	14 53
14 57	135 41.0	59.9	N17 46.9	1.0	17 Sat	136 58.1	2.0	N13 59.1	0.2	14 50
14 58	135 38.8	59.9	N18 10.7	1.0	18 SUN	137 45.2	2.0	N14 03.1	0.2	14 47
14 58	135 36.7	59.9	N18 34.1	1.0	19 Mon	138 32.3	2.0	N14 07.1	0.2	14 44
14 58	135 34.6	59.9	N18 57.1	0.9	20 Tu	139 19.3	2.0	N14 11.2	0.2	14 41
14 58	135 32.7	59.9	N19 19.7	0.9	21 Wed	140 06.2	1.9	N14 15.2	0.2	14 38
14 58	135 30.8	59.9	N19 41.9	0.9	22 Th	140 53.0	1.9	N14 19.3	0.2	14 35
14 58	135 29.1	59.9	N20 03.6	0.9	23 Fri	141 39.8	1.9	N14 23.3	0.2	14 31
14 58	135 27.5	59.9	N20 24.9	0.9	24 Sat	142 26.5	1.9	N14 27.4	0.2	14 28
14 58	135 26.0	59.9	N20 45.8	0.9	25 SUN	143 13.1	1.9	N14 31.4	0.2	14 25
14 58	135 24.7	15	N21 06.2	0.8	26 Mon	143 59.7	1.9	N14 35.5	0.2	14 22
14 58	135 23.6	0.0	N21 26.1	0.8	27 Tu	144 46.2	1.9	N14 39.6	0.2	14 19
14 59	135 22.6	0.0	N21 45.6	0.8	28 Wed	145 32.6	1.9	N14 43.7	0.2	14 16
14 59	135 21.9	0.0	N22 04.6	0.8	29 Th	146 18.9	1.9	N14 47.8	0.2	14 13
14 59	135 21.4	0.0	N22 23.1	0.8	30 Fri	147 05.2	1.9	N14 51.9	0.2	14 10
14 59	135 21.2	0.0	N22 41.1	0.7	31 Sat	147 51.4	1.9	N14 56.0	0.2	14 07

VENUS, Av. Mag. –4.4
SHA March 5 333; 10 328; 15 323; 20 318; 25 313; 30 308

JUPITER, Av. Mag. –2.1
SHA March 5 324; 10 323; 15 322; 20 321; 25 320; 30 319

MARS					SATURN					
Mer Pass h m	GHA ° '	Mean Var/hr 15°+	Dec ° '	Mean Var/hr '	Day	GHA ° '	Mean Var/hr 15°+	Dec ° '	Mean Var/hr '	Mer Pass h m
00 34	351 34.6	3.4	N 9 52.2	0.4	1 Th	311 13.0	2.6	S 8 38.4	0.0	03 15
00 28	352 56.0	3.4	N10 00.9	0.4	2 Fri	312 14.2	2.6	S 8 37.4	0.0	03 11
00 23	354 17.4	3.4	N10 09.5	0.4	3 Sat	313 15.6	2.6	S 8 36.4	0.0	03 06
00 17	355 39.0	3.4	N10 18.1	0.3	4 SUN	314 17.0	2.6	S 8 35.3	0.0	03 02
00 12	357 00.6	3.4	N10 26.5	0.3	5 Mon	315 18.5	2.6	S 8 34.2	0.0	02 58
00 06	358 22.3	3.4	N10 34.9	0.3	6 Tu	316 20.1	2.6	S 8 33.1	0.1	02 54
00 01	359 43.9	3.4	N10 43.1	0.3	7 Wed	317 21.7	2.6	S 8 31.9	0.0	02 50
23 50	1 05.5	3.4	N10 51.1	0.3	8 Th	318 23.5	2.6	S 8 30.7	0.0	02 46
23 45	2 27.0	3.4	N10 59.0	0.3	9 Fri	319 25.3	2.6	S 8 29.5	0.0	02 42
23 39	3 48.4	3.4	N11 06.7	0.3	10 Sat	320 27.2	2.6	S 8 28.3	0.1	02 38
23 34	5 09.6	3.4	N11 14.2	0.3	11 SUN	321 29.2	2.6	S 8 27.0	0.1	02 34
23 29	6 30.6	3.4	N11 21.5	0.3	12 Mon	322 31.2	2.6	S 8 25.7	0.1	02 29
23 23	7 51.3	3.4	N11 28.6	0.3	13 Tu	323 33.4	2.6	S 8 24.4	0.1	02 25
23 18	9 11.8	3.3	N11 35.5	0.3	14 Wed	324 35.6	2.6	S 8 23.1	0.1	02 21
23 13	10 31.9	3.3	N11 42.1	0.3	15 Th	325 37.8	2.6	S 8 21.7	0.1	02 17
23 07	11 51.6	3.3	N11 48.4	0.3	16 Fri	326 40.2	2.6	S 8 20.3	0.1	02 13
23 02	13 11.0	3.3	N11 54.5	0.2	17 Sat	327 42.6	2.6	S 8 18.9	0.1	02 09
22 57	14 29.9	3.3	N12 00.4	0.2	18 SUN	328 45.0	2.6	S 8 17.5	0.1	02 05
22 52	15 48.3	3.2	N12 05.9	0.2	19 Mon	329 47.6	2.6	S 8 16.1	0.1	02 00
22 47	17 06.2	3.2	N12 11.2	0.2	20 Tu	330 50.2	2.6	S 8 14.6	0.1	01 56
22 42	18 23.6	3.2	N12 16.1	0.2	21 Wed	331 52.8	2.6	S 8 13.1	0.1	01 52
22 37	19 40.5	3.2	N12 20.8	0.2	22 Th	332 55.5	2.6	S 8 11.6	0.1	01 48
22 31	20 56.7	3.2	N12 25.2	0.2	23 Fri	333 58.3	2.6	S 8 10.1	0.1	01 44
22 26	22 12.4	3.1	N12 29.2	0.2	24 Sat	335 01.1	2.6	S 8 08.6	0.1	01 40
22 22	23 27.4	3.1	N12 33.0	0.1	25 SUN	336 04.0	2.6	S 8 07.0	0.1	01 35
22 17	24 41.7	3.1	N12 36.4	0.1	26 Mon	337 06.9	2.6	S 8 05.5	0.1	01 31
22 12	25 55.4	3.0	N12 39.5	0.1	27 Tu	338 09.9	2.6	S 8 03.9	0.1	01 27
22 07	27 08.4	3.0	N12 42.3	0.1	28 Wed	339 12.9	2.6	S 8 02.3	0.1	01 23
22 02	28 20.7	3.0	N12 44.8	0.1	29 Th	340 16.0	2.6	S 8 00.7	0.1	01 19
21 58	29 32.3	2.9	N12 46.9	0.1	30 Fri	341 19.1	2.6	S 7 59.1	0.1	01 15
21 53	30 43.1	2.9	N12 48.8	0.1	31 Sat	342 22.3	2.6	S 7 57.5	0.1	01 10

MARS, Av. Mag. –1.0
SHA March 5 194; 10 196; 15 198; 20 199; 25 201; 30 202

SATURN, Av. Mag. +0.4
SHA March 5 152; 10 152; 15 153; 20 153; 25 153; 30 154

MARCH 2012

No.	Name	Mag	Transit h m	Dec ° '	SHA ° '
	Oh GMT March 1				
Ψ	ARIES	–	13 21	–	–
1	Alpheratz	2.1	13 30	N29 09.5	357 44.8
2	Ankaa	2.4	13 48	S42 14.4	353 16.9
3	Schedar	2.2	14 02	N56 36.3	349 42.1
4	Diphda	2.0	14 05	S17 55.2	348 57.1
5	Achernar	0.5	14 59	S57 10.6	335 27.9
6	POLARIS	2.0	16 07	N89 19.2	318 30.2
7	Hamal	2.0	15 29	N23 31.2	328 02.0
8	Acamar	3.2	16 19	S40 15.6	315 19.2
9	Menkar	2.5	16 23	N 4 08.1	314 16.2
10	Mirfak	1.8	16 46	N49 54.4	308 41.9
11	Aldebaran	0.9	17 57	N16 31.9	290 50.5
12	Rigel	0.1	18 35	S 8 11.5	281 12.9
13	Capella	0.1	18 38	N46 00.7	280 35.8
14	Bellatrix	1.6	18 46	N 6 21.5	278 33.0
15	Elnath	1.7	18 47	N28 37.0	278 13.7
16	Alnilam	1.7	18 57	S 1 11.9	275 47.2
17	Betelgeuse	0.1–1.2	19 16	N 7 24.4	271 02.2
18	Canopus	–0.7	19 44	S52 42.5	263 56.4
19	Sirius	–1.5	20 06	S16 44.3	258 34.4
20	Adhara	1.5	20 19	S28 59.7	255 13.1
21	Castor	1.6	20 55	N31 51.6	246 08.9
22	Procyon	0.4	21 00	N 5 11.4	245 00.5
23	Pollux	1.1	21 06	N27 59.7	243 28.6
24	Avior	1.9	21 42	S59 33.3	234 18.0
25	Suhail	2.2	22 28	S43 29.3	222 52.8
26	Miaplacidus	1.7	22 33	S69 46.4	221 39.2
27	Alphard	2.0	22 48	S 8 43.0	217 56.7
28	Regulus	1.4	23 28	N11 54.2	207 44.1
29	Dubhe	1.8	0 28	N61 41.0	193 52.1
30	Denebola	2.1	1 13	N14 30.0	182 34.2
31	Gienah	2.6	1 39	S17 36.8	175 52.9
32	Acrux	1.3	1 50	S63 10.1	173 09.6
33	Gacrux	1.6	1 55	S57 11.0	172 01.3
34	Mimosa	1.3	2 11	S59 45.4	167 52.4
35	Alioth	1.8	2 17	N55 53.4	166 20.9
36	Spica	1.0	2 49	S11 13.7	158 31.9
37	Alkaid	1.9	3 11	N49 14.9	152 59.1
38	Hadar	0.6	3 27	S60 25.8	148 48.7
39	Menkent	2.1	3 30	S36 25.8	148 08.3
40	Arcturus	0.0	3 39	N19 06.9	145 56.3
41	Rigil Kent	–0.3	4 03	S60 53.0	139 52.5
42	Zuben'ubi	2.8	4 14	S16 05.6	137 06.2
43	Kochab	2.1	4 13	N74 06.0	137 19.2
44	Alphecca	2.2	4 58	N26 40.2	126 11.6
45	Antares	1.0	5 52	S26 27.5	112 27.2
46	Atria	1.9	6 12	S69 02.7	107 29.8
47	Sabik	2.4	6 33	S15 44.4	102 13.5
48	Shaula	1.6	6 57	S37 06.6	96 23.1
49	Rasalhague	2.1	6 58	N12 32.9	96 07.3
50	Eltanin	2.2	7 19	N51 29.0	90 46.6
51	Kaus Aust.	1.9	7 47	S34 22.5	83 45.0
52	Vega	0.0	7 59	N38 47.5	80 39.7
53	Nunki	2.0	8 18	S26 16.8	75 59.5
54	Altair	0.8	9 13	N 8 54.0	62 09.3
55	Peacock	1.9	9 48	S56 41.5	53 21.0
56	Deneb	1.3	10 03	N45 19.3	49 32.4
57	Enif	2.4	11 06	N 9 55.8	33 48.3
58	Al Na'ir	1.7	11 30	S46 54.0	27 45.2
59	Fomalhaut	1.2	12 19	S29 33.4	15 25.3
60	Markab	2.5	12 27	N15 16.2	13 39.5

SUN

Yr	Day of Mth	Week	Transit h m	Semi-Diam	Lat 52°N Twilight h m	Sunrise h m	Sunset h m	Twilight h m
61	1	Th	12 12	16.2	06 11	06 45	17 40	18 14
62	2	Fri	12 12	16.2	06 09	06 43	17 42	18 16
63	3	Sat	12 12	16.2	06 07	06 41	17 44	18 18
64	4	Sun	12 12	16.2	06 05	06 38	17 46	18 19
65	5	Mon	12 11	16.1	06 03	06 36	17 47	18 21
66	6	Tu	12 11	16.1	06 00	06 34	17 49	18 23
67	7	Wed	12 11	16.1	05 58	06 32	17 51	18 25
68	8	Th	12 11	16.1	05 56	06 29	17 53	18 26
69	9	Fri	12 10	16.1	05 54	06 27	17 55	18 28
70	10	Sat	12 10	16.1	05 51	06 25	17 56	18 30
71	11	Sun	12 10	16.1	05 49	06 23	17 58	18 32
72	12	Mon	12 10	16.1	05 47	06 20	18 00	18 33
73	13	Tu	12 09	16.1	05 45	06 18	18 02	18 35
74	14	Wed	12 09	16.1	05 42	06 16	18 03	18 37
75	15	Th	12 09	16.1	05 40	06 13	18 05	18 39
76	16	Fri	12 08	16.1	05 38	06 11	18 07	18 40
77	17	Sat	12 08	16.1	05 35	06 09	18 09	18 42
78	18	Sun	12 08	16.1	05 33	06 07	18 10	18 44
79	19	Mon	12 08	16.1	05 31	06 04	18 12	18 46
80	20	Tu	12 07	16.1	05 28	06 02	18 14	18 48
81	21	Wed	12 07	16.1	05 26	06 00	18 15	18 49
82	22	Th	12 07	16.1	05 24	05 57	18 17	18 51
83	23	Fri	12 06	16.1	05 21	05 55	18 19	18 53
84	24	Sat	12 06	16.1	05 19	05 53	18 21	18 55
85	25	Sun	12 06	16.1	05 17	05 50	18 22	18 56
86	26	Mon	12 06	16.1	05 14	05 48	18 24	18 58
87	27	Tu	12 05	16.1	05 12	05 46	18 26	19 00
88	28	Wed	12 05	16.0	05 09	05 43	18 28	19 02
89	29	Th	12 05	16.0	05 07	05 41	18 29	19 03
90	30	Fri	12 04	16.0	05 05	05 39	18 31	19 05
91	31	Sat	12 04	16.0	05 02	05 36	18 33	19 07

Lat Corr to Sunrise, Sunset etc.

Lat °	Twilight h m	Sunrise h m	Sunset h m	Twilight h m
N70	–0 18	+0 08	–1 06	+0 19
68	–0 15	+0 06	–1 05	+0 15
66	–0 12	+0 05	–0 04	+0 13
64	–0 09	+0 04	–0 03	+0 10
62	–0 06	+0 03	–0 02	+0 07
N60	–0 04	+0 03	–0 02	+0 05
58	–0 03	+0 02	–0 02	+0 04
56	–0 01	+0 01	–0 01	+0 02
54	–0 01	+0 01	–0 01	+0 01
50	+0 01	0 00	0 00	–0 01
N45	+0 03	–0 01	+0 02	–0 02
40	+0 05	–0 02	+0 02	–0 04
35	+0 06	–0 03	+0 03	–0 05
30	+0 07	–0 04	+0 04	–0 06
20	+0 07	–0 05	+0 05	–0 06
N10	+0 07	–0 07	+0 06	–0 06
0	+0 06	–0 08	+0 07	–0 05
S10	+0 04	–0 09	+0 09	–1 03
20	+1 01	–0 11	+0 10	0 00
30	–0 03	–0 14	+0 12	+0 04
S35	–0 06	–0 15	+0 14	+0 07
40	–0 10	–0 17	+0 15	+0 11
45	–0 13	–0 18	+0 17	+0 14
S50	–0 18	–0 20	+0 19	+0 20

NOTES
The corrections to sunrise etc. are for middle of March.

MOON

Yr	Day of Mth	Week	Age days	Transit (Upper) h m	Diff m	Semi-diam '	Hor Par '	Lat 52°N Moonrise h m	Moonset h m
61	1	Th	09	18 35	51	15.0	54.9	10 13	02 05
62	2	Fri	10	19 26	51	15.1	55.4	11 06	02 56
63	3	Sat	11	20 17	51	15.3	56.2	12 08	03 39
64	4	Sun	12	21 08	52	15.5	57.0	13 17	04 16
65	5	Mon	13	22 00	51	15.8	57.9	14 31	04 46
66	6	Tu	14	22 51	51	16.0	58.7	15 49	05 13
67	7	Wed	15	23 42	52	16.2	59.5	17 10	05 37
68	8	Th	16	24 34	–	16.4	60.0	18 33	05 59
69	9	Fri	17	00 34	54	16.5	60.4	19 57	06 22
70	10	Sat	18	01 28	55	16.5	60.5	21 21	06 46
71	11	Sun	19	02 23	57	16.4	60.4	22 45	07 14
72	12	Mon	20	03 20	59	16.4	60.0	– –	07 48
73	13	Tu	21	04 19	60	16.2	59.5	00 03	08 30
74	14	Wed	22	05 19	58	16.1	58.9	01 14	09 22
75	15	Th	23	06 17	56	15.9	58.3	02 12	10 24
76	16	Fri	24	07 13	53	15.7	57.7	02 59	11 32
77	17	Sat	25	08 06	50	15.6	57.1	03 36	12 44
78	18	Sun	26	08 56	47	15.4	56.5	04 05	13 57
79	19	Mon	27	09 43	45	15.3	56.0	04 28	15 09
80	20	Tu	28	10 28	44	15.1	55.5	04 49	16 19
81	21	Wed	29	11 12	43	15.0	55.1	05 08	17 28
82	22	Th	30	11 55	43	14.9	54.7	05 27	18 36
83	23	Fri	01	12 38	43	14.8	54.4	05 46	19 43
84	24	Sat	02	13 21	45	14.8	54.2	06 06	20 49
85	25	Sun	03	14 06	46	14.7	54.1	06 29	21 54
86	26	Mon	04	14 52	47	14.7	54.0	06 57	22 57
87	27	Tu	05	15 39	49	14.8	54.1	07 30	23 55
88	28	Wed	06	16 28	49	14.8	54.4	08 10	– –
89	29	Th	07	17 17	50	14.9	54.8	08 58	00 47
90	30	Fri	08	18 07	50	15.1	55.3	09 55	01 33
91	31	Sat	09	18 57	50	15.3	56.1	10 59	02 11

Phases of the Moon

	Phase	d	h	m
☾	First Quarter	1	01	21
○	Full Moon	8	09	39
☽	Last Quarter	15	01	25
●	New Moon	22	14	37
☾	First Quarter	30	19	41

	d	h
Perigee	10	10
Apogee	26	06

GMT	SUN GHA ° ′	SUN Dec ° ′	ARIES GHA ° ′
Sunday, 1st April			
00	179 01.8	N 4 36.9	189 45.5
02	209 02.2	4 38.8	219 50.4
04	239 02.5	4 40.7	249 55.3
06	269 02.9	4 42.6	280 00.3
08	299 03.3	4 44.6	310 05.2
10	329 03.6	4 46.5	340 10.1
12	359 04.0	4 48.4	10 15.1
14	29 04.4	4 50.3	40 20.0
16	59 04.7	4 52.3	70 24.9
18	89 05.1	4 54.2	100 29.8
20	119 05.5	4 56.1	130 34.8
22	149 05.9	N 4 58.0	160 39.7
Monday, 2nd April			
00	179 06.2	N 4 59.9	190 44.6
02	209 06.6	5 01.9	220 49.6
04	239 07.0	5 03.8	250 54.5
06	269 07.3	5 05.7	280 59.4
08	299 07.7	5 07.6	311 04.3
10	329 08.1	5 09.5	341 09.3
12	359 08.4	5 11.5	11 14.2
14	29 08.8	5 13.4	41 19.1
16	59 09.2	5 15.3	71 24.1
18	89 09.5	5 17.2	101 29.0
20	119 09.9	5 19.1	131 33.9
22	149 10.3	N 5 21.0	161 38.8
Tuesday, 3rd April			
00	179 10.6	N 5 22.9	191 43.8
02	209 11.0	5 24.9	221 48.7
04	239 11.4	5 26.8	251 53.6
06	269 11.7	5 28.7	281 58.6
08	299 12.1	5 30.6	312 03.5
10	329 12.5	5 32.5	342 08.4
12	359 12.8	5 34.4	12 13.3
14	29 13.2	5 36.3	42 18.3
16	59 13.6	5 38.2	72 23.2
18	89 13.9	5 40.1	102 28.1
20	119 14.3	5 42.0	132 33.0
22	149 14.7	N 5 44.0	162 38.0
Wednesday, 4th April			
00	179 15.0	N 5 45.9	192 42.9
02	209 15.4	5 47.8	222 47.8
04	239 15.8	5 49.7	252 52.8
06	269 16.1	5 51.6	282 57.7
08	299 16.5	5 53.5	313 02.6
10	329 16.8	5 55.4	343 07.5
12	359 17.2	5 57.3	13 12.5
14	29 17.6	5 59.2	43 17.4
16	59 17.9	6 01.1	73 22.3
18	89 18.3	6 03.0	103 27.3
20	119 18.7	6 04.9	133 32.2
22	149 19.0	N 6 06.8	163 37.1
Thursday, 5th April			
00	179 19.4	N 6 08.7	193 42.0
02	209 19.7	6 10.6	223 47.0
04	239 20.1	6 12.5	253 51.9
06	269 20.5	6 14.3	283 56.8
08	299 20.8	6 16.2	314 01.8
10	329 21.2	6 18.1	344 06.7
12	359 21.5	6 20.0	14 11.6
14	29 21.9	6 21.9	44 16.5
16	59 22.3	6 23.8	74 21.5
18	89 22.6	6 25.7	104 26.4
20	119 23.0	6 27.6	134 31.3
22	149 23.3	N 6 29.5	164 36.2

GMT	SUN GHA ° ′	SUN Dec ° ′	ARIES GHA ° ′
Friday, 6th April			
00	179 23.7	N 6 31.4	194 41.2
02	209 24.0	6 33.2	224 46.1
04	239 24.4	6 35.1	254 51.0
06	269 24.7	6 37.0	284 56.0
08	299 25.1	6 38.9	315 00.9
10	329 25.5	6 40.8	345 05.8
12	359 25.8	6 42.7	15 10.7
14	29 26.2	6 44.6	45 15.7
16	59 26.5	6 46.4	75 20.6
18	89 26.9	6 48.3	105 25.5
20	119 27.2	6 50.2	135 30.5
22	149 27.6	N 6 52.1	165 35.4
Saturday, 7th April			
00	179 27.9	N 6 53.9	195 40.3
02	209 28.3	6 55.8	225 45.2
04	239 28.6	6 57.7	255 50.2
06	269 29.0	6 59.6	285 55.1
08	299 29.3	7 01.5	316 00.0
10	329 29.7	7 03.3	346 05.0
12	359 30.0	7 05.2	16 09.9
14	29 30.4	7 07.1	46 14.8
16	59 30.7	7 08.9	76 19.7
18	89 31.1	7 10.8	106 24.7
20	119 31.4	7 12.7	136 29.6
22	149 31.8	N 7 14.6	166 34.5
Sunday, 8th April			
00	179 32.1	N 7 16.4	196 39.5
02	209 32.5	7 18.3	226 44.4
04	239 32.8	7 20.2	256 49.3
06	269 33.2	7 22.0	286 54.2
08	299 33.5	7 23.9	316 59.2
10	329 33.9	7 25.7	347 04.1
12	359 34.2	7 27.6	17 09.0
14	29 34.5	7 29.5	47 13.9
16	59 34.9	7 31.3	77 18.9
18	89 35.2	7 33.2	107 23.8
20	119 35.6	7 35.1	137 28.7
22	149 35.9	N 7 36.9	167 33.7
Monday, 9th April			
00	179 36.3	N 7 38.8	197 38.6
02	209 36.6	7 40.6	227 43.5
04	239 36.9	7 42.5	257 48.4
06	269 37.3	7 44.3	287 53.4
08	299 37.6	7 46.2	317 58.3
10	329 38.0	7 48.0	348 03.2
12	359 38.3	7 49.9	18 08.2
14	29 38.6	7 51.7	48 13.1
16	59 39.0	7 53.6	78 18.0
18	89 39.3	7 55.4	108 22.9
20	119 39.7	7 57.3	138 27.9
22	149 40.0	N 7 59.1	168 32.8
Tuesday, 10th April			
00	179 40.3	N 8 01.0	198 37.7
02	209 40.7	8 02.8	228 42.7
04	239 41.0	8 04.7	258 47.6
06	269 41.3	8 06.5	288 52.5
08	299 41.7	8 08.4	318 57.4
10	329 42.0	8 10.2	349 02.4
12	359 42.3	8 12.1	19 07.3
14	29 42.7	8 13.9	49 12.2
16	59 43.0	8 15.7	79 17.2
18	89 43.3	8 17.6	109 22.1
20	119 43.7	8 19.4	139 27.0
22	149 44.0	N 8 21.2	169 31.9

GMT	SUN GHA ° ′	SUN Dec ° ′	ARIES GHA ° ′
Wednesday, 11th April			
00	179 44.3	N 8 23.1	199 36.9
02	209 44.7	8 24.9	229 41.8
04	239 45.0	8 26.7	259 46.7
06	269 45.3	8 28.6	289 51.7
08	299 45.6	8 30.4	319 56.6
10	329 46.0	8 32.2	350 01.5
12	359 46.3	8 34.1	20 06.4
14	29 46.6	8 35.9	50 11.4
16	59 46.9	8 37.7	80 16.3
18	89 47.3	8 39.6	110 21.2
20	119 47.6	8 41.4	140 26.2
22	149 47.9	N 8 43.2	170 31.1
Thursday, 12th April			
00	179 48.2	N 8 45.0	200 36.0
02	209 48.6	8 46.9	230 40.9
04	239 48.9	8 48.7	260 45.9
06	269 49.2	8 50.5	290 50.8
08	299 49.5	8 52.3	320 55.7
10	329 49.8	8 54.1	351 00.7
12	359 50.2	8 55.9	21 05.6
14	29 50.5	8 57.8	51 10.5
16	59 50.8	8 59.6	81 15.4
18	89 51.1	9 01.4	111 20.4
20	119 51.4	9 03.2	141 25.3
22	149 51.8	N 9 05.0	171 30.2
Friday, 13th April			
00	179 52.1	N 9 06.8	201 35.1
02	209 52.4	9 08.6	231 40.1
04	239 52.7	9 10.5	261 45.0
06	269 53.0	9 12.3	291 49.9
08	299 53.3	9 14.1	321 54.9
10	329 53.6	9 15.9	351 59.8
12	359 54.0	9 17.7	22 04.7
14	29 54.3	9 19.5	52 09.6
16	59 54.6	9 21.3	82 14.6
18	89 54.9	9 23.1	112 19.5
20	119 55.2	9 24.9	142 24.4
22	149 55.5	N 9 26.7	172 29.4
Saturday, 14th April			
00	179 55.8	N 9 28.5	202 34.3
02	209 56.1	9 30.3	232 39.2
04	239 56.4	9 32.1	262 44.1
06	269 56.7	9 33.9	292 49.1
08	299 57.1	9 35.7	322 54.0
10	329 57.4	9 37.5	352 58.9
12	359 57.7	9 39.3	23 03.9
14	29 58.0	9 41.0	53 08.8
16	59 58.3	9 42.8	83 13.7
18	89 58.6	9 44.6	113 18.6
20	119 58.9	9 46.4	143 23.6
22	149 59.2	N 9 48.2	173 28.5
Sunday, 15th April			
00	179 59.5	N 9 50.0	203 33.4
02	209 59.8	9 51.8	233 38.4
04	240 00.1	9 53.6	263 43.3
06	270 00.4	9 55.3	293 48.2
08	300 00.7	9 57.1	323 53.1
10	330 01.0	9 58.9	353 58.1
12	0 01.3	10 00.7	24 03.0
14	30 01.6	10 02.5	54 07.9
16	60 01.9	10 04.2	84 12.9
18	90 02.2	10 06.0	114 17.8
20	120 02.5	10 07.8	144 22.7
22	150 02.8	N10 09.6	174 27.6

GMT	SUN GHA ° ′	SUN Dec ° ′	ARIES GHA ° ′
Monday, 16th April			
00	180 03.1	N10 11.3	204 32.6
02	210 03.3	10 13.1	234 37.5
04	240 03.6	10 14.9	264 42.4
06	270 03.9	10 16.6	294 47.3
08	300 04.2	10 18.4	324 52.3
10	330 04.5	10 20.2	354 57.2
12	0 04.8	10 21.9	25 02.1
14	30 05.1	10 23.7	55 07.1
16	60 05.4	10 25.5	85 12.0
18	90 05.7	10 27.2	115 16.9
20	120 05.9	10 29.0	145 21.8
22	150 06.2	N10 30.7	175 26.8
Tuesday, 17th April			
00	180 06.5	N10 32.5	205 31.7
02	210 06.8	10 34.3	235 36.6
04	240 07.1	10 36.0	265 41.6
06	270 07.4	10 37.8	295 46.5
08	300 07.7	10 39.5	325 51.4
10	330 07.9	10 41.3	355 56.3
12	0 08.2	10 43.0	26 01.3
14	30 08.5	10 44.8	56 06.2
16	60 08.8	10 46.5	86 11.1
18	90 09.1	10 48.3	116 16.1
20	120 09.3	10 50.0	146 21.0
22	150 09.6	N10 51.7	176 25.9
Wednesday, 18th April			
00	180 09.9	N10 53.5	206 30.8
02	210 10.2	10 55.2	236 35.8
04	240 10.4	10 57.0	266 40.7
06	270 10.7	10 58.7	296 45.6
08	300 11.0	11 00.4	326 50.6
10	330 11.3	11 02.2	356 55.5
12	0 11.5	11 03.9	27 00.4
14	30 11.8	11 05.7	57 05.3
16	60 12.1	11 07.4	87 10.3
18	90 12.4	11 09.1	117 15.2
20	120 12.6	11 10.9	147 20.1
22	150 12.9	N11 12.6	177 25.0
Thursday, 19th April			
00	180 13.2	N11 14.3	207 30.0
02	210 13.4	11 16.0	237 34.9
04	240 13.7	11 17.8	267 39.8
06	270 14.0	11 19.5	297 44.8
08	300 14.2	11 21.2	327 49.7
10	330 14.5	11 22.9	357 54.6
12	0 14.8	11 24.6	27 59.5
14	30 15.0	11 26.4	58 04.5
16	60 15.3	11 28.1	88 09.4
18	90 15.5	11 29.8	118 14.3
20	120 15.8	11 31.5	148 19.3
22	150 16.1	N11 33.2	178 24.2
Friday, 20th April			
00	180 16.3	N11 34.9	208 29.1
02	210 16.6	11 36.6	238 34.0
04	240 16.8	11 38.4	268 39.0
06	270 17.1	11 40.1	298 43.9
08	300 17.4	11 41.8	328 48.8
10	330 17.6	11 43.5	358 53.8
12	0 17.9	11 45.2	28 58.7
14	30 18.1	11 46.9	59 03.6
16	60 18.4	11 48.6	89 08.5
18	90 18.6	11 50.3	119 13.5
20	120 18.9	11 52.0	149 18.4
22	150 19.1	N11 53.7	179 23.3

GMT	SUN GHA ° ′	SUN Dec ° ′	ARIES GHA ° ′
Saturday, 21st April			
00	180 19.4	N11 55.4	209 28.3
02	210 19.6	11 57.1	239 33.2
04	240 19.9	11 58.8	269 38.1
06	270 20.1	12 00.5	299 43.0
08	300 20.4	12 02.2	329 48.0
10	330 20.6	12 03.8	359 52.9
12	0 20.9	12 05.5	29 57.8
14	30 21.1	12 07.2	60 02.7
16	60 21.3	12 08.9	90 07.7
18	90 21.6	12 10.6	120 12.6
20	120 21.8	12 12.3	150 17.5
22	150 22.1	N12 13.9	180 22.5
Sunday, 22nd April			
00	180 22.3	N12 15.6	210 27.4
02	210 22.6	12 17.3	240 32.3
04	240 22.8	12 19.0	270 37.2
06	270 23.0	12 20.7	300 42.2
08	300 23.3	12 22.3	330 47.1
10	330 23.5	12 24.0	0 52.0
12	0 23.7	12 25.7	30 57.0
14	30 24.0	12 27.3	61 01.9
16	60 24.2	12 29.0	91 06.8
18	90 24.5	12 30.7	121 11.7
20	120 24.7	12 32.3	151 16.7
22	150 24.9	N12 34.0	181 21.6
Monday, 23rd April			
00	180 25.1	N12 35.7	211 26.5
02	210 25.4	12 37.3	241 31.5
04	240 25.6	12 39.0	271 36.4
06	270 25.8	12 40.6	301 41.3
08	300 26.1	12 42.3	331 46.2
10	330 26.3	12 44.0	1 51.2
12	0 26.5	12 45.6	31 56.1
14	30 26.7	12 47.3	62 01.0
16	60 27.0	12 48.9	92 06.0
18	90 27.2	12 50.6	122 10.9
20	120 27.4	12 52.2	152 15.8
22	150 27.6	N12 53.9	182 20.7
Tuesday, 24th April			
00	180 27.9	N12 55.5	212 25.7
02	210 28.1	12 57.1	242 30.6
04	240 28.3	12 58.8	272 35.5
06	270 28.5	13 00.4	302 40.5
08	300 28.7	13 02.1	332 45.4
10	330 29.0	13 03.7	2 50.3
12	0 29.2	13 05.3	32 55.2
14	30 29.4	13 07.0	63 00.2
16	60 29.6	13 08.6	93 05.1
18	90 29.8	13 10.2	123 10.0
20	120 30.0	13 11.9	153 14.9
22	150 30.3	N13 13.5	183 19.9
Wednesday, 25th April			
00	180 30.5	N13 15.1	213 24.8
02	210 30.7	13 16.7	243 29.7
04	240 30.9	13 18.4	273 34.7
06	270 31.1	13 20.0	303 39.6
08	300 31.3	13 21.6	333 44.5
10	330 31.5	13 23.2	3 49.4
12	0 31.7	13 24.9	33 54.4
14	30 31.9	13 26.5	63 59.3
16	60 32.1	13 28.1	94 04.2
18	90 32.3	13 29.7	124 09.2
20	120 32.5	13 31.3	154 14.1
22	150 32.7	N13 32.9	184 19.0

GMT	SUN GHA ° ′	SUN Dec ° ′	ARIES GHA ° ′
Thursday, 26th April			
00	180 32.9	N13 34.5	214 23.9
02	210 33.2	13 36.1	244 28.9
04	240 33.4	13 37.7	274 33.8
06	270 33.6	13 39.3	304 38.7
08	300 33.7	13 40.9	334 43.7
10	330 33.9	13 42.5	4 48.6
12	0 34.1	13 44.1	34 53.5
14	30 34.3	13 45.7	64 58.4
16	60 34.5	13 47.3	95 03.4
18	90 34.7	13 48.9	125 08.3
20	120 34.9	13 50.5	155 13.2
22	150 35.1	N13 52.1	185 18.2
Friday, 27th April			
00	180 35.3	N13 53.7	215 23.1
02	210 35.5	13 55.3	245 28.0
04	240 35.7	13 56.9	275 32.9
06	270 35.9	13 58.5	305 37.9
08	300 36.1	14 00.0	335 42.8
10	330 36.3	14 01.6	5 47.7
12	0 36.4	14 03.2	35 52.7
14	30 36.6	14 04.8	65 57.6
16	60 36.8	14 06.4	96 02.5
18	90 37.0	14 07.9	126 07.4
20	120 37.2	14 09.5	156 12.4
22	150 37.4	N14 11.1	186 17.3
Saturday, 28th April			
00	180 37.5	N14 12.7	216 22.2
02	210 37.7	14 14.2	246 27.2
04	240 37.9	14 15.8	276 32.1
06	270 38.1	14 17.4	306 37.0
08	300 38.3	14 18.9	336 41.9
10	330 38.4	14 20.5	6 46.9
12	0 38.6	14 22.0	36 51.8
14	30 38.8	14 23.6	66 56.7
16	60 39.0	14 25.2	97 01.6
18	90 39.1	14 26.7	127 06.6
20	120 39.3	14 28.3	157 11.5
22	150 39.5	N14 29.8	187 16.4
Sunday, 29th April			
00	180 39.7	N14 31.4	217 21.4
02	210 39.8	14 32.9	247 26.3
04	240 40.0	14 34.5	277 31.2
06	270 40.2	14 36.0	307 36.1
08	300 40.3	14 37.5	337 41.1
10	330 40.5	14 39.1	7 46.0
12	0 40.7	14 40.6	37 50.9
14	30 40.8	14 42.2	67 55.9
16	60 41.0	14 43.7	98 00.8
18	90 41.2	14 45.2	128 05.7
20	120 41.3	14 46.8	158 10.6
22	150 41.5	N14 48.3	188 15.6
Monday, 30th April			
00	180 41.7	N14 49.8	218 20.5
02	210 41.8	14 51.4	248 25.4
04	240 42.0	14 52.9	278 30.4
06	270 42.1	14 54.4	308 35.3
08	300 42.3	14 55.9	338 40.2
10	330 42.4	14 57.5	8 45.1
12	0 42.6	14 59.0	38 50.1
14	30 42.8	15 00.5	68 55.0
16	60 42.9	15 02.0	98 59.9
18	90 43.1	15 03.5	129 04.9
20	120 43.2	15 05.0	159 09.8
22	150 43.4	N15 06.6	189 14.7

APRIL 2012

MOON

Day	GMT hr	GHA ° '	Mean Var/hr 14°+	Dec ° '	Mean Var/hr '
1 Sun	0	73 09.3	29.8	N17 16.1	7.9
	6	160 08.3	29.8	N16 29.0	8.5
	12	247 07.4	29.9	N15 38.4	9.0
	18	334 06.6	29.9	N14 44.4	9.6
2 Mon	0	61 05.9	29.8	N13 47.3	10.1
	6	148 05.0	29.9	N12 47.1	10.6
	12	235 04.1	29.8	N11 44.0	11.0
	18	322 03.1	29.8	N10 38.2	11.4
3 Tu	0	49 01.7	29.7	N 9 29.7	11.8
	6	136 00.0	29.7	N 8 18.8	12.2
	12	222 57.8	29.5	N 7 05.7	12.5
	18	309 54.9	29.4	N 5 50.5	12.9
4 Wed	0	36 51.3	29.3	N 4 33.6	13.1
	6	123 46.9	29.1	N 3 15.2	13.3
	12	210 41.3	28.9	N 1 55.5	13.5
	18	297 34.6	28.6	N 0 34.8	13.5
5 Th	0	24 26.6	28.4	S 0 46.5	13.6
	6	111 17.1	28.1	S 2 08.2	13.7
	12	198 05.9	27.8	S 3 30.0	13.6
	18	284 53.0	27.5	S 4 51.4	13.5
6 Fri	0	11 38.1	27.1	S 6 12.2	13.3
	6	98 21.3	26.8	S 7 32.0	13.1
	12	185 02.3	26.5	S 8 50.3	12.7
	18	271 41.2	26.1	S10 06.8	12.3
7 Sat	0	358 17.8	25.7	S11 21.1	11.9
	6	84 52.1	25.3	S12 32.9	11.4
	12	171 24.1	25.0	S13 41.7	10.9
	18	257 54.0	24.6	S14 47.1	10.2
8 Sun	0	344 21.7	24.3	S15 48.9	9.6
	6	70 47.3	23.9	S16 46.6	8.9
	12	157 11.1	23.7	S17 39.9	8.0
	18	243 33.3	23.5	S18 28.6	7.2
9 Mon	0	329 54.0	23.2	S19 12.5	6.4
	6	56 13.6	23.1	S19 51.1	5.5
	12	142 32.4	23.1	S20 24.5	4.6
	18	228 50.8	23.0	S20 52.4	3.6
10 Tu	0	315 09.0	23.0	S21 14.7	2.7
	6	41 27.5	23.2	S21 31.5	1.8
	12	127 46.6	23.4	S21 42.6	0.8
	18	214 06.7	23.6	S21 48.1	0.0
11 Wed	0	300 28.3	23.9	S21 48.2	0.9
	6	26 51.6	24.2	S21 42.8	1.8
	12	113 16.9	24.7	S21 32.1	2.7
	18	199 44.6	25.1	S21 16.4	3.5
12 Th	0	286 14.8	25.5	S20 55.8	4.3
	6	12 47.8	26.1	S20 30.5	5.0
	12	99 23.8	26.5	S20 00.8	5.7
	18	186 02.9	27.1	S19 26.9	6.4
13 Fri	0	272 45.1	27.6	S18 49.1	7.0
	6	359 30.5	28.1	S18 07.6	7.6
	12	86 19.1	28.6	S17 22.7	8.1
	18	173 10.9	29.2	S16 34.7	8.6
14 Sat	0	260 05.7	29.7	S15 43.8	9.0
	6	347 03.7	30.2	S14 50.2	9.3
	12	74 04.5	30.7	S13 54.3	9.7
	18	161 08.2	31.1	S12 56.2	10.0
15 Sun	0	248 14.5	31.5	S11 56.2	10.3
	6	335 23.3	31.9	S10 54.5	10.5
	12	62 34.5	32.2	S 9 51.3	10.8
	18	149 47.8	32.6	S 8 46.8	10.9
16 Mon	0	237 03.1	32.9	S 7 41.2	11.1
	6	324 20.2	33.1	S 6 34.7	11.2
	12	51 38.9	33.4	S 5 27.5	11.3
	18	138 59.1	33.6	S 4 19.8	11.3
17 Tu	0	226 20.6	33.8	S 3 11.7	11.4
	6	313 43.1	33.9	S 2 03.4	11.4
	12	41 06.5	34.0	S 0 55.1	11.4
	18	128 30.6	34.1	N 0 13.1	11.3
18 Wed	0	215 55.3	34.1	N 1 21.1	11.3
	6	303 20.3	34.3	N 2 28.6	11.1
	12	30 45.6	34.3	N 3 35.6	11.0
	18	118 10.9	34.2	N 4 41.8	10.9
19 Th	0	205 36.1	34.1	N 5 47.2	10.7
	6	293 01.1	34.1	N 6 51.6	10.5
	12	20 25.7	34.0	N 7 54.8	10.3
	18	107 49.8	33.9	N 8 56.8	10.1
20 Fri	0	195 13.3	33.8	N 9 57.3	9.8
	6	282 36.0	33.6	N10 56.2	9.6
	12	9 57.9	33.5	N11 53.4	9.2
	18	97 18.8	33.3	N12 48.8	8.9
21 Sat	0	184 38.8	33.1	N13 42.1	8.5
	6	271 57.6	32.9	N14 33.4	8.2
	12	359 15.3	32.7	N15 22.3	7.8
	18	86 31.8	32.5	N16 08.9	7.3
22 Sun	0	173 47.0	32.3	N16 52.9	6.8
	6	261 01.0	32.1	N17 34.3	6.4
	12	348 13.8	31.9	N18 12.9	5.9
	18	75 25.3	31.7	N18 48.6	5.4
23 Mon	0	162 35.6	31.5	N19 21.3	4.9
	6	249 44.7	31.3	N19 50.8	4.3
	12	336 52.6	31.1	N20 17.2	3.8
	18	63 59.5	30.9	N20 40.2	3.2
24 Tu	0	151 05.4	30.8	N20 59.8	2.6
	6	238 10.3	30.7	N21 15.9	2.1
	12	325 14.5	30.6	N21 28.5	1.5
	18	52 17.9	30.5	N21 37.4	0.8
25 Wed	0	139 20.6	30.4	N21 42.8	0.2
	6	226 22.9	30.3	N21 44.4	0.4
	12	313 24.8	30.3	N21 42.3	1.0
	18	40 26.4	30.2	N21 36.4	1.7
26 Th	0	127 27.8	30.3	N21 26.8	2.3
	6	214 29.1	30.2	N21 13.5	2.9
	12	301 30.5	30.2	N20 56.4	3.5
	18	28 31.9	30.3	N20 35.6	4.2
27 Fri	0	115 33.6	30.3	N20 11.2	4.7
	6	202 35.5	30.4	N19 43.1	5.4
	12	289 37.7	30.4	N19 11.5	5.9
	18	16 40.3	30.5	N18 36.3	6.5
28 Sat	0	103 43.3	30.5	N17 57.8	7.0
	6	190 46.6	30.6	N17 15.8	7.6
	12	277 50.3	30.7	N16 30.6	8.1
	18	4 54.3	30.7	N15 42.2	8.6
29 Sun	0	91 58.7	30.7	N14 50.7	9.1
	6	179 03.3	30.7	N13 56.2	9.6
	12	266 08.1	30.8	N12 58.9	10.0
	18	353 12.9	30.8	N11 58.9	10.5
30 Mon	0	80 17.8	30.8	N10 56.2	10.9
	6	167 22.5	30.8	N 9 51.1	11.3
	12	254 26.9	30.7	N 8 43.7	11.6
	18	341 31.0	30.6	N 7 34.2	11.9

PLANETS

VENUS Mer Pass h m	VENUS GHA ° '	VENUS Mean Var/hr 15°+	VENUS Dec ° '	VENUS Mean Var/hr '	Day	JUPITER GHA ° '	JUPITER Mean Var/hr 15°+	JUPITER Dec ° '	JUPITER Mean Var/hr '	JUPITER Mer Pass h m
14 59	135 21.3	0.0	N22 58.7	0.7	1 SUN	148 37.6	1.9	N15 00.0	0.2	14 04
14 59	135 21.6	0.0	N23 15.7	0.7	2 Mon	149 23.7	1.9	N15 04.1	0.2	14 01
14 58	135 22.4	0.0	N23 32.3	0.7	3 Tu	150 09.7	1.9	N15 08.2	0.2	13 58
14 58	135 23.4	0.1	N23 48.3	0.7	4 Wed	150 55.7	1.9	N15 12.3	0.2	13 55
14 58	135 24.9	0.1	N24 03.9	0.6	5 Th	151 41.7	1.9	N15 16.4	0.2	13 51
14 58	135 26.8	0.1	N24 18.9	0.6	6 Fri	152 27.5	1.9	N15 20.5	0.2	13 48
14 58	135 29.1	0.1	N24 33.4	0.6	7 Sat	153 13.4	1.9	N15 24.6	0.2	13 45
14 58	135 31.9	0.1	N24 47.4	0.6	8 SUN	153 59.1	1.9	N15 28.7	0.2	13 42
14 58	135 35.2	0.2	N25 00.9	0.5	9 Mon	154 44.9	1.9	N15 32.8	0.2	13 39
14 57	135 39.0	0.2	N25 13.8	0.5	10 Tu	155 30.5	1.9	N15 36.8	0.2	13 36
14 57	135 43.5	0.2	N25 26.3	0.5	11 Wed	156 16.2	1.9	N15 40.9	0.2	13 33
14 57	135 48.5	0.2	N25 38.2	0.5	12 Th	157 01.7	1.9	N15 45.0	0.2	13 30
14 56	135 54.2	0.3	N25 49.6	0.5	13 Fri	157 47.3	1.9	N15 49.0	0.2	13 27
14 56	136 00.6	0.3	N26 00.4	0.4	14 Sat	158 32.8	1.9	N15 53.1	0.2	13 24
14 55	136 07.6	0.3	N26 10.8	0.4	15 SUN	159 18.2	1.9	N15 57.1	0.2	13 21
14 55	136 15.5	0.4	N26 20.6	0.4	16 Mon	160 03.6	1.9	N16 01.2	0.2	13 18
14 54	136 24.2	0.4	N26 29.9	0.4	17 Tu	160 48.9	1.9	N16 05.2	0.2	13 15
14 53	136 33.7	0.4	N26 38.6	0.3	18 Wed	161 34.2	1.9	N16 09.2	0.2	13 12
14 53	136 44.2	0.5	N26 46.9	0.3	19 Th	162 19.5	1.9	N16 13.2	0.2	13 09
14 52	136 55.5	0.5	N26 54.6	0.3	20 Fri	163 04.8	1.9	N16 17.2	0.2	13 06
14 51	137 08.0	0.6	N27 01.8	0.3	21 Sat	163 50.0	1.9	N16 21.2	0.2	13 03
14 50	137 21.4	0.6	N27 08.5	0.3	22 SUN	164 35.1	1.9	N16 25.2	0.2	13 00
14 49	137 36.0	0.7	N27 14.7	0.2	23 Mon	165 20.2	1.9	N16 29.2	0.2	12 57
14 48	137 51.8	0.7	N27 20.4	0.2	24 Tu	166 05.3	1.9	N16 33.1	0.2	12 54
14 47	138 08.8	0.8	N27 25.5	0.2	25 Wed	166 50.4	1.9	N16 37.1	0.2	12 51
14 45	138 27.1	0.8	N27 30.2	0.2	26 Th	167 35.4	1.9	N16 41.0	0.2	12 48
14 44	138 46.8	0.9	N27 34.3	0.2	27 Fri	168 20.4	1.9	N16 44.9	0.2	12 45
14 43	139 07.9	0.9	N27 37.9	0.1	28 Sat	169 05.4	1.9	N16 48.9	0.2	12 42
14 41	139 30.5	1.0	N27 41.1	0.1	29 SUN	169 50.4	1.9	N16 52.7	0.2	12 39
14 39	139 54.7	1.1	N27 43.7	0.1	30 Mon	170 35.3	1.9	N16 56.6	0.2	12 36

VENUS, Av. Mag. –4.6
SHA April 5 302; 10 297; 15 293; 20 288; 25 285; 30 282

JUPITER, Av. Mag. –2.0
SHA April 5 318; 10 317; 15 316; 20 315; 25 313; 30 312

MARS Mer Pass h m	MARS GHA ° '	MARS Mean Var/hr 15°+	MARS Dec ° '	MARS Mean Var/hr '	Day	SATURN GHA ° '	SATURN Mean Var/hr 15°+	SATURN Dec ° '	SATURN Mean Var/hr '	SATURN Mer Pass h m
21 48	31 53.2	2.9	N12 50.3	0.1	1 SUN	343 25.4	2.6	S 7 55.9	0.1	01 06
21 44	33 02.6	2.9	N12 51.5	0.0	2 Mon	344 28.7	2.6	S 7 54.3	0.1	01 02
21 39	34 11.3	2.8	N12 52.5	0.0	3 Tu	345 31.9	2.6	S 7 52.6	0.1	00 58
21 35	35 19.1	2.8	N12 53.1	0.0	4 Wed	346 35.2	2.6	S 7 51.0	0.1	00 53
21 30	36 26.3	2.8	N12 53.4	0.0	5 Th	347 38.5	2.6	S 7 49.3	0.1	00 49
21 26	37 32.7	2.7	N12 53.4	0.0	6 Fri	348 41.8	2.6	S 7 47.7	0.1	00 45
21 22	38 38.3	2.7	N12 53.2	0.0	7 Sat	349 45.2	2.6	S 7 46.0	0.1	00 41
21 17	39 43.3	2.7	N12 52.6	0.0	8 SUN	350 48.6	2.6	S 7 44.3	0.1	00 37
21 13	40 47.4	2.6	N12 51.8	0.0	9 Mon	351 52.0	2.6	S 7 42.7	0.1	00 32
21 09	41 50.9	2.6	N12 50.7	0.1	10 Tu	352 55.4	2.6	S 7 41.0	0.1	00 28
21 05	42 53.6	2.6	N12 49.3	0.1	11 Wed	353 58.8	2.6	S 7 39.3	0.1	00 24
21 01	43 55.6	2.6	N12 47.6	0.1	12 Th	355 02.2	2.6	S 7 37.7	0.1	00 20
20 57	44 56.9	2.5	N12 45.7	0.1	13 Fri	356 05.7	2.6	S 7 36.0	0.1	00 16
20 53	45 57.4	2.5	N12 43.5	0.1	14 Sat	357 09.2	2.6	S 7 34.3	0.1	00 11
20 49	46 57.3	2.5	N12 41.0	0.1	15 SUN	358 12.6	2.6	S 7 32.7	0.1	00 07
20 45	47 56.4	2.4	N12 38.3	0.1	16 Mon	359 16.1	2.6	S 7 31.0	0.1	00 03
20 41	48 54.9	2.4	N12 35.4	0.1	17 Tu	0 19.6	2.6	S 7 29.3	0.1	23 54
20 37	49 52.6	2.4	N12 32.2	0.1	18 Wed	1 23.0	2.6	S 7 27.7	0.1	23 50
20 33	50 49.7	2.4	N12 28.7	0.2	19 Th	2 26.5	2.6	S 7 26.0	0.1	23 46
20 30	51 46.1	2.3	N12 25.0	0.2	20 Fri	3 30.0	2.6	S 7 24.4	0.1	23 42
20 26	52 41.9	2.3	N12 21.1	0.2	21 Sat	4 33.4	2.6	S 7 22.8	0.1	23 38
20 22	53 37.0	2.3	N12 17.0	0.2	22 SUN	5 36.9	2.6	S 7 21.2	0.1	23 33
20 19	54 31.5	2.2	N12 12.6	0.2	23 Mon	6 40.3	2.6	S 7 19.5	0.1	23 29
20 15	55 25.3	2.2	N12 08.0	0.2	24 Tu	7 43.7	2.6	S 7 17.9	0.1	23 25
20 12	56 18.5	2.2	N12 03.1	0.2	25 Wed	8 47.1	2.6	S 7 16.4	0.1	23 21
20 08	57 11.1	2.2	N11 58.1	0.2	26 Th	9 50.5	2.6	S 7 14.8	0.1	23 17
20 05	58 03.1	2.1	N11 52.8	0.2	27 Fri	10 53.9	2.6	S 7 13.2	0.1	23 12
20 02	58 54.5	2.1	N11 47.4	0.2	28 Sat	11 57.2	2.6	S 7 11.7	0.1	23 08
19 58	59 45.4	2.1	N11 41.7	0.2	29 SUN	13 00.5	2.6	S 7 10.1	0.1	23 04
19 55	60 35.7	2.1	N11 35.8	0.3	30 Mon	14 03.8	2.6	S 7 08.6	0.1	23 00

MARS, Av. Mag. –0.4
SHA April 5 203; 10 203; 15 203; 20 203; 25 203; 30 202

SATURN, Av. Mag. +0.3
SHA April 5 154; 10 154; 15 155; 20 155; 25 155; 30 156

STARS

No.	Name	Mag	Transit h m	Dec ° '	SHA ° '
Oh GMT April 1					
Ψ	ARIES	–	11 19	–	–
1	Alpheratz	2.1	11 28	N29 09.4	357 44.7
2	Ankaa	2.4	11 46	S42 14.3	353 16.9
3	Schedar	2.2	12 00	N56 36.2	349 42.0
4	Diphda	2.0	12 03	S17 55.2	348 57.0
5	Achernar	0.5	12 57	S57 10.5	335 27.9
6	POLARIS	2.0	14 04	N89 19.1	318 36.3
7	Hamal	2.0	13 27	N23 31.2	328 02.0
8	Acamar	3.2	14 17	S40 15.5	315 19.3
9	Menkar	2.5	14 22	N 4 08.2	314 16.2
10	Mirfak	1.8	14 44	N49 54.3	308 42.0
11	Aldebaran	0.9	15 55	N16 31.9	290 50.6
12	Rigel	0.1	16 33	S 8 11.5	281 13.0
13	Capella	0.1	16 36	N46 00.6	280 35.9
14	Bellatrix	1.6	16 44	N 6 21.5	278 33.1
15	Elnath	1.7	16 45	N28 37.0	278 13.9
16	Alnilam	1.7	16 55	S 1 11.9	275 47.4
17	Betelgeuse	0.1–1.2	17 14	N 7 24.4	271 02.3
18	Canopus	–0.7	17 42	S52 42.5	263 56.7
19	Sirius	–1.5	18 04	S16 44.3	258 34.5
20	Adhara	1.5	18 17	S28 59.7	255 13.3
21	Castor	1.6	18 53	N31 51.6	246 09.0
22	Procyon	0.4	18 58	N 5 11.4	245 00.6
23	Pollux	1.1	19 04	N27 59.7	243 28.7
24	Avior	1.9	19 41	S59 33.4	234 18.3
25	Suhail	2.2	20 26	S43 29.4	222 52.9
26	Miaplacidus	1.7	20 31	S69 46.5	221 39.6
27	Alphard	2.0	20 46	S 8 43.0	217 56.8
28	Regulus	1.4	21 26	N11 54.2	207 44.2
29	Dubhe	1.8	22 22	N61 41.1	193 52.2
30	Denebola	2.1	23 07	N14 30.0	182 34.2
31	Gienah	2.6	23 34	S17 36.9	175 52.8
32	Acrux	1.3	23 44	S63 10.3	173 09.6
33	Gacrux	1.6	23 49	S57 11.2	172 01.3
34	Mimosa	1.3	0 09	S59 45.6	167 52.3
35	Alioth	1.8	0 16	N55 53.5	166 20.8
36	Spica	1.0	0 47	S11 13.7	158 31.8
37	Alkaid	1.9	1 09	N49 15.0	152 59.0
38	Hadar	0.6	1 26	S60 26.0	148 48.4
39	Menkent	2.1	1 28	S36 25.9	148 08.2
40	Arcturus	0.0	1 37	N19 06.9	145 56.1
41	Rigil Kent	–0.3	2 01	S60 53.1	139 52.2
42	Zuben'ubi	2.8	2 12	S16 05.7	137 06.0
43	Kochab	2.1	2 11	N74 06.2	137 18.8
44	Alphecca	2.2	2 56	N26 40.2	126 11.4
45	Antares	1.0	3 51	S26 27.5	112 27.0
46	Atria	1.9	4 10	S69 02.7	107 29.2
47	Sabik	2.4	4 31	S15 44.4	102 13.3
48	Shaula	1.6	4 55	S37 06.6	96 22.8
49	Rasalhague	2.1	4 56	N12 33.0	96 07.0
50	Eltanin	2.2	5 17	N51 29.0	90 46.3
51	Kaus Aust.	1.9	5 45	S34 22.5	83 44.8
52	Vega	0.0	5 57	N38 47.5	80 39.4
53	Nunki	2.0	6 16	S26 16.7	75 59.2
54	Altair	0.8	7 11	N 8 54.0	62 09.0
55	Peacock	1.9	7 46	S56 41.4	53 20.6
56	Deneb	1.3	8 02	N45 19.3	49 32.2
57	Enif	2.4	9 04	N 9 55.9	33 48.1
58	Al Na'ir	1.7	9 28	S46 53.8	27 45.0
59	Fomalhaut	1.2	10 18	S29 33.3	15 25.1
60	Markab	2.5	10 25	N15 16.2	13 39.4

SUN

Lat 52°N

Yr	Day of Mth	Week	Transit h m	Semi-Diam	Twilight h m	Sunrise h m	Sunset h m	Twilight h m
92	1	Sun	12 04	16.0	05 00	05 34	18 34	19 09
93	2	Mon	12 03	16.0	04 57	05 32	18 36	19 11
94	3	Tu	12 03	16.0	04 55	05 30	18 38	19 12
95	4	Wed	12 03	16.0	04 53	05 27	18 40	19 14
96	5	Th	12 03	16.0	04 50	05 25	18 41	19 16
97	6	Fri	12 02	16.0	04 48	05 23	18 43	19 18
98	7	Sat	12 02	16.0	04 46	05 20	18 45	19 20
99	8	Sun	12 02	16.0	04 43	05 18	18 46	19 22
100	9	Mon	12 01	16.0	04 41	05 16	18 48	19 23
101	10	Tu	12 01	16.0	04 38	05 14	18 50	19 25
102	11	Wed	12 01	16.0	04 36	05 11	18 51	19 27
103	12	Th	12 01	16.0	04 34	05 09	18 53	19 29
104	13	Fri	12 00	16.0	04 31	05 07	18 55	19 31
105	14	Sat	12 00	16.0	04 29	05 05	18 57	19 33
106	15	Sun	12 00	16.0	04 27	05 03	18 58	19 34
107	16	Mon	12 00	16.0	04 24	05 00	19 00	19 36
108	17	Tu	11 59	16.0	04 22	04 58	19 02	19 38
109	18	Wed	11 59	16.0	04 20	04 56	19 03	19 40
110	19	Th	11 59	15.9	04 17	04 54	19 05	19 42
111	20	Fri	11 59	15.9	04 15	04 52	19 07	19 44
112	21	Sat	11 59	15.9	04 13	04 50	19 09	19 46
113	22	Sun	11 58	15.9	04 11	04 48	19 10	19 48
114	23	Mon	11 58	15.9	04 08	04 46	19 12	19 49
115	24	Tu	11 58	15.9	04 06	04 44	19 14	19 51
116	25	Wed	11 58	15.9	04 04	04 42	19 15	19 53
117	26	Th	11 58	15.9	04 02	04 40	19 17	19 55
118	27	Fri	11 58	15.9	04 00	04 38	19 19	19 57
119	28	Sat	11 57	15.9	03 57	04 36	19 20	19 59
120	29	Sun	11 57	15.9	03 55	04 34	19 22	20 01
121	30	Mon	11 57	15.9	03 53	04 32	19 24	20 03

Lat Corr to Sunrise, Sunset etc.

Lat °	Twilight h m	Sunrise h m	Sunset h m	Twilight h m
N70	–1 50	–1 05	+1 09	+1 50
68	–1 26	–0 53	+0 55	+1 27
66	–1 07	–0 42	+0 44	+1 08
64	–0 52	–0 33	+0 35	+0 53
62	–0 40	–0 26	+0 27	+0 41
N60	–0 30	–0 19	+0 20	+0 30
58	–0 21	–0 13	+0 14	+0 21
56	–0 13	–0 08	+0 09	+0 13
54	–0 06	–0 04	+0 04	+0 06
50	+0 05	+0 04	–0 04	–0 05
N45	+0 17	+0 12	–0 11	–0 17
40	+0 26	+0 19	–0 18	–0 25
35	+0 33	+0 25	–0 24	–0 33
30	+0 40	+0 30	–0 29	–0 40
20	+0 51	+0 38	–0 38	–0 50
N10	+0 59	+0 45	–0 42	–0 59
0	+1 07	+0 52	–0 53	–1 07
S10	+1 12	+0 59	–0 59	–1 13
20	+1 18	+1 05	–1 06	–1 18
30	+1 24	+1 13	–1 15	–1 24
S35	+1 27	+1 17	–1 19	–1 27
40	+1 30	+1 22	–1 24	–1 30
45	+1 34	+1 28	–1 30	–1 34
S50	+1 38	+1 35	–1 36	–1 38

NOTES

The corrections to sunrise etc. are for middle of April.

MOON

Yr	Day of Mth	Week	Age days	Transit (Upper) h m	Diff m	Semi-diam '	Hor Par '	Lat 52°N Moonrise h m	Lat 52°N Moonset h m
92	1	Sun	10	19 47	50	15.5	56.9	12 09	02 44
93	2	Mon	11	20 37	50	15.8	57.8	13 23	03 11
94	3	Tu	12	21 27	52	16.0	58.8	14 41	03 36
95	4	Wed	13	22 19	53	16.3	59.7	16 01	03 59
96	5	Th	14	23 12	55	16.5	60.4	17 25	04 21
97	6	Fri	15	24 07	–	16.6	61.0	18 50	04 45
98	7	Sat	16	00 07	58	16.7	61.2	20 16	05 12
99	8	Sun	17	01 05	61	16.6	61.1	21 40	05 44
100	9	Mon	18	02 06	61	16.5	60.7	22 57	06 24
101	10	Tu	19	03 07	61	16.4	60.1	– –	07 14
102	11	Wed	20	04 08	59	16.2	59.3	00 03	08 15
103	12	Th	21	05 07	55	15.9	58.5	00 55	09 23
104	13	Fri	22	06 02	52	15.7	57.7	01 36	10 35
105	14	Sat	23	06 54	48	15.5	56.9	02 08	11 48
106	15	Sun	24	07 42	45	15.3	56.2	02 33	13 00
107	16	Mon	25	08 27	44	15.1	55.6	02 55	14 11
108	17	Tu	26	09 11	42	15.0	55.1	03 14	15 19
109	18	Wed	27	09 53	43	14.9	54.7	03 33	16 27
110	19	Th	28	10 36	43	14.8	54.4	03 51	17 33
111	20	Fri	29	11 19	44	14.8	54.1	04 11	18 40
112	21	Sat	00	12 03	46	14.7	54.0	04 34	19 45
113	22	Sun	01	12 49	47	14.7	54.0	05 00	20 48
114	23	Mon	02	13 36	48	14.7	54.0	05 31	21 48
115	24	Tu	03	14 24	49	14.8	54.1	06 09	22 42
116	25	Wed	04	15 13	49	14.8	54.4	06 54	23 29
117	26	Th	05	16 02	49	14.9	54.8	07 48	– –
118	27	Fri	06	16 51	49	15.1	55.3	08 48	00 09
119	28	Sat	07	17 40	48	15.2	55.9	09 55	00 43
120	29	Sun	08	18 28	48	15.5	56.7	11 05	01 12
121	30	Mon	09	19 16	50	15.7	57.6	12 19	01 37

Phases of the Moon

	d	h	m
○ Full Moon	6	19	19
☽ Last Quarter	13	10	50
● New Moon	21	07	18
☾ First Quarter	29	09	57

	d	h
Perigee	7	17
Apogee	22	14

 MAY 2012

GMT	SUN GHA ° ′	SUN Dec ° ′	ARIES GHA ° ′
	Tuesday, 1st May		
00	180 43.5	**N15 08.1**	219 19.6
02	210 43.7	**15 09.6**	249 24.6
04	240 43.8	**15 11.1**	279 29.5
06	270 44.0	**15 12.6**	309 34.4
08	300 44.1	**15 14.1**	339 39.4
10	330 44.3	**15 15.6**	9 44.3
12	0 44.4	**15 17.1**	39 49.2
14	30 44.5	**15 18.6**	69 54.1
16	60 44.7	**15 20.1**	99 59.1
18	90 44.8	**15 21.6**	130 04.0
20	120 45.0	**15 23.1**	160 08.9
22	150 45.1	**N15 24.5**	190 13.8
	Wednesday, 2nd May		
00	180 45.2	**N15 26.0**	220 18.8
02	210 45.4	**15 27.5**	250 23.7
04	240 45.5	**15 29.0**	280 28.6
06	270 45.7	**15 30.5**	310 33.6
08	300 45.8	**15 32.0**	340 38.5
10	330 45.9	**15 33.4**	10 43.4
12	0 46.1	**15 34.9**	40 48.3
14	30 46.2	**15 36.4**	70 53.3
16	60 46.3	**15 37.9**	100 58.2
18	90 46.5	**15 39.3**	131 03.1
20	120 46.6	**15 40.8**	161 08.1
22	150 46.7	**N15 42.3**	191 13.0
	Thursday, 3rd May		
00	180 46.8	**N15 43.8**	221 17.9
02	210 47.0	**15 45.2**	251 22.8
04	240 47.1	**15 46.7**	281 27.8
06	270 47.2	**15 48.1**	311 32.7
08	300 47.4	**15 49.6**	341 37.6
10	330 47.5	**15 51.1**	11 42.6
12	0 47.6	**15 52.5**	41 47.5
14	30 47.7	**15 54.0**	71 52.4
16	60 47.8	**15 55.4**	101 57.3
18	90 48.0	**15 56.9**	132 02.3
20	120 48.1	**15 58.3**	162 07.2
22	150 48.2	**N15 59.8**	192 12.1
	Friday, 4th May		
00	180 48.3	**N16 01.2**	222 17.1
02	210 48.4	**16 02.7**	252 22.0
04	240 48.5	**16 04.1**	282 26.9
06	270 48.7	**16 05.5**	312 31.8
08	300 48.8	**16 07.0**	342 36.8
10	330 48.9	**16 08.4**	12 41.7
12	0 49.0	**16 09.8**	42 46.6
14	30 49.1	**16 11.3**	72 51.5
16	60 49.2	**16 12.7**	102 56.5
18	90 49.3	**16 14.1**	133 01.4
20	120 49.4	**16 15.6**	163 06.3
22	150 49.5	**N16 17.0**	193 11.3
	Saturday, 5th May		
00	180 49.6	**N16 18.4**	223 16.2
02	210 49.7	**16 19.8**	253 21.1
04	240 49.9	**16 21.2**	283 26.0
06	270 50.0	**16 22.7**	313 31.0
08	300 50.1	**16 24.1**	343 35.9
10	330 50.2	**16 25.5**	13 40.8
12	0 50.3	**16 26.9**	43 45.8
14	30 50.4	**16 28.3**	73 50.7
16	60 50.5	**16 29.7**	103 55.6
18	90 50.6	**16 31.1**	134 00.5
20	120 50.6	**16 32.5**	164 05.5
22	150 50.7	**N16 33.9**	194 10.4

GMT	SUN GHA ° ′	SUN Dec ° ′	ARIES GHA ° ′
	Sunday, 6th May		
00	180 50.8	**N16 35.3**	224 15.3
02	210 50.9	**16 36.7**	254 20.3
04	240 51.0	**16 38.1**	284 25.2
06	270 51.1	**16 39.5**	314 30.1
08	300 51.2	**16 40.9**	344 35.0
10	330 51.3	**16 42.3**	14 40.0
12	0 51.4	**16 43.7**	44 44.9
14	30 51.5	**16 45.1**	74 49.8
16	60 51.6	**16 46.4**	104 54.8
18	90 51.6	**16 47.8**	134 59.7
20	120 51.7	**16 49.2**	165 04.6
22	150 51.8	**N16 50.6**	195 09.5
	Monday, 7th May		
00	180 51.9	**N16 52.0**	225 14.5
02	210 52.0	**16 53.3**	255 19.4
04	240 52.0	**16 54.7**	285 24.3
06	270 52.1	**16 56.1**	315 29.3
08	300 52.2	**16 57.4**	345 34.2
10	330 52.3	**16 58.8**	15 39.1
12	0 52.4	**17 00.2**	45 44.0
14	30 52.4	**17 01.5**	75 49.0
16	60 52.5	**17 02.9**	105 53.9
18	90 52.6	**17 04.3**	135 58.8
20	120 52.7	**17 05.6**	166 03.8
22	150 52.7	**N17 07.0**	196 08.7
	Tuesday, 8th May		
00	180 52.8	**N17 08.3**	226 13.6
02	210 52.9	**17 09.7**	256 18.5
04	240 52.9	**17 11.0**	286 23.5
06	270 53.0	**17 12.4**	316 28.4
08	300 53.1	**17 13.7**	346 33.3
10	330 53.1	**17 15.1**	16 38.2
12	0 53.2	**17 16.4**	46 43.2
14	30 53.3	**17 17.7**	76 48.1
16	60 53.3	**17 19.1**	106 53.0
18	90 53.4	**17 20.4**	136 58.0
20	120 53.4	**17 21.7**	167 02.9
22	150 53.5	**N17 23.1**	197 07.8
	Wednesday, 9th May		
00	180 53.6	**N17 24.4**	227 12.7
02	210 53.6	**17 25.7**	257 17.7
04	240 53.7	**17 27.1**	287 22.6
06	270 53.7	**17 28.4**	317 27.5
08	300 53.8	**17 29.7**	347 32.5
10	330 53.8	**17 31.0**	17 37.4
12	0 53.9	**17 32.3**	47 42.3
14	30 53.9	**17 33.6**	77 47.2
16	60 54.0	**17 35.0**	107 52.2
18	90 54.0	**17 36.3**	137 57.1
20	120 54.1	**17 37.6**	168 02.0
22	150 54.1	**N17 38.9**	198 07.0
	Thursday, 10th May		
00	180 54.2	**N17 40.2**	228 11.9
02	210 54.2	**17 41.5**	258 16.8
04	240 54.3	**17 42.8**	288 21.7
06	270 54.3	**17 44.1**	318 26.7
08	300 54.3	**17 45.4**	348 31.6
10	330 54.4	**17 46.7**	18 36.5
12	0 54.4	**17 48.0**	48 41.5
14	30 54.5	**17 49.3**	78 46.4
16	60 54.5	**17 50.6**	108 51.3
18	90 54.5	**17 51.8**	138 56.2
20	120 54.6	**17 53.1**	169 01.2
22	150 54.6	**N17 54.4**	199 06.1

GMT	SUN GHA ° ′	SUN Dec ° ′	ARIES GHA ° ′
	Friday, 11th May		
00	180 54.6	**N17 55.7**	229 11.0
02	210 54.7	**17 57.0**	259 16.0
04	240 54.7	**17 58.2**	289 20.9
06	270 54.7	**17 59.5**	319 25.8
08	300 54.8	**18 00.8**	349 30.7
10	330 54.8	**18 02.1**	19 35.7
12	0 54.8	**18 03.3**	49 40.6
14	30 54.9	**18 04.6**	79 45.5
16	60 54.9	**18 05.8**	109 50.5
18	90 54.9	**18 07.1**	139 55.4
20	120 54.9	**18 08.4**	170 00.3
22	150 54.9	**N18 09.6**	200 05.2
	Saturday, 12th May		
00	180 55.0	**N18 10.9**	230 10.2
02	210 55.0	**18 12.1**	260 15.1
04	240 55.0	**18 13.4**	290 20.0
06	270 55.0	**18 14.6**	320 25.0
08	300 55.0	**18 15.9**	350 29.9
10	330 55.1	**18 17.1**	20 34.8
12	0 55.1	**18 18.4**	50 39.7
14	30 55.1	**18 19.6**	80 44.7
16	60 55.1	**18 20.8**	110 49.6
18	90 55.1	**18 22.1**	140 54.5
20	120 55.1	**18 23.3**	170 59.5
22	150 55.1	**N18 24.5**	201 04.4
	Sunday, 13th May		
00	180 55.1	**N18 25.8**	231 09.3
02	210 55.2	**18 27.0**	261 14.2
04	240 55.2	**18 28.2**	291 19.2
06	270 55.2	**18 29.4**	321 24.1
08	300 55.2	**18 30.7**	351 29.0
10	330 55.2	**18 31.9**	21 33.9
12	0 55.2	**18 33.1**	51 38.9
14	30 55.2	**18 34.3**	81 43.8
16	60 55.2	**18 35.5**	111 48.7
18	90 55.2	**18 36.7**	141 53.7
20	120 55.2	**18 37.9**	171 58.6
22	150 55.2	**N18 39.1**	202 03.5
	Monday, 14th May		
00	180 55.2	**N18 40.3**	232 08.4
02	210 55.2	**18 41.5**	262 13.4
04	240 55.2	**18 42.7**	292 18.3
06	270 55.2	**18 43.9**	322 23.2
08	300 55.1	**18 45.1**	352 28.2
10	330 55.1	**18 46.3**	22 33.1
12	0 55.1	**18 47.5**	52 38.0
14	30 55.1	**18 48.7**	82 42.9
16	60 55.1	**18 49.9**	112 47.9
18	90 55.1	**18 51.1**	142 52.8
20	120 55.1	**18 52.3**	172 57.7
22	150 55.1	**N18 53.4**	203 02.7
	Tuesday, 15th May		
00	180 55.0	**N18 54.6**	233 07.6
02	210 55.0	**18 55.8**	263 12.5
04	240 55.0	**18 57.0**	293 17.4
06	270 55.0	**18 58.1**	323 22.4
08	300 55.0	**18 59.3**	353 27.3
10	330 55.0	**19 00.5**	23 32.2
12	0 54.9	**19 01.6**	53 37.2
14	30 54.9	**19 02.8**	83 42.1
16	60 54.9	**19 03.9**	113 47.0
18	90 54.9	**19 05.1**	143 51.9
20	120 54.8	**19 06.2**	173 56.9
22	150 54.8	**N19 07.4**	204 01.8

GMT	SUN GHA ° ′	SUN Dec ° ′	ARIES GHA ° ′
	Wednesday, 16th May		
00	180 54.8	**N19 08.5**	234 06.7
02	210 54.8	**19 09.7**	264 11.6
04	240 54.7	**19 10.8**	294 16.6
06	270 54.7	**19 12.0**	324 21.5
08	300 54.7	**19 13.1**	354 26.4
10	330 54.6	**19 14.3**	24 31.4
12	0 54.6	**19 15.4**	54 36.3
14	30 54.6	**19 16.5**	84 41.2
16	60 54.5	**19 17.7**	114 46.1
18	90 54.5	**19 18.8**	144 51.1
20	120 54.5	**19 19.9**	174 56.0
22	150 54.4	**N19 21.0**	205 00.9
	Thursday, 17th May		
00	180 54.4	**N19 22.2**	235 05.9
02	210 54.3	**19 23.3**	265 10.8
04	240 54.3	**19 24.4**	295 15.7
06	270 54.2	**19 25.5**	325 20.6
08	300 54.2	**19 26.6**	355 25.6
10	330 54.2	**19 27.7**	25 30.5
12	0 54.1	**19 28.9**	55 35.4
14	30 54.1	**19 30.0**	85 40.4
16	60 54.0	**19 31.1**	115 45.3
18	90 54.0	**19 32.2**	145 50.2
20	120 53.9	**19 33.3**	175 55.1
22	150 53.9	**N19 34.4**	206 00.1
	Friday, 18th May		
00	180 53.8	**N19 35.5**	236 05.0
02	210 53.8	**19 36.6**	266 09.9
04	240 53.7	**19 37.6**	296 14.9
06	270 53.7	**19 38.7**	326 19.8
08	300 53.6	**19 39.8**	356 24.7
10	330 53.5	**19 40.9**	26 29.6
12	0 53.5	**19 42.0**	56 34.6
14	30 53.4	**19 43.1**	86 39.5
16	60 53.4	**19 44.1**	116 44.4
18	90 53.3	**19 45.2**	146 49.3
20	120 53.3	**19 46.3**	176 54.3
22	150 53.2	**N19 47.3**	206 59.2
	Saturday, 19th May		
00	180 53.1	**N19 48.4**	237 04.1
02	210 53.1	**19 49.5**	267 09.1
04	240 53.0	**19 50.5**	297 14.0
06	270 52.9	**19 51.6**	327 18.9
08	300 52.9	**19 52.7**	357 23.8
10	330 52.8	**19 53.7**	27 28.8
12	0 52.7	**19 54.8**	57 33.7
14	30 52.7	**19 55.8**	87 38.6
16	60 52.6	**19 56.9**	117 43.6
18	90 52.5	**19 57.9**	147 48.5
20	120 52.4	**19 59.0**	177 53.4
22	150 52.4	**N20 00.0**	207 58.3
	Sunday, 20th May		
00	180 52.3	**N20 01.0**	238 03.3
02	210 52.2	**20 02.1**	268 08.2
04	240 52.1	**20 03.1**	298 13.1
06	270 52.1	**20 04.1**	328 18.1
08	300 52.0	**20 05.2**	358 23.0
10	330 51.9	**20 06.2**	28 27.9
12	0 51.8	**20 07.2**	58 32.8
14	30 51.7	**20 08.2**	88 37.8
16	60 51.7	**20 09.3**	118 42.7
18	90 51.6	**20 10.3**	148 47.6
20	120 51.5	**20 11.3**	178 52.6
22	150 51.4	**N20 12.3**	208 57.5

GMT	SUN GHA ° ′	SUN Dec ° ′	ARIES GHA ° ′
	Monday, 21st May		
00	180 51.3	**N20 13.3**	239 02.4
02	210 51.2	**20 14.3**	269 07.3
04	240 51.2	**20 15.3**	299 12.3
06	270 51.1	**20 16.3**	329 17.2
08	300 51.0	**20 17.3**	359 22.1
10	330 50.9	**20 18.3**	29 27.1
12	0 50.8	**20 19.3**	59 32.0
14	30 50.7	**20 20.3**	89 36.9
16	60 50.6	**20 21.3**	119 41.8
18	90 50.5	**20 22.3**	149 46.8
20	120 50.4	**20 23.3**	179 51.7
22	150 50.3	**N20 24.3**	209 56.6
	Tuesday, 22nd May		
00	180 50.2	**N20 25.2**	240 01.6
02	210 50.1	**20 26.2**	270 06.5
04	240 50.0	**20 27.2**	300 11.4
06	270 49.9	**20 28.2**	330 16.3
08	300 49.8	**20 29.1**	0 21.3
10	330 49.7	**20 30.1**	30 26.2
12	0 49.6	**20 31.1**	60 31.1
14	30 49.5	**20 32.0**	90 36.1
16	60 49.4	**20 33.0**	120 41.0
18	90 49.3	**20 34.0**	150 45.9
20	120 49.2	**20 34.9**	180 50.8
22	150 49.1	**N20 35.9**	210 55.8
	Wednesday, 23rd May		
00	180 49.0	**N20 36.8**	241 00.7
02	210 48.9	**20 37.8**	271 05.6
04	240 48.8	**20 38.7**	301 10.5
06	270 48.7	**20 39.7**	331 15.5
08	300 48.6	**20 40.6**	1 20.4
10	330 48.4	**20 41.5**	31 25.3
12	0 48.3	**20 42.5**	61 30.3
14	30 48.2	**20 43.4**	91 35.2
16	60 48.1	**20 44.3**	121 40.1
18	90 48.0	**20 45.3**	151 45.0
20	120 47.9	**20 46.2**	181 50.0
22	150 47.8	**N20 47.1**	211 54.9
	Thursday, 24th May		
00	180 47.6	**N20 48.1**	241 59.8
02	210 47.5	**20 49.0**	272 04.8
04	240 47.4	**20 49.9**	302 09.7
06	270 47.3	**20 50.8**	332 14.6
08	300 47.2	**20 51.7**	2 19.5
10	330 47.0	**20 52.6**	32 24.5
12	0 46.9	**20 53.5**	62 29.4
14	30 46.8	**20 54.4**	92 34.3
16	60 46.7	**20 55.3**	122 39.3
18	90 46.5	**20 56.2**	152 44.2
20	120 46.4	**20 57.1**	182 49.1
22	150 46.3	**N20 58.0**	212 54.0
	Friday, 25th May		
00	180 46.2	**N20 58.9**	242 59.0
02	210 46.0	**20 59.8**	273 03.9
04	240 45.9	**21 00.7**	303 08.8
06	270 45.8	**21 01.6**	333 13.8
08	300 45.6	**21 02.5**	3 18.7
10	330 45.5	**21 03.3**	33 23.6
12	0 45.4	**21 04.2**	63 28.5
14	30 45.2	**21 05.1**	93 33.5
16	60 45.1	**21 06.0**	123 38.4
18	90 45.0	**21 06.8**	153 43.3
20	120 44.8	**21 07.7**	183 48.3
22	150 44.7	**N21 08.6**	213 53.2

GMT	SUN GHA ° ′	SUN Dec ° ′	ARIES GHA ° ′
	Saturday, 26th May		
00	180 44.6	**N21 09.4**	243 58.1
02	210 44.4	**21 10.3**	274 03.0
04	240 44.3	**21 11.1**	304 08.0
06	270 44.1	**21 12.0**	334 12.9
08	300 44.0	**21 12.8**	4 17.8
10	330 43.9	**21 13.7**	34 22.8
12	0 43.7	**21 14.5**	64 27.7
14	30 43.6	**21 15.4**	94 32.6
16	60 43.4	**21 16.2**	124 37.5
18	90 43.3	**21 17.1**	154 42.5
20	120 43.1	**21 17.9**	184 47.4
22	150 43.0	**N21 18.7**	214 52.3
	Sunday, 27th May		
00	180 42.8	**N21 19.6**	244 57.3
02	210 42.7	**21 20.4**	275 02.2
04	240 42.6	**21 21.2**	305 07.1
06	270 42.4	**21 22.0**	335 12.0
08	300 42.3	**21 22.9**	5 17.0
10	330 42.1	**21 23.7**	35 21.9
12	0 41.9	**21 24.5**	65 26.8
14	30 41.8	**21 25.3**	95 31.7
16	60 41.6	**21 26.1**	125 36.7
18	90 41.5	**21 26.9**	155 41.6
20	120 41.3	**21 27.7**	185 46.5
22	150 41.2	**N21 28.5**	215 51.5
	Monday, 28th May		
00	180 41.0	**N21 29.3**	245 56.4
02	210 40.9	**21 30.1**	276 01.3
04	240 40.7	**21 30.9**	306 06.2
06	270 40.5	**21 31.7**	336 11.2
08	300 40.4	**21 32.5**	6 16.1
10	330 40.2	**21 33.3**	36 21.0
12	0 40.1	**21 34.1**	66 26.0
14	30 39.9	**21 34.9**	96 30.9
16	60 39.7	**21 35.6**	126 35.8
18	90 39.6	**21 36.4**	156 40.7
20	120 39.4	**21 37.2**	186 45.7
22	150 39.3	**N21 38.0**	216 50.6
	Tuesday, 29th May		
00	180 39.1	**N21 38.7**	246 55.5
02	210 38.9	**21 39.5**	277 00.5
04	240 38.8	**21 40.3**	307 05.4
06	270 38.6	**21 41.0**	337 10.3
08	300 38.4	**21 41.8**	7 15.2
10	330 38.2	**21 42.5**	37 20.2
12	0 38.1	**21 43.3**	67 25.1
14	30 37.9	**21 44.0**	97 30.0
16	60 37.7	**21 44.8**	127 35.0
18	90 37.6	**21 45.5**	157 39.9
20	120 37.4	**21 46.3**	187 44.8
22	150 37.2	**N21 47.0**	217 49.7
	Wednesday, 30th May		
00	180 37.0	**N21 47.8**	247 54.7
02	210 36.9	**21 48.5**	277 59.6
04	240 36.7	**21 49.2**	308 04.5
06	270 36.5	**21 50.0**	338 09.5
08	300 36.3	**21 50.7**	8 14.4
10	330 36.2	**21 51.4**	38 19.3
12	0 36.0	**21 52.1**	68 24.2
14	30 35.8	**21 52.9**	98 29.2
16	60 35.6	**21 53.6**	128 34.1
18	90 35.4	**21 54.3**	158 39.0
20	120 35.3	**21 55.0**	188 43.9
22	150 35.1	**N21 55.7**	218 48.9

GMT	SUN GHA ° ′	SUN Dec ° ′	ARIES GHA ° ′
	Thursday, 31st May		
00	180 34.9	**N21 56.4**	248 53.8
02	210 34.7	**21 57.1**	278 58.7
04	240 34.5	**21 57.8**	309 03.7
06	270 34.4	**N21 58.5**	339 08.6
08	300 34.2	**N21 59.2**	9 13.5
10	330 34.0	**21 59.9**	39 18.4
12	0 33.8	**22 00.6**	69 23.4
14	30 33.6	**N22 01.3**	99 28.3
16	60 33.4	**N22 02.0**	129 33.2
18	90 33.2	**22 02.6**	159 38.2
20	120 33.0	**22 03.3**	189 43.1
22	150 32.9	**N22 04.0**	219 48.0

Day	GMT hr	GHA ° ′	Mean Var/hr 14°+	Dec ° ′	Mean Var/hr ′
1 Tu	0	68 34.5	30.4	N 6 22.6	12.3
	6	155 37.3	30.3	N 5 09.3	12.5
	12	242 39.2	30.2	N 3 54.4	12.7
	18	329 40.0	29.9	N 2 38.1	13.0
2 Wed	0	56 39.6	29.7	N 1 20.7	13.1
	6	143 37.8	29.4	N 0 02.4	13.2
	12	230 34.4	29.1	S 1 16.6	13.3
	18	317 29.1	28.8	S 2 35.9	13.2
3 Th	0	44 21.9	28.4	S 3 55.4	13.2
	6	131 12.6	28.0	S 5 14.6	13.1
	12	218 00.9	27.6	S 6 33.2	12.9
	18	304 46.7	27.2	S 7 50.9	12.8
4 Fri	0	31 29.9	26.7	S 9 07.4	12.5
	6	118 10.4	26.2	S10 22.2	12.1
	12	204 48.0	25.7	S11 35.0	11.7
	18	291 22.8	25.3	S12 45.5	11.3
5 Sat	0	17 54.7	24.8	S13 53.1	10.7
	6	104 23.7	24.4	S14 57.5	10.1
	12	190 49.9	23.9	S15 58.3	9.4
	18	277 13.4	23.4	S16 55.2	8.7
6 Sun	0	3 34.4	23.1	S17 47.7	7.9
	6	89 53.0	22.7	S18 35.6	7.1
	12	176 09.7	22.5	S19 18.5	6.2
	18	262 24.6	22.3	S19 56.2	5.3
7 Mon	0	348 38.2	22.1	S20 28.5	4.3
	6	74 50.8	22.0	S20 55.1	3.4
	12	161 02.9	22.0	S21 15.9	2.4
	18	247 15.0	22.1	S21 30.8	1.4
8 Tu	0	333 27.5	22.2	S21 39.9	0.5
	6	59 40.9	22.5	S21 43.1	0.5
	12	145 55.7	22.7	S21 40.5	1.5
	18	232 12.2	23.2	S21 32.2	2.4
9 Wed	0	318 30.9	23.6	S21 18.3	3.2
	6	44 52.1	24.0	S20 59.2	4.1
	12	131 16.2	24.6	S20 35.0	5.0
	18	217 43.3	25.2	S20 05.9	5.7
10 Th	0	304 13.8	25.7	S19 32.4	6.4
	6	30 47.7	26.3	S18 54.5	7.0
	12	117 25.2	26.9	S18 12.7	7.6
	18	204 06.3	27.5	S17 27.3	8.2
11 Fri	0	290 51.1	28.1	S16 38.5	8.7
	6	17 39.4	28.7	S15 46.7	9.2
	12	104 31.2	29.3	S14 52.2	9.5
	18	191 26.5	29.8	S13 55.1	9.9
12 Sat	0	278 25.0	30.3	S12 56.0	10.2
	6	5 26.7	30.8	S11 54.9	10.5
	12	92 31.4	31.3	S10 52.1	10.7
	18	179 38.8	31.7	S 9 47.9	10.9
13 Sun	0	266 48.9	32.1	S 8 42.5	11.1
	6	354 01.4	32.5	S 7 36.1	11.2
	12	81 16.0	32.8	S 6 28.9	11.3
	18	168 32.7	33.1	S 5 21.1	11.3
14 Mon	0	255 51.1	33.4	S 4 12.9	11.4
	6	343 11.1	33.6	S 3 04.5	11.4
	12	70 32.5	33.8	S 1 55.9	11.4
	18	157 55.0	33.9	S 0 47.5	11.3
15 Tu	0	245 18.5	34.1	N 0 20.7	11.3
	6	332 42.7	34.1	N 1 28.6	11.3
	12	60 07.5	34.2	N 2 35.9	11.1
	18	147 32.7	34.2	N 3 42.6	11.0
16 Wed	0	234 58.1	34.3	N 4 48.5	10.8
	6	322 23.5	34.2	N 5 53.5	10.7
	12	49 48.8	34.1	N 6 57.5	10.4
	18	137 13.8	34.1	N 8 00.2	10.2
17 Th	0	224 38.4	33.9	N 9 01.6	10.0
	6	312 02.4	33.8	N10 01.6	9.8
	12	39 25.6	33.7	N11 00.1	9.5
	18	126 48.0	33.5	N11 56.8	9.2
18 Fri	0	214 09.5	33.4	N12 51.6	8.8
	6	301 29.9	33.1	N13 44.5	8.5
	12	28 49.1	33.0	N14 35.3	8.1
	18	116 07.2	32.7	N15 23.9	7.7
19 Sat	0	203 24.0	32.6	N16 10.1	7.2
	6	290 39.5	32.4	N16 53.8	6.8
	12	17 53.6	32.1	N17 34.9	6.4
	18	105 06.5	31.9	N18 13.2	5.9
20 Sun	0	192 17.9	31.7	N18 48.7	5.4
	6	279 28.0	31.5	N19 21.2	4.9
	12	6 36.9	31.3	N19 50.5	4.3
	18	Eclipse of the Sun occurs today			
21 Mon	0	180 50.9	30.8	N20 39.5	3.2
	6	267 56.3	30.7	N20 59.0	2.6
	12	355 00.7	30.6	N21 14.9	2.0
	18	82 04.2	30.5	N21 27.3	1.5
22 Tu	0	169 06.9	30.4	N21 36.1	0.8
	6	256 09.0	30.2	N21 41.2	0.2
	12	343 10.6	30.2	N21 42.6	0.4
	18	70 11.9	30.2	N21 40.2	1.1
23 Wed	0	157 12.9	30.1	N21 34.1	1.7
	6	244 13.8	30.1	N21 24.2	2.3
	12	331 14.8	30.2	N21 10.6	2.9
	18	58 15.9	30.2	N20 53.3	3.5
24 Th	0	145 17.3	30.3	N20 32.3	4.2
	6	232 19.1	30.4	N20 07.6	4.8
	12	319 21.3	30.4	N19 39.4	5.4
	18	46 24.1	30.6	N19 07.7	5.9
25 Fri	0	133 27.5	30.6	N18 32.5	6.5
	6	220 31.6	30.8	N17 54.0	7.0
	12	307 36.3	30.9	N17 12.3	7.5
	18	34 41.6	31.0	N16 27.4	8.1
26 Sat	0	121 47.7	31.2	N15 39.4	8.5
	6	208 54.3	31.2	N14 48.6	9.0
	12	296 01.6	31.3	N13 55.0	9.4
	18	23 09.3	31.3	N12 58.7	9.9
27 Sun	0	110 17.5	31.5	N11 59.8	10.2
	6	197 26.0	31.4	N10 58.6	10.6
	12	284 34.8	31.4	N 9 55.1	11.0
	18	11 43.6	31.4	N 8 49.5	11.3
28 Mon	0	98 52.3	31.5	N 7 42.0	11.6
	6	186 00.9	31.4	N 6 32.7	11.8
	12	273 09.1	31.2	N 5 21.8	12.1
	18	0 16.7	31.2	N 4 09.4	12.3
29 Tu	0	87 23.6	31.0	N 2 55.8	12.4
	6	174 29.6	30.8	N 1 41.2	12.6
	12	261 34.4	30.6	N 0 25.7	12.7
	18	348 37.9	30.3	S 0 50.5	12.7
30 Wed	0	75 39.9	30.0	S 2 07.0	12.8
	6	162 40.2	29.7	S 3 23.7	12.8
	12	249 38.5	29.3	S 4 40.3	12.7
	18	336 34.6	29.0	S 5 56.5	12.6
31 Th	0	63 28.4	28.5	S 7 12.1	12.5
	6	150 19.7	28.0	S 8 26.7	12.2
	12	237 08.3	27.6	S 9 40.1	12.0
	18	323 54.1	27.0	S10 51.8	11.6

VENUS						JUPITER				
Mer Pass h m	GHA ° ′	Mean Var/hr 15°+	Dec ° ′	Mean Var/hr ′	Day	GHA ° ′	Mean Var/hr 15°+	Dec ° ′	Mean Var/hr ′	Mer Pass h m
14 38	140 20.5	1.2	N27 45.9	0.1	1 Tu	171 20.2	1.9	N17 00.5	0.2	12 33
14 36	140 48.1	1.2	N27 47.5	0.0	2 Wed	172 05.1	1.9	N17 04.3	0.2	12 30
14 34	141 17.4	1.3	N27 48.6	0.0	3 Th	172 49.9	1.9	N17 08.2	0.2	12 27
14 31	141 48.5	1.4	N27 49.3	0.0	4 Fri	173 34.8	1.9	N17 12.0	0.2	12 24
14 29	142 21.6	1.5	N27 49.4	0.0	5 Sat	174 19.6	1.9	N17 15.8	0.2	12 21
14 27	142 56.7	1.5	N27 49.0	0.0	6 SUN	175 04.4	1.9	N17 19.6	0.2	12 18
14 24	143 33.9	1.6	N27 48.2	0.1	7 Mon	175 49.2	1.9	N17 23.3	0.2	12 15
14 21	144 13.2	1.7	N27 46.8	0.1	8 Tu	176 34.0	1.9	N17 27.1	0.2	12 12
14 19	144 54.7	1.8	N27 44.9	0.1	9 Wed	177 18.7	1.9	N17 30.8	0.2	12 09
14 16	145 38.5	1.9	N27 42.5	0.1	10 Th	178 03.5	1.9	N17 34.5	0.2	12 06
14 12	146 24.6	2.0	N27 39.5	0.1	11 Fri	178 48.2	1.9	N17 38.2	0.2	12 03
14 09	147 13.1	2.1	N27 36.0	0.2	12 Sat	179 32.9	1.9	N17 41.9	0.1	12 00
14 06	148 04.1	2.2	N27 32.0	0.2	13 SUN	180 17.6	1.9	N17 45.5	0.1	11 57
14 02	148 57.5	2.3	N27 27.4	0.2	14 Mon	181 02.3	1.9	N17 49.1	0.2	11 54
13 58	149 53.5	2.4	N27 22.2	0.2	15 Tu	181 47.0	1.9	N17 52.7	0.1	11 51
13 54	150 52.0	2.6	N27 16.5	0.3	16 Wed	182 31.7	1.9	N17 56.3	0.2	11 48
13 50	151 53.2	2.7	N27 10.2	0.3	17 Th	183 16.3	1.9	N17 59.9	0.1	11 45
13 46	152 57.0	2.8	N27 03.2	0.3	18 Fri	184 01.0	1.9	N18 03.5	0.1	11 42
13 41	154 03.5	2.9	N26 55.7	0.3	19 Sat	184 45.7	1.9	N18 07.0	0.1	11 40
13 36	155 12.7	3.0	N26 47.5	0.4	20 SUN	185 30.3	1.9	N18 10.5	0.1	11 37
13 32	156 24.4	3.1	N26 38.6	0.4	21 Mon	186 15.0	1.9	N18 14.0	0.1	11 34
13 27	157 38.8	3.2	N26 29.1	0.4	22 Tu	186 59.7	1.9	N18 17.4	0.1	11 31
13 21	158 55.8	3.3	N26 19.0	0.5	23 Wed	187 44.3	1.9	N18 20.9	0.1	11 28
13 16	160 15.3	3.4	N26 08.1	0.5	24 Th	188 29.0	1.9	N18 24.3	0.1	11 25
13 10	161 37.3	3.5	N25 56.6	0.5	25 Fri	189 13.7	1.9	N18 27.6	0.1	11 22
13 05	163 01.6	3.6	N25 44.4	0.5	26 Sat	189 58.3	1.9	N18 31.0	0.1	11 19
12 59	164 28.1	3.7	N25 31.5	0.6	27 SUN	190 43.0	1.9	N18 34.3	0.1	11 16
12 53	165 56.8	3.8	N25 17.9	0.6	28 Mon	191 27.7	1.9	N18 37.7	0.1	11 13
12 47	167 27.3	3.9	N25 03.7	0.6	29 Tu	192 12.4	1.9	N18 41.0	0.1	11 10
12 41	168 59.7	3.9	N24 48.8	0.6	30 Wed	192 57.1	1.9	N18 44.2	0.1	11 07
12 34	170 33.6	4.0	N24 33.3	0.7	31 Th	193 41.9	1.9	N18 47.5	0.1	11 04

VENUS, Av. Mag. –4.5
SHA May 5 279; 10 277; 15 277; 20 277; 25 279; 30 281

JUPITER, Av. Mag. –2.0
SHA May 5 311; 10 310; 15 309; 20 307; 25 306; 30 305

MARS						SATURN				
Mer Pass h m	GHA ° ′	Mean Var/hr 15°+	Dec ° ′	Mean Var/hr ′	Day	GHA ° ′	Mean Var/hr 15°+	Dec ° ′	Mean Var/hr ′	Mer Pass h m
19 52	61 25.4	2.0	N11 29.8	0.3	1 Tu	15 07.0	2.6	S 7 07.1	0.1	22 56
19 48	62 14.6	2.0	N11 23.5	0.3	2 Wed	16 10.3	2.6	S 7 05.6	0.1	22 51
19 45	63 03.2	2.0	N11 17.1	0.3	3 Th	17 13.4	2.6	S 7 04.2	0.1	22 47
19 42	63 51.4	2.0	N11 10.5	0.3	4 Fri	18 16.6	2.6	S 7 02.7	0.1	22 43
19 39	64 39.1	2.0	N11 03.7	0.3	5 Sat	19 19.7	2.6	S 7 01.3	0.1	22 39
19 36	65 26.2	1.9	N10 56.7	0.3	6 SUN	20 22.8	2.6	S 6 59.9	0.1	22 35
19 33	66 12.9	1.9	N10 49.6	0.3	7 Mon	21 25.8	2.6	S 6 58.5	0.1	22 30
19 30	66 59.1	1.9	N10 42.2	0.3	8 Tu	22 28.8	2.6	S 6 57.2	0.1	22 26
19 27	67 44.9	1.9	N10 34.8	0.3	9 Wed	23 31.7	2.6	S 6 55.8	0.1	22 22
19 24	68 30.2	1.9	N10 27.1	0.3	10 Th	24 34.6	2.6	S 6 54.5	0.1	22 18
19 21	69 15.1	1.8	N10 19.3	0.3	11 Fri	25 37.4	2.6	S 6 53.2	0.1	22 14
19 18	69 59.5	1.8	N10 11.3	0.3	12 Sat	26 40.2	2.6	S 6 51.9	0.0	22 09
19 15	70 43.5	1.8	N10 03.2	0.3	13 SUN	27 42.9	2.6	S 6 50.7	0.0	22 05
19 12	71 27.1	1.8	N 9 54.9	0.3	14 Mon	28 45.6	2.6	S 6 49.5	0.0	22 01
19 09	72 10.3	1.8	N 9 46.5	0.4	15 Tu	29 48.2	2.6	S 6 48.3	0.0	21 57
19 06	72 53.1	1.8	N 9 37.9	0.4	16 Wed	30 50.8	2.6	S 6 47.1	0.0	21 53
19 03	73 35.6	1.8	N 9 29.1	0.4	17 Th	31 53.3	2.6	S 6 46.0	0.0	21 49
19 01	74 17.6	1.7	N 9 20.3	0.4	18 Fri	32 55.7	2.6	S 6 44.9	0.0	21 45
18 58	74 59.3	1.7	N 9 11.2	0.4	19 Sat	33 58.1	2.6	S 6 43.8	0.0	21 40
18 55	75 40.6	1.7	N 9 02.0	0.4	20 SUN	35 00.4	2.6	S 6 42.7	0.0	21 36
18 52	76 21.5	1.7	N 8 52.7	0.4	21 Mon	36 02.7	2.6	S 6 41.7	0.0	21 32
18 50	77 02.1	1.7	N 8 43.3	0.4	22 Tu	37 04.9	2.6	S 6 40.7	0.0	21 28
18 47	77 42.3	1.7	N 8 33.7	0.4	23 Wed	38 07.0	2.6	S 6 39.8	0.0	21 24
18 44	78 22.3	1.6	N 8 23.9	0.4	24 Th	39 09.0	2.6	S 6 38.8	0.0	21 20
18 42	79 01.8	1.6	N 8 14.1	0.4	25 Fri	40 11.0	2.6	S 6 37.9	0.0	21 16
18 39	79 41.1	1.6	N 8 04.1	0.4	26 Sat	41 12.9	2.6	S 6 37.1	0.0	21 12
18 37	80 20.1	1.6	N 7 54.0	0.4	27 SUN	42 14.7	2.6	S 6 36.2	0.0	21 07
18 34	80 58.7	1.6	N 7 43.7	0.4	28 Mon	43 16.4	2.6	S 6 35.4	0.0	21 03
18 32	81 37.1	1.6	N 7 33.4	0.4	29 Tu	44 18.1	2.6	S 6 34.7	0.0	20 59
18 29	82 15.1	1.6	N 7 22.9	0.4	30 Wed	45 19.7	2.6	S 6 33.9	0.0	20 55
18 27	82 52.9	1.6	N 7 12.3	0.4	31 Th	46 21.2	2.6	S 6 33.2	0.0	20 51

MARS, Av. Mag. +0.2
SHA May 5 201; 10 200; 15 199; 20 198; 25 196; 30 194

SATURN, Av. Mag. +0.4
SHA May 5 156; 10 156; 15 157; 20 157; 25 157; 30 157

MAY 2012

STARS

No.	Name	Mag	Transit h m	Dec ° '	SHA ° '
	Oh GMT May 1				
Ψ	ARIES	–	9 21	–	–
1	Alpheratz	2.1	9 30	N29 09.4	357 44.5
2	Ankaa	2.4	9 48	S42 14.1	353 16.7
3	Schedar	2.2	10 02	N56 36.1	349 41.8
4	Diphda	2.0	10 05	S17 55.1	348 56.9
5	Achernar	0.5	10 59	S57 10.3	335 27.8
6	POLARIS	2.0	12 06	N89 19.0	318 34.7
7	Hamal	2.0	11 29	N23 31.1	328 02.0
8	Acamar	3.2	12 19	S40 15.3	315 19.3
9	Menkar	2.5	12 24	N 4 08.2	314 16.2
10	Mirfak	1.8	12 46	N49 54.2	308 42.0
11	Aldebaran	0.9	13 57	N16 31.9	290 50.6
12	Rigel	0.1	14 35	S 8 11.4	281 13.1
13	Capella	0.1	14 38	N46 00.6	280 36.0
14	Bellatrix	1.6	14 46	N 6 21.5	278 33.1
15	Elnath	1.7	14 47	N28 37.0	278 13.9
16	Alnilam	1.7	14 57	S 1 11.8	275 47.4
17	Betelgeuse	0.1–1.2	15 16	N 7 24.4	271 02.4
18	Canopus	–0.7	15 44	S52 42.5	263 56.9
19	Sirius	–1.5	16 06	S16 44.3	258 34.6
20	Adhara	1.5	16 19	S28 59.6	255 13.4
21	Castor	1.6	16 55	N31 51.6	246 09.2
22	Procyon	0.4	17 00	N 5 11.4	245 00.7
23	Pollux	1.1	17 06	N27 59.7	243 28.9
24	Avior	1.9	17 43	S59 33.4	234 18.6
25	Suhail	2.2	18 28	S43 29.4	222 53.1
26	Miaplacidus	1.7	18 33	S69 46.5	221 40.0
27	Alphard	2.0	18 48	S 8 43.0	217 56.9
28	Regulus	1.4	19 29	N11 54.3	207 44.3
29	Dubhe	1.8	20 24	N61 41.2	193 52.4
30	Denebola	2.1	21 09	N14 30.1	182 34.3
31	Gienah	2.6	21 36	S17 36.9	175 52.9
32	Acrux	1.3	21 46	S63 10.4	173 09.7
33	Gacrux	1.6	21 51	S57 11.3	172 01.4
34	Mimosa	1.3	22 08	S59 45.7	167 52.4
35	Alioth	1.8	22 14	N55 53.6	166 20.9
36	Spica	1.0	22 45	S11 13.7	158 31.8
37	Alkaid	1.9	23 07	N49 15.1	152 59.0
38	Hadar	0.6	23 24	S60 26.1	148 48.4
39	Menkent	2.1	23 26	S36 26.0	148 08.1
40	Arcturus	0.0	23 35	N19 07.0	145 56.1
41	Rigil Kent	–0.3	0 03	S60 53.3	139 52.1
42	Zuben'ubi	2.8	0 14	S16 05.7	137 05.9
43	Kochab	2.1	0 13	N74 06.3	137 18.8
44	Alphecca	2.2	0 58	N26 40.3	126 11.3
45	Antares	1.0	1 53	S26 27.5	112 26.8
46	Atria	1.9	2 12	S69 02.8	107 28.8
47	Sabik	2.4	2 33	S15 44.4	102 13.1
48	Shaula	1.6	2 57	S37 06.6	96 22.6
49	Rasalhague	2.1	2 58	N12 33.1	96 06.9
50	Eltanin	2.2	3 19	N51 29.1	90 46.0
51	Kaus Aust.	1.9	3 47	S34 22.5	83 44.5
52	Vega	0.0	3 59	N38 47.6	80 39.2
53	Nunki	2.0	4 18	S26 16.7	75 59.0
54	Altair	0.8	5 13	N 8 54.1	62 08.8
55	Peacock	1.9	5 48	S56 41.3	53 20.2
56	Deneb	1.3	6 04	N45 19.3	49 31.9
57	Enif	2.4	7 06	N 9 55.9	33 47.9
58	Al Na'ir	1.7	7 30	S46 53.7	27 44.7
59	Fomalhaut	1.2	8 20	S29 33.2	15 24.9
60	Markab	2.5	8 27	N15 16.3	13 39.2

SUN AND MOON

SUN

Yr	Day of Mth	Week	Transit h m	Semi-Diam	Lat 52°N Twilight h m	Sunrise h m	Sunset h m	Twilight h m
122	1	Tu	11 57	15.9	03 51	04 30	19 25	20 05
123	2	Wed	11 57	15.9	03 49	04 28	19 27	20 06
124	3	Th	11 57	15.9	03 47	04 26	19 29	20 08
125	4	Fri	11 57	15.9	03 45	04 24	19 30	20 10
126	5	Sat	11 57	15.9	03 43	04 22	19 32	20 12
127	6	Sun	11 57	15.9	03 41	04 20	19 34	20 14
128	7	Mon	11 57	15.9	03 39	04 19	19 35	20 16
129	8	Tu	11 56	15.9	03 37	04 17	19 37	20 18
130	9	Wed	11 56	15.9	03 35	04 15	19 39	20 20
131	10	Th	11 56	15.9	03 33	04 14	19 40	20 21
132	11	Fri	11 56	15.9	03 31	04 12	19 42	20 23
133	12	Sat	11 56	15.9	03 29	04 10	19 43	20 25
134	13	Sun	11 56	15.9	03 27	04 09	19 45	20 27
135	14	Mon	11 56	15.8	03 25	04 07	19 46	20 29
136	15	Tu	11 56	15.8	03 23	04 06	19 48	20 31
137	16	Wed	11 56	15.8	03 22	04 04	19 50	20 32
138	17	Th	11 56	15.8	03 20	04 03	19 51	20 34
139	18	Fri	11 56	15.8	03 18	04 01	19 53	20 36
140	19	Sat	11 56	15.8	03 17	04 00	19 54	20 38
141	20	Sun	11 57	15.8	03 15	03 59	19 55	20 39
142	21	Mon	11 57	15.8	03 13	03 57	19 57	20 41
143	22	Tu	11 57	15.8	03 12	03 56	19 58	20 43
144	23	Wed	11 57	15.8	03 10	03 55	20 00	20 44
145	24	Th	11 57	15.8	03 09	03 54	20 01	20 46
146	25	Fri	11 57	15.8	03 08	03 52	20 02	20 47
147	26	Sat	11 57	15.8	03 06	03 51	20 04	20 49
148	27	Sun	11 57	15.8	03 05	03 50	20 05	20 50
149	28	Mon	11 57	15.8	03 04	03 49	20 06	20 52
150	29	Tu	11 57	15.8	03 03	03 48	20 07	20 53
151	30	Wed	11 58	15.8	03 01	03 47	20 09	20 55
152	31	Th	11 58	15.8	03 00	03 47	20 10	20 56

Lat Corr to Sunrise, Sunset etc.

Lat °	Twilight h m	Sunrise h m	Sunset h m	Twilight h m
N70	TAN	–3 17	+3 32	TAN
68	TAN	–2 16	+2 20	TAN
66	TAN	–1 42	+1 45	TAN
64	–2 12	–1 17	+1 20	+2 13
62	–1 31	–0 58	+1 01	+1 31
N60	–1 04	–0 43	+0 44	+1 03
58	–0 42	–0 30	+0 30	+0 45
56	–0 27	–0 18	+0 19	+0 27
54	–0 12	–0 08	+0 09	+0 12
50	+0 11	+0 08	–0 08	–0 11
N45	+0 33	+0 25	–0 24	–0 32
40	+0 50	+0 39	–0 38	–0 49
35	+1 04	+0 50	–0 50	–1 03
30	+1 17	+1 00	–1 00	–1 16
20	+1 35	+1 17	–1 17	–1 35
N10	+1 51	+1 32	–1 32	–1 51
0	+2 06	+1 46	–1 46	–2 06
S10	+2 19	+1 59	–2 00	–2 19
20	+2 32	+2 13	–2 15	–2 32
30	+2 46	+2 30	–2 31	–2 47
S35	+2 54	+2 39	–2 41	–2 54
40	+3 03	+2 50	–2 52	–3 03
45	+3 13	+3 03	–3 04	–3 13
S50	+3 24	+3 18	–3 20	–3 24

NOTES
The corrections to sunrise etc. are for middle of May. TAN means Twilight all night.

MOON

Yr	Day of Mth	Week	Age days	Transit (Upper) h m	Diff m	Semi-diam '	Hor Par '	Lat 52°N Moonrise h m	Moonset h m
122	1	Tu	10	20 06	50	16.0	58.6	13 35	01 59
123	2	Wed	11	20 56	53	16.2	59.5	14 55	02 21
124	3	Th	12	21 49	56	16.4	60.3	16 17	02 44
125	4	Fri	13	22 45	60	16.6	61.0	17 42	03 09
126	5	Sat	14	23 45	62	16.7	61.4	19 08	03 38
127	6	Sun	15	24 47	–	16.7	61.4	20 30	04 14
128	7	Mon	16	00 47	64	16.7	61.1	21 44	05 00
129	8	Tu	17	01 51	62	16.5	60.5	22 44	05 57
130	9	Wed	18	02 53	59	16.3	59.8	23 32	07 05
131	10	Th	19	03 52	55	16.0	58.8	– –	08 19
132	11	Fri	20	04 47	50	15.8	57.9	00 08	09 34
133	12	Sat	21	05 37	48	15.5	57.0	00 36	10 49
134	13	Sun	22	06 25	44	15.3	56.2	01 00	12 01
135	14	Mon	23	07 09	43	15.1	55.5	01 20	13 10
136	15	Tu	24	07 52	43	15.0	54.9	01 39	14 18
137	16	Wed	25	08 35	43	14.8	54.5	01 58	15 25
138	17	Th	26	09 18	43	14.8	54.2	02 17	16 31
139	18	Fri	27	10 01	45	14.7	54.0	02 38	17 36
140	19	Sat	28	10 46	47	14.7	54.0	03 03	18 40
141	20	Sun	29	11 33	48	14.7	54.0	03 33	19 41
142	21	Mon	01	12 21	49	14.7	54.1	04 08	20 37
143	22	Tu	02	13 10	49	14.8	54.3	04 52	21 27
144	23	Wed	03	13 59	49	14.9	54.6	05 43	22 10
145	24	Th	04	14 48	49	15.0	55.0	06 41	22 45
146	25	Fri	05	15 37	47	15.1	55.5	07 46	23 15
147	26	Sat	06	16 24	48	15.3	56.0	08 54	23 41
148	27	Sun	07	17 12	47	15.4	56.7	10 05	– –
149	28	Mon	08	17 59	48	15.7	57.5	11 19	00 04
150	29	Tu	09	18 47	50	15.9	58.3	12 34	00 25
151	30	Wed	10	19 37	53	16.1	59.1	13 52	00 46
152	31	Th	11	20 30	56	16.3	59.9	15 13	01 09

Phases of the Moon

		d	h	m
○	Full Moon	6	03	35
☽	Last Quarter	12	21	47
●	New Moon	20	23	47
☾	First Quarter	28	20	16

	d	h
Perigee	6	04
Apogee	19	16

Friday, 1st June

GMT	SUN GHA ° ′	SUN Dec ° ′	ARIES GHA ° ′
00	180 32.7	N22 04.7	249 52.9
02	210 32.5	22 05.3	279 57.9
04	240 32.3	22 06.0	310 02.8
06	270 32.1	22 06.7	340 07.7
08	300 31.9	22 07.3	10 12.7
10	330 31.7	22 08.0	40 17.6
12	0 31.5	22 08.7	70 22.5
14	30 31.3	22 09.3	100 27.4
16	60 31.1	22 10.0	130 32.4
18	90 30.9	22 10.6	160 37.3
20	120 30.7	22 11.3	190 42.2
22	150 30.5	N22 11.9	220 47.2

Saturday, 2nd June

GMT	SUN GHA ° ′	SUN Dec ° ′	ARIES GHA ° ′
00	180 30.3	N22 12.5	250 52.1
02	210 30.1	22 13.2	280 57.0
04	240 29.9	22 13.8	311 01.9
06	270 29.7	22 14.5	341 06.9
08	300 29.5	22 15.1	11 11.8
10	330 29.3	22 15.7	41 16.7
12	0 29.1	22 16.3	71 21.7
14	30 28.9	22 17.0	101 26.6
16	60 28.7	22 17.6	131 31.5
18	90 28.5	22 18.2	161 36.4
20	120 28.3	22 18.8	191 41.4
22	150 28.1	N22 19.4	221 46.3

Sunday, 3rd June

GMT	SUN GHA ° ′	SUN Dec ° ′	ARIES GHA ° ′
00	180 27.9	N22 20.0	251 51.2
02	210 27.7	22 20.6	281 56.1
04	240 27.5	22 21.2	312 01.1
06	270 27.3	22 21.8	342 06.0
08	300 27.1	22 22.4	12 10.9
10	330 26.9	22 23.0	42 15.9
12	0 26.7	22 23.6	72 20.8
14	30 26.5	22 24.2	102 25.7
16	60 26.2	22 24.8	132 30.6
18	90 26.0	22 25.4	162 35.6
20	120 25.8	22 26.0	192 40.5
22	150 25.6	N22 26.6	222 45.4

Monday, 4th June

GMT	SUN GHA ° ′	SUN Dec ° ′	ARIES GHA ° ′
00	180 25.4	N22 27.1	252 50.4
02	210 25.2	22 27.7	282 55.3
04	240 25.0	22 28.3	313 00.2
06	270 24.8	22 28.8	343 05.1
08	300 24.5	22 29.4	13 10.1
10	330 24.3	22 30.0	43 15.0
12	0 24.1	22 30.5	73 19.9
14	30 23.9	22 31.1	103 24.9
16	60 23.7	22 31.6	133 29.8
18	90 23.5	22 32.2	163 34.7
20	120 23.2	22 32.7	193 39.6
22	150 23.0	N22 33.3	223 44.6

Tuesday, 5th June

GMT	SUN GHA ° ′	SUN Dec ° ′	ARIES GHA ° ′
00	180 22.8	N22 33.8	253 49.5
02	210 22.6	22 34.4	283 54.4
04	240 22.4	22 34.9	313 59.4
06	270 22.1	22 35.5	344 04.3
08	300 21.9	22 36.0	14 09.2
10	330 21.7	22 36.5	44 14.1
12	0 21.5	22 37.0	74 19.1
14	30 21.3	22 37.6	104 24.0
16	60 21.0	22 38.1	134 28.9
18	90 20.8	22 38.6	164 33.9
20	120 20.6	22 39.1	194 38.8
22	150 20.4	N22 39.6	224 43.7

Wednesday, 6th June

GMT	SUN GHA ° ′	SUN Dec ° ′	ARIES GHA ° ′
00	180 20.1	N22 40.1	254 48.6
02	210 19.9	22 40.7	284 53.6
04	240 19.7	22 41.2	314 58.5
06	270 19.4	22 41.7	345 03.4
08	300 19.2	22 42.2	15 08.4
10	330 19.0	22 42.7	45 13.3
12	0 18.8	22 43.2	75 18.2
14	30 18.5	22 43.6	105 23.1
16	60 18.3	22 44.1	135 28.1
18	90 18.1	22 44.6	165 33.0
20	120 17.8	22 45.1	195 37.9
22	150 17.6	N22 45.6	225 42.9

Thursday, 7th June

GMT	SUN GHA ° ′	SUN Dec ° ′	ARIES GHA ° ′
00	180 17.4	N22 46.1	255 47.8
02	210 17.1	22 46.5	285 52.7
04	240 16.9	22 47.0	315 57.6
06	270 16.7	22 47.5	346 02.6
08	300 16.4	22 47.9	16 07.5
10	330 16.2	22 48.4	46 12.4
12	0 16.0	22 48.9	76 17.4
14	30 15.7	22 49.3	106 22.3
16	60 15.5	22 49.8	136 27.2
18	90 15.3	22 50.2	166 32.1
20	120 15.0	22 50.7	196 37.1
22	150 14.8	N22 51.1	226 42.0

Friday, 8th June

GMT	SUN GHA ° ′	SUN Dec ° ′	ARIES GHA ° ′
00	180 14.5	N22 51.6	256 46.9
02	210 14.3	22 52.0	286 51.9
04	240 14.1	22 52.5	316 56.8
06	270 13.8	22 52.9	347 01.7
08	300 13.6	22 53.3	17 06.6
10	330 13.4	22 53.8	47 11.6
12	0 13.1	22 54.2	77 16.5
14	30 12.9	22 54.6	107 21.4
16	60 12.6	22 55.0	137 26.4
18	90 12.4	22 55.4	167 31.3
20	120 12.1	22 55.9	197 36.2
22	150 11.9	N22 56.3	227 41.1

Saturday, 9th June

GMT	SUN GHA ° ′	SUN Dec ° ′	ARIES GHA ° ′
00	180 11.7	N22 56.7	257 46.1
02	210 11.4	22 57.1	287 51.0
04	240 11.2	22 57.5	317 55.9
06	270 10.9	22 57.9	348 00.8
08	300 10.7	22 58.3	18 05.8
10	330 10.4	22 58.7	48 10.7
12	0 10.2	22 59.1	78 15.6
14	30 09.9	22 59.5	108 20.6
16	60 09.7	22 59.9	138 25.5
18	90 09.4	23 00.3	168 30.4
20	120 09.2	23 00.6	198 35.3
22	150 08.9	N23 01.0	228 40.3

Sunday, 10th June

GMT	SUN GHA ° ′	SUN Dec ° ′	ARIES GHA ° ′
00	180 08.7	N23 01.4	258 45.2
02	210 08.4	23 01.8	288 50.1
04	240 08.2	23 02.1	318 55.1
06	270 07.9	23 02.5	349 00.0
08	300 07.7	23 02.9	19 04.9
10	330 07.4	23 03.2	49 09.8
12	0 07.2	23 03.6	79 14.8
14	30 06.9	23 04.0	109 19.7
16	60 06.7	23 04.3	139 24.6
18	90 06.4	23 04.7	169 29.6
20	120 06.2	23 05.0	199 34.5
22	150 05.9	N23 05.4	229 39.4

Monday, 11th June

GMT	SUN GHA ° ′	SUN Dec ° ′	ARIES GHA ° ′
00	180 05.7	N23 05.7	259 44.3
02	210 05.4	23 06.0	289 49.3
04	240 05.2	23 06.4	319 54.2
06	270 04.9	23 06.7	349 59.1
08	300 04.6	23 07.0	20 04.1
10	330 04.4	23 07.4	50 09.0
12	0 04.1	23 07.7	80 13.9
14	30 03.9	23 08.0	110 18.8
16	60 03.6	23 08.3	140 23.8
18	90 03.4	23 08.7	170 28.7
20	120 03.1	23 09.0	200 33.6
22	150 02.8	N23 09.3	230 38.6

Tuesday, 12th June

GMT	SUN GHA ° ′	SUN Dec ° ′	ARIES GHA ° ′
00	180 02.6	N23 09.6	260 43.5
02	210 02.3	23 09.9	290 48.4
04	240 02.1	23 10.2	320 53.3
06	270 01.8	23 10.5	350 58.3
08	300 01.6	23 10.8	21 03.2
10	330 01.3	23 11.1	51 08.1
12	0 01.0	23 11.4	81 13.0
14	30 00.8	23 11.7	111 18.0
16	60 00.5	23 12.0	141 22.9
18	90 00.3	23 12.3	171 27.8
20	120 00.0	23 12.5	201 32.8
22	149 59.7	N23 12.8	231 37.7

Wednesday, 13th June

GMT	SUN GHA ° ′	SUN Dec ° ′	ARIES GHA ° ′
00	179 59.5	N23 13.1	261 42.6
02	209 59.2	23 13.4	291 47.5
04	239 58.9	23 13.6	321 52.5
06	269 58.7	23 13.9	351 57.4
08	299 58.4	23 14.2	22 02.3
10	329 58.2	23 14.4	52 07.3
12	359 57.9	23 14.7	82 12.2
14	29 57.6	23 14.9	112 17.1
16	59 57.4	23 15.2	142 22.0
18	89 57.1	23 15.4	172 27.0
20	119 56.8	23 15.7	202 31.9
22	149 56.6	N23 15.9	232 36.8

Thursday, 14th June

GMT	SUN GHA ° ′	SUN Dec ° ′	ARIES GHA ° ′
00	179 56.3	N23 16.2	262 41.8
02	209 56.0	23 16.4	292 46.7
04	239 55.8	23 16.6	322 51.6
06	269 55.5	23 16.9	352 56.5
08	299 55.2	23 17.1	23 01.5
10	329 55.0	23 17.3	53 06.4
12	359 54.7	23 17.5	83 11.3
14	29 54.4	23 17.8	113 16.3
16	59 54.2	23 18.0	143 21.2
18	89 53.9	23 18.2	173 26.1
20	119 53.6	23 18.4	203 31.0
22	149 53.4	N23 18.6	233 36.0

Friday, 15th June

GMT	SUN GHA ° ′	SUN Dec ° ′	ARIES GHA ° ′
00	179 53.1	N23 18.8	263 40.9
02	209 52.8	23 19.0	293 45.8
04	239 52.6	23 19.2	323 50.7
06	269 52.3	23 19.4	353 55.7
08	299 52.0	23 19.6	24 00.6
10	329 51.7	23 19.8	54 05.5
12	359 51.5	23 20.0	84 10.5
14	29 51.2	23 20.2	114 15.4
16	59 50.9	23 20.4	144 20.3
18	89 50.7	23 20.6	174 25.2
20	119 50.4	23 20.7	204 30.2
22	149 50.1	N23 20.9	234 35.1

Saturday, 16th June

GMT	SUN GHA ° ′	SUN Dec ° ′	ARIES GHA ° ′
00	179 49.9	N23 21.1	264 40.0
02	209 49.6	23 21.3	294 45.0
04	239 49.3	23 21.4	324 49.9
06	269 49.0	23 21.6	354 54.8
08	299 48.8	23 21.7	24 59.7
10	329 48.5	23 21.9	55 04.7
12	359 48.2	23 22.1	85 09.6
14	29 48.0	23 22.2	115 14.5
16	59 47.7	23 22.4	145 19.5
18	89 47.4	23 22.5	175 24.4
20	119 47.1	23 22.7	205 29.3
22	149 46.9	N23 22.8	235 34.2

Sunday, 17th June

GMT	SUN GHA ° ′	SUN Dec ° ′	ARIES GHA ° ′
00	179 46.6	N23 22.9	265 39.2
02	209 46.3	23 23.1	295 44.1
04	239 46.1	23 23.2	325 49.0
06	269 45.8	23 23.3	355 54.0
08	299 45.5	23 23.5	25 58.9
10	329 45.2	23 23.6	56 03.8
12	359 45.0	23 23.7	86 08.7
14	29 44.7	23 23.8	116 13.7
16	59 44.4	23 23.9	146 18.6
18	89 44.1	23 24.0	176 23.5
20	119 43.9	23 24.1	206 28.5
22	149 43.6	N23 24.3	236 33.4

Monday, 18th June

GMT	SUN GHA ° ′	SUN Dec ° ′	ARIES GHA ° ′
00	179 43.3	N23 24.4	266 38.3
02	209 43.0	23 24.5	296 43.2
04	239 42.8	23 24.6	326 48.2
06	269 42.5	23 24.7	356 53.1
08	299 42.2	23 24.7	26 58.0
10	329 41.9	23 24.8	57 03.0
12	359 41.7	23 24.9	87 07.9
14	29 41.4	23 25.0	117 12.8
16	59 41.1	23 25.1	147 17.7
18	89 40.9	23 25.2	177 22.7
20	119 40.6	23 25.2	207 27.6
22	149 40.3	N23 25.3	237 32.5

Tuesday, 19th June

GMT	SUN GHA ° ′	SUN Dec ° ′	ARIES GHA ° ′
00	179 40.0	N23 25.4	267 37.5
02	209 39.8	23 25.4	297 42.4
04	239 39.5	23 25.5	327 47.3
06	269 39.2	23 25.6	357 52.2
08	299 38.9	23 25.6	27 57.2
10	329 38.7	23 25.7	58 02.1
12	359 38.4	23 25.7	88 07.0
14	29 38.1	23 25.8	118 12.0
16	59 37.8	23 25.8	148 16.9
18	89 37.6	23 25.9	178 21.8
20	119 37.3	23 25.9	208 26.7
22	149 37.0	N23 25.9	238 31.7

Wednesday, 20th June

GMT	SUN GHA ° ′	SUN Dec ° ′	ARIES GHA ° ′
00	179 36.7	N23 26.0	268 36.6
02	209 36.5	23 26.0	298 41.5
04	239 36.2	23 26.0	328 46.4
06	269 35.9	23 26.1	358 51.4
08	299 35.6	23 26.1	28 56.3
10	329 35.4	23 26.1	59 01.2
12	359 35.1	23 26.1	89 06.2
14	29 34.8	23 26.1	119 11.1
16	59 34.5	23 26.2	149 16.0
18	89 34.3	23 26.2	179 20.9
20	119 34.0	23 26.2	209 25.9
22	149 33.7	N23 26.2	239 30.8

Thursday, 21st June

GMT	SUN GHA ° ′	SUN Dec ° ′	ARIES GHA ° ′
00	179 33.4	N23 26.2	269 35.7
02	209 33.2	23 26.2	299 40.7
04	239 32.9	23 26.2	329 45.6
06	269 32.6	23 26.2	359 50.5
08	299 32.4	23 26.1	29 55.4
10	329 32.1	23 26.1	60 00.4
12	359 31.8	23 26.1	90 05.3
14	29 31.5	23 26.1	120 10.2
16	59 31.3	23 26.1	150 15.2
18	89 31.0	23 26.0	180 20.1
20	119 30.7	23 26.0	210 25.0
22	149 30.4	N23 26.0	240 29.9

Friday, 22nd June

GMT	SUN GHA ° ′	SUN Dec ° ′	ARIES GHA ° ′
00	179 30.2	N23 25.9	270 34.9
02	209 29.9	23 25.9	300 39.8
04	239 29.6	23 25.9	330 44.7
06	269 29.3	23 25.8	0 49.7
08	299 29.1	23 25.8	30 54.6
10	329 28.8	23 25.7	60 59.5
12	359 28.5	23 25.7	91 04.4
14	29 28.3	23 25.6	121 09.4
16	59 28.0	23 25.6	151 14.3
18	89 27.7	23 25.5	181 19.2
20	119 27.4	23 25.4	211 24.2
22	149 27.2	N23 25.4	241 29.1

Saturday, 23rd June

GMT	SUN GHA ° ′	SUN Dec ° ′	ARIES GHA ° ′
00	179 26.9	N23 25.3	271 34.0
02	209 26.6	23 25.2	301 38.9
04	239 26.4	23 25.2	331 43.9
06	269 26.1	23 25.1	1 48.8
08	299 25.8	23 25.0	31 53.7
10	329 25.5	23 24.9	61 58.7
12	359 25.3	23 24.8	92 03.6
14	29 25.0	23 24.8	122 08.5
16	59 24.7	23 24.7	152 13.4
18	89 24.5	23 24.6	182 18.4
20	119 24.2	23 24.5	212 23.3
22	149 23.9	N23 24.4	242 28.2

Sunday, 24th June

GMT	SUN GHA ° ′	SUN Dec ° ′	ARIES GHA ° ′
00	179 23.7	N23 24.3	272 33.2
02	209 23.4	23 24.2	302 38.1
04	239 23.1	23 24.0	332 43.0
06	269 22.9	23 23.9	2 47.9
08	299 22.6	23 23.8	32 52.9
10	329 22.3	23 23.7	62 57.8
12	359 22.1	23 23.6	93 02.7
14	29 21.8	23 23.5	123 07.6
16	59 21.5	23 23.3	153 12.6
18	89 21.3	23 23.2	183 17.5
20	119 21.0	23 23.1	213 22.4
22	149 20.7	N23 22.9	243 27.4

Monday, 25th June

GMT	SUN GHA ° ′	SUN Dec ° ′	ARIES GHA ° ′
00	179 20.5	N23 22.8	273 32.3
02	209 20.2	23 22.7	303 37.2
04	239 19.9	23 22.5	333 42.1
06	269 19.7	23 22.4	3 47.1
08	299 19.4	23 22.2	33 52.0
10	329 19.1	23 22.1	63 56.9
12	359 18.9	23 21.9	94 01.9
14	29 18.6	23 21.8	124 06.8
16	59 18.3	23 21.6	154 11.7
18	89 18.1	23 21.4	184 16.6
20	119 17.8	23 21.3	214 21.6
22	149 17.5	N23 21.1	244 26.5

Tuesday, 26th June

GMT	SUN GHA ° ′	SUN Dec ° ′	ARIES GHA ° ′
00	179 17.3	N23 20.9	274 31.4
02	209 17.0	23 20.8	304 36.4
04	239 16.8	23 20.6	334 41.3
06	269 16.5	23 20.4	4 46.2
08	299 16.2	23 20.2	34 51.1
10	329 16.0	23 20.0	64 56.1
12	359 15.7	23 19.8	95 01.0
14	29 15.4	23 19.7	125 05.9
16	59 15.2	23 19.5	155 10.9
18	89 14.9	23 19.3	185 15.8
20	119 14.7	23 19.1	215 20.7
22	149 14.4	N23 18.9	245 25.6

Wednesday, 27th June

GMT	SUN GHA ° ′	SUN Dec ° ′	ARIES GHA ° ′
00	179 14.1	N23 18.7	275 30.6
02	209 13.9	23 18.4	305 35.5
04	239 13.6	23 18.2	335 40.4
06	269 13.4	23 18.0	5 45.4
08	299 13.1	23 17.8	35 50.3
10	329 12.9	23 17.6	65 55.2
12	359 12.6	23 17.4	96 00.1
14	29 12.3	23 17.1	126 05.1
16	59 12.1	23 16.9	156 10.0
18	89 11.8	23 16.7	186 14.9
20	119 11.6	23 16.4	216 19.8
22	149 11.3	N23 16.2	246 24.8

Thursday, 28th June

GMT	SUN GHA ° ′	SUN Dec ° ′	ARIES GHA ° ′
00	179 11.1	N23 16.0	276 29.7
02	209 10.8	23 15.7	306 34.6
04	239 10.6	23 15.5	336 39.6
06	269 10.3	23 15.2	6 44.5
08	299 10.1	23 15.0	36 49.4
10	329 09.8	23 14.7	66 54.3
12	359 09.5	23 14.5	96 59.3
14	29 09.3	23 14.2	127 04.2
16	59 09.0	23 13.9	157 09.1
18	89 08.8	23 13.7	187 14.1
20	119 08.5	23 13.4	217 19.0
22	149 08.3	N23 13.1	247 23.9

Friday, 29th June

GMT	SUN GHA ° ′	SUN Dec ° ′	ARIES GHA ° ′
00	179 08.0	N23 12.9	277 28.8
02	209 07.8	23 12.6	307 33.8
04	239 07.5	23 12.3	337 38.7
06	269 07.3	23 12.0	7 43.6
08	299 07.0	23 11.7	37 48.6
10	329 06.8	23 11.5	67 53.5
12	359 06.5	23 11.2	97 58.4
14	29 06.3	23 10.9	128 03.3
16	59 06.1	23 10.6	158 08.3
18	89 05.8	23 10.3	188 13.2
20	119 05.6	23 10.0	218 18.1
22	149 05.3	N23 09.7	248 23.1

Saturday, 30th June

GMT	SUN GHA ° ′	SUN Dec ° ′	ARIES GHA ° ′
00	179 05.1	N23 09.4	278 28.0
02	209 04.8	23 09.1	308 32.9
04	239 04.6	23 08.7	338 37.8
06	269 04.3	23 08.4	8 42.8
08	299 04.1	23 08.1	38 47.7
10	329 03.9	23 07.8	68 52.6
12	359 03.6	23 07.5	98 57.6
14	29 03.4	23 07.1	129 02.5
16	59 03.1	23 06.8	159 07.4
18	89 02.9	23 06.5	189 12.3
20	119 02.6	23 06.1	219 17.3
22	149 02.4	N23 05.8	249 22.2

JUNE 2012

MOON

Day	GMT hr	GHA ° '	Mean Var/hr 14°+	Dec ° '	Mean Var/hr '
1 Fri	0	50 36.9	26.6	S12 01.6	11.2
	6	137 16.6	26.0	S13 09.1	10.7
	12	223 53.2	25.5	S14 13.9	10.3
	18	310 26.6	25.0	S15 15.7	9.7
2 Sat	0	36 56.8	24.5	S16 14.2	9.0
	6	123 24.0	24.0	S17 08.8	8.4
	12	209 48.2	23.5	S17 59.4	7.6
	18	296 09.7	23.1	S18 45.5	6.9
3 Sun	0	22 28.5	22.7	S19 26.8	5.9
	6	108 45.1	22.4	S20 03.0	5.1
	12	194 59.7	22.2	S20 34.0	4.1
	18	281 12.8	21.9	S20 59.4	3.2
4 Mon	0	7 24.7	21.9	S21 19.0	2.2
	6	93 36.0	21.9	S21 32.9	1.2
	12	179 47.1	21.9	S21 40.8	0.3
	18	265 58.5	22.0	S21 42.8	0.8
5 Tu	0	352 10.7	22.3	S21 38.9	1.7
	6	78 24.2	22.6	S21 29.2	2.7
	12	164 39.5	23.0	S21 13.8	3.5
	18	250 56.9	23.3	S20 52.9	4.5
6 Wed	0	337 16.9	23.9	S20 26.8	5.3
	6	63 39.7	24.3	S19 55.6	6.0
	12	150 05.7	24.9	S19 19.7	6.8
	18	236 34.9	25.5	S18 39.4	7.4
7 Th	0	323 07.7	26.1	S17 55.0	8.1
	6	49 44.1	26.7	S17 06.9	8.7
	12	136 24.2	27.3	S16 15.3	9.2
	18	223 07.9	28.0	S15 20.6	9.7
8 Fri	0	309 55.3	28.5	S14 23.1	10.0
	6	36 46.2	29.1	S13 23.1	10.3
	12	123 40.5	29.7	S12 21.0	10.7
	18	210 38.1	30.2	S11 17.1	10.9
9 Sat	0	297 38.9	30.6	S10 11.6	11.2
	6	24 42.7	31.2	S 9 04.7	11.3
	12	111 49.3	31.6	S 7 56.8	11.4
	18	198 58.4	31.9	S 6 48.1	11.6
10 Sun	0	286 09.9	32.3	S 5 38.8	11.7
	6	13 23.6	32.6	S 4 29.1	11.6
	12	100 39.3	32.9	S 3 19.1	11.7
	18	187 56.7	33.2	S 2 09.2	11.6
11 Mon	0	275 15.6	33.4	S 0 59.4	11.5
	6	2 35.8	33.6	N 0 10.0	11.5
	12	89 57.1	33.8	N 1 19.0	11.4
	18	177 19.4	33.9	N 2 27.4	11.3
12 Tu	0	264 42.3	33.9	N 3 35.0	11.1
	6	352 05.7	34.0	N 4 41.7	10.9
	12	79 29.4	33.9	N 5 47.4	10.8
	18	166 53.3	34.0	N 6 52.0	10.5
13 Wed	0	254 17.1	34.0	N 7 55.2	10.2
	6	341 40.7	33.9	N 8 57.1	10.0
	12	69 03.9	33.8	N 9 57.5	9.8
	18	156 26.6	33.6	N10 56.2	9.5
14 Th	0	243 48.6	33.5	N11 53.1	9.2
	6	331 09.8	33.4	N12 48.2	8.8
	12	58 30.1	33.2	N13 41.3	8.5
	18	145 49.4	33.0	N14 32.2	8.1
15 Fri	0	233 07.5	32.8	N15 20.9	7.7
	6	320 24.5	32.6	N16 07.3	7.3
	12	47 40.1	32.4	N16 51.1	6.8
	18	134 54.5	32.1	N17 32.4	6.4
16 Sat	0	222 07.5	31.9	N18 10.9	5.9
	6	309 19.1	31.7	N18 46.5	5.4
	12	36 29.4	31.5	N19 19.2	4.9
	18	123 38.3	31.2	N19 48.8	4.3

Day	GMT hr	GHA ° '	Mean Var/hr 14°+	Dec ° '	Mean Var/hr '
17 Sun	0	210 45.9	31.0	N20 15.3	3.8
	6	297 52.2	30.9	N20 38.4	3.2
	12	24 57.3	30.6	N20 58.1	2.7
	18	112 01.3	30.4	N21 14.4	2.1
18 Mon	0	199 04.3	30.3	N21 27.1	1.5
	6	286 06.4	30.2	N21 36.2	0.8
	12	13 07.7	30.2	N21 41.6	0.2
	18	100 08.4	30.0	N21 43.2	0.4
19 Tu	0	187 08.6	30.0	N21 41.1	1.0
	6	274 08.4	30.0	N21 35.2	1.7
	12	1 08.0	29.9	N21 25.5	2.3
	18	88 07.5	30.0	N21 12.0	2.9
20 Wed	0	175 07.2	30.0	N20 54.7	3.5
	6	262 07.0	30.1	N20 33.6	4.1
	12	349 07.2	30.1	N20 08.9	4.8
	18	76 07.8	30.2	N19 40.5	5.4
21 Th	0	163 09.0	30.3	N19 08.6	5.9
	6	250 10.8	30.5	N18 33.2	6.5
	12	337 13.3	30.6	N17 54.4	7.1
	18	64 16.6	30.7	N17 12.4	7.6
22 Fri	0	151 20.6	30.8	N16 27.1	8.1
	6	238 25.5	31.0	N15 38.9	8.6
	12	325 31.2	31.1	N14 47.7	9.0
	18	52 37.6	31.2	N13 53.8	9.5
23 Sat	0	139 44.8	31.3	N12 57.3	9.9
	6	226 52.7	31.4	N11 58.3	10.3
	12	314 01.1	31.5	N10 57.0	10.6
	18	41 10.0	31.5	N 9 53.6	10.9
24 Sun	0	128 19.4	31.6	N 8 48.2	11.2
	6	215 29.0	31.6	N 7 41.0	11.5
	12	302 38.7	31.6	N 6 32.1	11.8
	18	29 48.5	31.6	N 5 21.9	11.9
25 Mon	0	116 58.0	31.5	N 4 10.3	12.1
	6	204 07.2	31.5	N 2 57.7	12.2
	12	291 15.9	31.3	N 1 44.2	12.4
	18	18 23.9	31.2	N 0 30.0	12.4
26 Tu	0	105 30.9	31.0	S 0 44.7	12.5
	6	192 36.9	30.8	S 1 59.6	12.5
	12	279 41.6	30.5	S 3 14.5	12.4
	18	6 44.8	30.2	S 4 29.3	12.4
27 Wed	0	93 46.3	29.9	S 5 43.6	12.3
	6	180 45.9	29.6	S 6 57.2	12.1
	12	267 43.4	29.2	S 8 09.9	11.9
	18	354 38.7	28.8	S 9 21.3	11.7
28 Th	0	81 31.5	28.4	S10 31.3	11.4
	6	168 21.7	27.9	S11 39.5	11.0
	12	255 09.2	27.4	S12 45.5	10.6
	18	341 53.9	26.9	S13 49.2	10.2
29 Fri	0	68 35.7	26.4	S14 50.2	9.6
	6	155 14.5	25.9	S15 48.1	9.1
	12	241 50.3	25.5	S16 42.7	8.4
	18	328 23.2	24.9	S17 33.6	7.8
30 Sat	0	54 53.2	24.5	S18 20.5	7.0
	6	141 20.5	24.1	S19 03.1	6.3
	12	227 45.2	23.7	S19 41.2	5.4
	18	314 07.5	23.4	S20 14.5	4.6

PLANETS

VENUS					JUPITER					
Mer Pass h m	GHA ° '	Mean Var/hr 15°+	Dec ° '	Mean Var/hr '	Day	GHA ° '	Mean Var/hr 15°+	Dec ° '	Mean Var/hr '	Mer Pass h m
12 28	172 08.8	4.0	N24 17.3	0.7	1 Fri	194 26.6	1.9	N18 50.7	0.1	11 01
12 22	173 45.2	4.1	N24 00.7	0.7	2 Sat	195 11.4	1.9	N18 53.9	0.1	10 58
12 15	175 22.5	4.1	N23 43.7	0.7	3 SUN	195 56.1	1.9	N18 57.0	0.1	10 55
12 09	177 00.3	4.1	N23 26.3	0.7	4 Mon	196 40.9	1.9	N19 00.1	0.1	10 52
12 02	178 38.5	4.1	N23 08.5	0.8	5 Tu	197 25.7	1.9	N19 03.3	0.1	10 49
11 56	180 16.6	4.1	N22 50.2	0.7	6 Wed	198 10.6	1.9	N19 06.3	0.1	10 46
11 49	181 54.1	4.1	N22 32.5	0.8	7 Th	198 55.4	1.9	N19 09.4	0.1	10 43
11 43	183 31.8	4.0	N22 14.2	0.8	8 Fri	199 40.3	1.9	N19 12.4	0.1	10 40
11 36	185 08.8	4.0	N21 55.9	0.8	9 Sat	200 25.2	1.9	N19 15.4	0.1	10 37
11 30	186 44.9	4.0	N21 37.7	0.8	10 SUN	201 10.1	1.9	N19 18.4	0.1	10 34
11 24	188 19.9	3.9	N21 19.7	0.7	11 Mon	201 55.0	1.9	N19 21.3	0.1	10 31
11 18	189 53.4	3.8	N21 02.0	0.7	12 Tu	202 40.0	1.9	N19 24.3	0.1	10 28
11 11	191 25.5	3.8	N20 44.6	0.7	13 Wed	203 25.0	1.9	N19 27.2	0.1	10 25
11 06	192 55.8	3.7	N20 27.6	0.7	14 Th	204 10.0	1.9	N19 30.0	0.1	10 22
11 00	194 24.2	3.6	N20 11.1	0.7	15 Fri	204 55.0	1.9	N19 32.9	0.1	10 19
10 54	195 50.6	3.5	N19 55.3	0.6	16 Sat	205 40.1	1.9	N19 35.7	0.1	10 16
10 48	197 14.9	3.4	N19 40.0	0.6	17 SUN	206 25.2	1.9	N19 38.4	0.1	10 13
10 43	198 37.1	3.3	N19 25.5	0.6	18 Mon	207 10.4	1.9	N19 41.2	0.1	10 10
10 38	199 56.9	3.2	N19 11.7	0.5	19 Tu	207 55.6	1.9	N19 43.9	0.1	10 07
10 33	201 14.4	3.1	N18 58.7	0.5	20 Wed	208 40.8	1.9	N19 46.6	0.1	10 04
10 28	202 29.5	3.0	N18 46.5	0.5	21 Th	209 26.1	1.9	N19 49.3	0.1	10 01
10 23	203 42.2	2.9	N18 35.2	0.4	22 Fri	210 11.4	1.9	N19 51.9	0.1	09 58
10 19	204 52.5	2.8	N18 24.7	0.4	23 Sat	210 56.7	1.9	N19 54.5	0.1	09 55
10 14	206 00.4	2.7	N18 15.0	0.4	24 SUN	211 42.1	1.9	N19 57.1	0.1	09 52
10 10	207 05.9	2.6	N18 06.3	0.3	25 Mon	212 27.5	1.9	N19 59.6	0.1	09 49
10 06	208 09.0	2.5	N17 58.4	0.3	26 Tu	213 13.0	1.9	N20 02.2	0.1	09 46
10 02	209 09.7	2.4	N17 51.3	0.3	27 Wed	213 58.6	1.9	N20 04.7	0.1	09 43
09 58	210 08.1	2.3	N17 45.1	0.2	28 Th	214 44.1	1.9	N20 07.1	0.1	09 40
09 54	211 04.2	2.2	N17 39.7	0.2	29 Fri	215 29.8	1.9	N20 09.5	0.1	09 37
09 51	211 58.1	2.2	N17 35.1	0.2	30 Sat	216 15.5	1.9	N20 11.9	0.1	09 34

VENUS, Av. Mag. –4.3
SHA June 5 285; 10 288; 15 291; 20 293; 25 294; 30 294

JUPITER, Av. Mag. –2.0
SHA June 5 304; 10 302; 15 301; 20 300; 25 299; 30 298

MARS					SATURN					
Mer Pass h m	GHA ° '	Mean Var/hr 15°+	Dec ° '	Mean Var/hr '	Day	GHA ° '	Mean Var/hr 15°+	Dec ° '	Mean Var/hr '	Mer Pass h m
18 24	83 30.4	1.6	N 7 01.6	0.5	1 Fri	47 22.6	2.6	S 6 32.6	0.0	20 47
18 22	84 07.6	1.5	N 6 50.8	0.5	2 Sat	48 24.0	2.5	S 6 31.9	0.0	20 43
18 19	84 44.6	1.5	N 6 39.8	0.5	3 SUN	49 25.2	2.6	S 6 31.3	0.0	20 39
18 17	85 21.3	1.5	N 6 28.8	0.5	4 Mon	50 26.4	2.5	S 6 30.8	0.0	20 35
18 14	85 57.8	1.5	N 6 17.6	0.5	5 Tu	51 27.5	2.5	S 6 30.3	0.0	20 31
18 12	86 34.0	1.5	N 6 06.3	0.5	6 Wed	52 28.5	2.5	S 6 29.8	0.0	20 27
18 10	87 10.0	1.5	N 5 55.0	0.5	7 Th	53 29.5	2.5	S 6 29.3	0.0	20 23
18 07	87 45.7	1.5	N 5 43.5	0.5	8 Fri	54 30.3	2.5	S 6 28.9	0.0	20 19
18 05	88 21.2	1.5	N 5 31.9	0.5	9 Sat	55 31.1	2.5	S 6 28.5	0.0	20 15
18 02	88 56.5	1.5	N 5 20.3	0.5	10 SUN	56 31.8	2.5	S 6 28.2	0.0	20 10
18 00	89 31.5	1.5	N 5 08.5	0.5	11 Mon	57 32.4	2.5	S 6 27.9	0.0	20 06
17 58	90 06.3	1.4	N 4 56.6	0.5	12 Tu	58 32.9	2.5	S 6 27.6	0.0	20 02
17 56	90 40.9	1.4	N 4 44.6	0.5	13 Wed	59 33.3	2.5	S 6 27.3	0.0	19 58
17 53	91 15.3	1.4	N 4 32.6	0.5	14 Th	60 33.6	2.5	S 6 27.1	0.0	19 54
17 51	91 49.5	1.4	N 4 20.4	0.5	15 Fri	61 33.8	2.5	S 6 27.0	0.0	19 50
17 49	92 23.4	1.4	N 4 08.2	0.5	16 Sat	62 34.0	2.5	S 6 26.9	0.0	19 46
17 47	92 57.2	1.4	N 3 55.8	0.5	17 SUN	63 34.0	2.5	S 6 26.8	0.0	19 42
17 44	93 30.7	1.4	N 3 43.4	0.5	18 Mon	64 34.0	2.5	S 6 26.7	0.0	19 38
17 42	94 04.1	1.4	N 3 30.9	0.5	19 Tu	65 33.9	2.5	S 6 26.7	0.0	19 34
17 40	94 37.2	1.4	N 3 18.3	0.5	20 Wed	66 33.6	2.5	S 6 26.7	0.0	19 31
17 38	95 10.2	1.4	N 3 05.6	0.5	21 Th	67 33.3	2.5	S 6 26.8	0.0	19 27
17 36	95 42.9	1.4	N 2 52.9	0.5	22 Fri	68 32.9	2.5	S 6 26.9	0.0	19 23
17 33	96 15.5	1.3	N 2 40.0	0.5	23 Sat	69 32.4	2.5	S 6 27.0	0.0	19 19
17 31	96 47.9	1.3	N 2 27.1	0.5	24 SUN	70 31.8	2.5	S 6 27.2	0.0	19 15
17 29	97 20.1	1.3	N 2 14.1	0.5	25 Mon	71 31.1	2.5	S 6 27.4	0.0	19 11
17 27	97 52.1	1.3	N 2 01.0	0.5	26 Tu	72 30.3	2.5	S 6 27.7	0.0	19 07
17 25	98 23.9	1.3	N 1 47.9	0.6	27 Wed	73 29.5	2.5	S 6 28.0	0.0	19 03
17 23	98 55.6	1.3	N 1 34.7	0.6	28 Th	74 28.5	2.5	S 6 28.3	0.0	18 59
17 21	99 27.1	1.3	N 1 21.4	0.6	29 Fri	75 27.4	2.5	S 6 28.7	0.0	18 55
17 19	99 58.4	1.3	N 1 08.1	0.6	30 Sat	76 26.3	2.5	S 6 29.1	0.0	18 51

MARS, Av. Mag. +0.7
SHA June 5 192; 10 190; 15 188; 20 186; 25 184; 30 182

SATURN, Av. Mag. +0.6
SHA June 5 158; 10 158; 15 158; 20 158; 25 158; 30 158

STARS

No.	Name	Mag	Transit h m	Dec ° '	SHA ° '
	Oh GMT June 1				
Ψ	ARIES	–	7 19	–	–
1	Alpheratz	2.1	7 28	N29 09.5	357 44.3
2	Ankaa	2.4	7 46	S42 14.0	353 16.5
3	Schedar	2.2	8 00	N56 36.1	349 41.5
4	Diphda	2.0	8 03	S17 54.9	348 56.7
5	Achernar	0.5	8 57	S57 10.1	335 27.6
6	POLARIS	2.0	10 05	N89 18.8	318 26.3
7	Hamal	2.0	9 27	N23 31.2	328 01.8
8	Acamar	3.2	10 17	S40 15.2	315 19.2
9	Menkar	2.5	10 22	N 4 08.3	314 16.0
10	Mirfak	1.8	10 44	N49 54.1	308 41.8
11	Aldebaran	0.9	11 55	N16 31.9	290 50.5
12	Rigel	0.1	12 34	S 8 11.3	281 13.1
13	Capella	0.1	12 36	N46 00.5	280 36.0
14	Bellatrix	1.6	12 44	N 6 21.5	278 33.1
15	Elnath	1.7	12 45	N28 36.9	278 13.9
16	Alnilam	1.7	12 55	S 1 11.8	275 47.4
17	Betelgeuse	0.1–1.2	13 14	N 7 24.4	271 02.4
18	Canopus	–0.7	13 42	S52 42.3	263 57.0
19	Sirius	–1.5	14 04	S16 44.2	258 34.7
20	Adhara	1.5	14 17	S28 59.5	255 13.4
21	Castor	1.6	14 53	N31 51.6	246 09.2
22	Procyon	0.4	14 58	N 5 11.4	245 00.8
23	Pollux	1.1	15 04	N27 59.7	243 28.9
24	Avior	1.9	15 41	S59 33.3	234 18.8
25	Suhail	2.2	16 26	S43 29.3	222 53.2
26	Miaplacidus	1.7	16 31	S69 46.5	221 40.4
27	Alphard	2.0	16 46	S 8 43.0	217 57.0
28	Regulus	1.4	17 27	N11 54.3	207 44.4
29	Dubhe	1.8	18 22	N61 41.2	193 52.7
30	Denebola	2.1	19 07	N14 30.1	182 34.4
31	Gienah	2.6	19 34	S17 36.9	175 53.0
32	Acrux	1.3	19 45	S63 10.5	173 09.9
33	Gacrux	1.6	19 49	S57 11.3	172 01.5
34	Mimosa	1.3	20 06	S59 45.8	167 52.5
35	Alioth	1.8	20 12	N55 53.7	166 21.1
36	Spica	1.0	20 43	S11 13.7	158 31.8
37	Alkaid	1.9	21 05	N49 15.2	152 59.2
38	Hadar	0.6	21 22	S60 26.2	148 48.5
39	Menkent	2.1	21 24	S36 26.1	148 08.1
40	Arcturus	0.0	21 33	N19 07.1	145 56.1
41	Rigil Kent	–0.3	21 57	S60 53.4	139 52.2
42	Zuben'ubi	2.8	22 08	S16 05.7	137 05.9
43	Kochab	2.1	22 08	N74 06.5	137 19.1
44	Alphecca	2.2	22 52	N26 40.5	126 11.3
45	Antares	1.0	23 47	S26 27.5	112 26.7
46	Atria	1.9	0 11	S69 03.0	107 28.6
47	Sabik	2.4	0 32	S15 44.3	102 13.0
48	Shaula	1.6	0 55	S37 06.7	96 22.4
49	Rasalhague	2.1	0 56	N12 33.2	96 06.8
50	Eltanin	2.2	1 17	N51 29.3	90 45.9
51	Kaus Aust.	1.9	1 45	S34 22.5	83 44.3
52	Vega	0.0	1 58	N38 47.8	80 39.0
53	Nunki	2.0	2 16	S26 16.7	75 58.8
54	Altair	0.8	3 11	N 8 54.2	62 08.6
55	Peacock	1.9	3 47	S56 41.3	53 19.9
56	Deneb	1.3	4 02	N45 19.5	49 31.6
57	Enif	2.4	5 04	N 9 56.0	33 47.6
58	Al Na'ir	1.7	5 29	S46 53.7	27 44.4
59	Fomalhaut	1.2	6 18	S29 33.1	15 24.7
60	Markab	2.5	6 25	N15 16.4	13 39.0

SUN

Lat 52°N

Yr	Day of Mth	Week	Transit h m	Semi-Diam	Twilight h m	Sunrise h m	Sunset h m	Twilight h m
153	1	Fri	11 58	15.8	02 59	03 46	20 11	20 57
154	2	Sat	11 58	15.8	02 58	03 45	20 12	20 59
155	3	Sun	11 58	15.8	02 57	03 44	20 13	21 00
156	4	Mon	11 58	15.8	02 56	03 44	20 14	21 01
157	5	Tu	11 59	15.8	02 56	03 43	20 15	21 02
158	6	Wed	11 59	15.8	02 55	03 42	20 16	21 03
159	7	Th	11 59	15.8	02 54	03 42	20 17	21 04
160	8	Fri	11 59	15.8	02 53	03 41	20 17	21 05
161	9	Sat	11 59	15.8	02 53	03 41	20 18	21 06
162	10	Sun	12 00	15.8	02 52	03 40	20 19	21 07
163	11	Mon	12 00	15.8	02 52	03 40	20 20	21 08
164	12	Tu	12 00	15.8	02 51	03 40	20 20	21 09
165	13	Wed	12 00	15.8	02 51	03 40	20 21	21 09
166	14	Th	12 00	15.8	02 51	03 40	20 21	21 10
167	15	Fri	12 01	15.8	02 51	03 39	20 22	21 11
168	16	Sat	12 01	15.8	02 51	03 39	20 22	21 11
169	17	Sun	12 01	15.8	02 51	03 39	20 23	21 12
170	18	Mon	12 01	15.8	02 51	03 39	20 23	21 12
171	19	Tu	12 01	15.8	02 51	03 39	20 23	21 12
172	20	Wed	12 02	15.8	02 51	03 40	20 24	21 13
173	21	Th	12 02	15.8	02 51	03 40	20 24	21 13
174	22	Fri	12 02	15.8	02 51	03 40	20 24	21 13
175	23	Sat	12 02	15.8	02 51	03 40	20 24	21 13
176	24	Sun	12 03	15.8	02 52	03 41	20 24	21 13
177	25	Mon	12 03	15.8	02 52	03 41	20 24	21 13
178	26	Tu	12 03	15.8	02 53	03 42	20 24	21 13
179	27	Wed	12 03	15.8	02 53	03 42	20 24	21 13
180	28	Th	12 03	15.8	02 54	03 43	20 24	21 12
181	29	Fri	12 04	15.8	02 55	03 43	20 24	21 12
182	30	Sat	12 04	15.8	02 56	03 44	20 23	21 12

Lat Corr to Sunrise, Sunset etc.

Lat °	Twilight h m	Sunrise h m	Sunset h m	Twilight h m
N70	SAH	SAH	SAH	SAH
68	SAH	SAH	SAH	SAH
66	SAH	SAH	SAH	SAH
64	TAN	–2 06	+2 07	TAN
62	TAN	–1 29	+1 30	TAN
N60	–1 57	–1 03	+1 04	+1 58
58	–1 09	–0 43	+0 43	+1 11
56	–0 39	–0 26	+2 26	+0 40
54	–0 17	–0 12	+0 12	+0 18
50	+0 16	+0 11	–0 11	–0 15
N45	+0 45	+0 33	–0 34	–0 46
40	+1 07	+0 51	–0 51	–1 07
35	+1 25	+1 06	–1 06	–1 25
30	+1 40	+1 19	–1 20	–1 40
20	+2 05	+1 41	–1 41	–2 05
N10	+2 26	+2 00	–2 00	–2 26
0	+2 44	+2 18	–2 18	–2 44
S10	+3 01	+2 35	–2 36	–3 01
20	+3 18	+2 54	–2 54	–3 18
30	+3 37	+3 15	–3 15	–3 37
S35	+3 48	+3 27	–3 28	–3 48
40	+4 00	+3 41	–3 42	– 3 59
45	+4 13	+3 58	–3 58	–4 13
S50	+4 29	+4 19	–4 19	–4 29

NOTES
The corrections to sunrise etc. are for middle of June. SAH means Sun above Horizon. TAN means Twilight all night.

MOON

Yr	Day of Mth	Week	Age days	Transit (Upper) h m	Diff m	Semi-diam '	Hor Par '	Lat 52°N Moonrise h m	Lat 52°N Moonset h m
153	1	Fri	12	21 26	60	16.5	60.6	16 37	01 35
154	2	Sat	13	22 26	63	16.6	61.0	18 00	02 06
155	3	Sun	14	23 29	64	16.7	61.2	19 18	02 46
156	4	Mon	15	24 33	–	16.6	61.0	20 26	03 37
157	5	Tu	16	00 33	62	16.5	60.6	21 21	04 40
158	6	Wed	17	01 35	58	16.3	59.9	22 04	05 53
159	7	Th	18	02 33	55	16.1	59.0	22 36	07 11
160	8	Fri	19	03 28	50	15.8	58.1	23 03	08 29
161	9	Sat	20	04 18	47	15.6	57.1	23 25	09 44
162	10	Sun	21	05 05	44	15.3	56.3	23 45	10 57
163	11	Mon	22	05 49	44	15.1	55.5	– –	12 07
164	12	Tu	23	06 33	42	15.0	54.9	00 04	13 15
165	13	Wed	24	07 15	44	14.8	54.5	00 23	14 21
166	14	Th	25	07 59	44	14.8	54.2	00 44	15 27
167	15	Fri	26	08 43	46	14.7	54.1	01 07	16 31
168	16	Sat	27	09 29	48	14.7	54.0	01 35	17 34
169	17	Sun	28	10 17	49	14.8	54.2	02 08	18 32
170	18	Mon	29	11 06	49	14.8	54.4	02 49	19 24
171	19	Tu	30	11 55	50	14.9	54.6	03 37	20 09
172	20	Wed	01	12 45	49	15.0	55.0	04 34	20 48
173	21	Th	02	13 34	49	15.1	55.4	05 37	21 20
174	22	Fri	03	14 23	47	15.2	55.9	06 45	21 46
175	23	Sat	04	15 10	47	15.4	56.4	07 56	22 10
176	24	Sun	05	15 57	47	15.5	57.0	09 08	22 31
177	25	Mon	06	16 44	48	15.7	57.6	10 22	22 52
178	26	Tu	07	17 32	50	15.9	58.2	11 38	23 14
179	27	Wed	08	18 22	53	16.0	58.8	12 55	23 37
180	28	Th	09	19 15	57	16.2	59.4	14 15	– –
181	29	Fri	10	20 12	59	16.3	60.0	15 36	00 05
182	30	Sat	11	21 11	62	16.4	60.3	16 54	00 40

Phases of the Moon

Phase	d	h	m
○ Full Moon	4	11	12
☽ Last Quarter	11	10	41
● New Moon	19	15	02
☾ First Quarter	27	03	30

	d	h
Perigee	3	13
Apogee	16	01

GMT	SUN GHA ° ′	SUN Dec ° ′	ARIES GHA ° ′
Sunday, 1st July			
00	179 02.2	N23 05.5	279 27.1
02	209 01.9	23 05.1	309 32.1
04	239 01.7	23 04.8	339 37.0
06	269 01.5	23 04.4	9 41.9
08	299 01.2	23 04.1	39 46.8
10	329 01.0	23 03.7	69 51.8
12	359 00.7	23 03.3	99 56.7
14	29 00.5	23 03.0	130 01.6
16	59 00.3	23 02.6	160 06.5
18	89 00.0	23 02.3	190 11.5
20	118 59.8	23 01.9	220 16.4
22	148 59.6	N23 01.5	250 21.3
Monday, 2nd July			
00	178 59.3	N23 01.1	280 26.3
02	208 59.1	23 00.8	310 31.2
04	238 58.9	23 00.4	340 36.1
06	268 58.6	23 00.0	10 41.0
08	298 58.4	22 59.6	40 46.0
10	328 58.2	22 59.2	70 50.9
12	358 57.9	22 58.8	100 55.8
14	28 57.7	22 58.4	131 00.8
16	58 57.5	22 58.0	161 05.7
18	88 57.3	22 57.6	191 10.6
20	118 57.0	22 57.2	221 15.5
22	148 56.8	N22 56.8	251 20.5
Tuesday, 3rd July			
00	178 56.6	N22 56.4	281 25.4
02	208 56.3	22 56.0	311 30.3
04	238 56.1	22 55.6	341 35.3
06	268 55.9	22 55.2	11 40.2
08	298 55.7	22 54.8	41 45.1
10	328 55.4	22 54.3	71 50.0
12	358 55.2	22 53.9	101 55.0
14	28 55.0	22 53.5	131 59.9
16	58 54.8	22 53.1	162 04.8
18	88 54.6	22 52.6	192 09.8
20	118 54.3	22 52.2	222 14.7
22	148 54.1	N22 51.8	252 19.6
Wednesday, 4th July			
00	178 53.9	N22 51.3	282 24.5
02	208 53.7	22 50.9	312 29.5
04	238 53.5	22 50.4	342 34.4
06	268 53.2	22 50.0	12 39.3
08	298 53.0	22 49.5	42 44.3
10	328 52.8	22 49.1	72 49.2
12	358 52.6	22 48.6	102 54.1
14	28 52.4	22 48.2	132 59.0
16	58 52.1	22 47.7	163 04.0
18	88 51.9	22 47.2	193 08.9
20	118 51.7	22 46.8	223 13.8
22	148 51.5	N22 46.3	253 18.8
Thursday, 5th July			
00	178 51.3	N22 45.8	283 23.7
02	208 51.1	22 45.3	313 28.6
04	238 50.9	22 44.9	343 33.5
06	268 50.7	22 44.4	13 38.5
08	298 50.4	22 43.9	43 43.4
10	328 50.2	22 43.4	73 48.3
12	358 50.0	22 42.9	103 53.3
14	28 49.8	22 42.4	133 58.2
16	58 49.6	22 41.9	164 03.1
18	88 49.4	22 41.4	194 08.0
20	118 49.2	22 40.9	224 13.0
22	148 49.0	N22 40.4	254 17.9
Friday, 6th July			
00	178 48.8	N22 39.9	284 22.8
02	208 48.6	22 39.4	314 27.8
04	238 48.4	22 38.9	344 32.7
06	268 48.2	22 38.4	14 37.6
08	298 48.0	22 37.8	44 42.5
10	328 47.7	22 37.3	74 47.5
12	358 47.5	22 36.8	104 52.4
14	28 47.3	22 36.3	134 57.3
16	58 47.1	22 35.7	165 02.3
18	88 46.9	22 35.2	195 07.2
20	118 46.7	22 34.7	225 12.1
22	148 46.5	N22 34.1	255 17.0
Saturday, 7th July			
00	178 46.3	N22 33.6	285 22.0
02	208 46.1	22 33.1	315 26.9
04	238 45.9	22 32.5	345 31.8
06	268 45.7	22 32.0	15 36.7
08	298 45.6	22 31.4	45 41.7
10	328 45.4	22 30.9	75 46.6
12	358 45.2	22 30.3	105 51.5
14	28 45.0	22 29.7	135 56.5
16	58 44.8	22 29.2	166 01.4
18	88 44.6	22 28.6	196 06.3
20	118 44.4	22 28.1	226 11.2
22	148 44.2	N22 27.5	256 16.2
Sunday, 8th July			
00	178 44.0	N22 26.9	286 21.1
02	208 43.8	22 26.3	316 26.0
04	238 43.6	22 25.8	346 31.0
06	268 43.4	22 25.2	16 35.9
08	298 43.2	22 24.6	46 40.8
10	328 43.1	22 24.0	76 45.7
12	358 42.9	22 23.4	106 50.7
14	28 42.7	22 22.8	136 55.6
16	58 42.5	22 22.2	167 00.5
18	88 42.3	22 21.6	197 05.5
20	118 42.1	22 21.0	227 10.4
22	148 41.9	N22 20.4	257 15.3
Monday, 9th July			
00	178 41.8	N22 19.8	287 20.2
02	208 41.6	22 19.2	317 25.2
04	238 41.4	22 18.6	347 30.1
06	268 41.2	22 18.0	17 35.0
08	298 41.0	22 17.4	47 40.0
10	328 40.9	22 16.8	77 44.9
12	358 40.7	22 16.2	107 49.8
14	28 40.5	22 15.5	137 54.7
16	58 40.3	22 14.9	167 59.7
18	88 40.1	22 14.3	198 04.6
20	118 40.0	22 13.6	228 09.5
22	148 39.8	N22 13.0	258 14.5
Tuesday, 10th July			
00	178 39.6	N22 12.4	288 19.4
02	208 39.4	22 11.7	318 24.3
04	238 39.3	22 11.1	348 29.2
06	268 39.1	22 10.4	18 34.2
08	298 38.9	22 09.8	48 39.1
10	328 38.8	22 09.1	78 44.0
12	358 38.6	22 08.5	108 48.9
14	28 38.4	22 07.8	138 53.9
16	58 38.2	22 07.2	168 58.8
18	88 38.1	22 06.5	199 03.7
20	118 37.9	22 05.9	229 08.7
22	148 37.7	N22 05.2	259 13.6
Wednesday, 11th July			
00	178 37.6	N22 04.5	289 18.5
02	208 37.4	22 03.9	319 23.4
04	238 37.3	22 03.2	349 28.4
06	268 37.1	22 02.5	19 33.3
08	298 36.9	22 01.8	49 38.2
10	328 36.8	22 01.1	79 43.2
12	358 36.6	22 00.5	109 48.1
14	28 36.4	21 59.8	139 53.0
16	58 36.3	21 59.1	169 57.9
18	88 36.1	21 58.4	200 02.9
20	118 36.0	21 57.7	230 07.8
22	148 35.8	N21 57.0	260 12.7
Thursday, 12th July			
00	178 35.6	N21 56.3	290 17.7
02	208 35.5	21 55.6	320 22.6
04	238 35.3	21 54.9	350 27.5
06	268 35.2	21 54.2	20 32.4
08	298 35.0	21 53.5	50 37.4
10	328 34.9	21 52.8	80 42.3
12	358 34.7	21 52.0	110 47.2
14	28 34.6	21 51.3	140 52.2
16	58 34.4	21 50.6	170 57.1
18	88 34.3	21 49.9	201 02.0
20	118 34.1	21 49.2	231 06.9
22	148 34.0	N21 48.4	261 11.9
Friday, 13th July			
00	178 33.8	N21 47.7	291 16.8
02	208 33.7	21 47.0	321 21.7
04	238 33.5	21 46.2	351 26.7
06	268 33.4	21 45.5	21 31.6
08	298 33.2	21 44.7	51 36.5
10	328 33.1	21 44.0	81 41.4
12	358 33.0	21 43.2	111 46.4
14	28 32.8	21 42.5	141 51.3
16	58 32.7	21 41.7	171 56.2
18	88 32.5	21 41.0	202 01.1
20	118 32.4	21 40.2	232 06.1
22	148 32.3	N21 39.5	262 11.0
Saturday, 14th July			
00	178 32.1	N21 38.7	292 15.9
02	208 32.0	21 37.9	322 20.9
04	238 31.9	21 37.2	352 25.8
06	268 31.7	21 36.4	22 30.7
08	298 31.6	21 35.6	52 35.6
10	328 31.5	21 34.9	82 40.6
12	358 31.3	21 34.1	112 45.5
14	28 31.2	21 33.3	142 50.4
16	58 31.1	21 32.5	172 55.4
18	88 30.9	21 31.7	203 00.3
20	118 30.8	21 30.9	233 05.2
22	148 30.7	N21 30.2	263 10.1
Sunday, 15th July			
00	178 30.5	N21 29.4	293 15.1
02	208 30.4	21 28.6	323 20.0
04	238 30.3	21 27.8	353 24.9
06	268 30.2	21 27.0	23 29.9
08	298 30.0	21 26.2	53 34.8
10	328 29.9	21 25.4	83 39.7
12	358 29.8	21 24.6	113 44.6
14	28 29.7	21 23.7	143 49.6
16	58 29.6	21 22.9	173 54.5
18	88 29.4	21 22.1	203 59.4
20	118 29.3	21 21.3	234 04.4
22	148 29.2	N21 20.5	264 09.3
Monday, 16th July			
00	178 29.1	N21 19.7	294 14.2
02	208 29.0	21 18.8	324 19.1
04	238 28.8	21 18.0	354 24.1
06	268 28.7	21 17.2	24 29.0
08	298 28.6	21 16.3	54 33.9
10	328 28.5	21 15.5	84 38.9
12	358 28.4	21 14.7	114 43.8
14	28 28.3	21 13.8	144 48.7
16	58 28.2	21 13.0	174 53.6
18	88 28.1	21 12.1	204 58.6
20	118 28.0	21 11.3	235 03.5
22	148 27.9	N21 10.4	265 08.4
Tuesday, 17th July			
00	178 27.7	N21 09.6	295 13.4
02	208 27.6	21 08.7	325 18.3
04	238 27.5	21 07.9	355 23.2
06	268 27.4	21 07.0	25 28.1
08	298 27.3	21 06.1	55 33.1
10	328 27.2	21 05.3	85 38.0
12	358 27.1	21 04.4	115 42.9
14	28 27.0	21 03.5	145 47.9
16	58 26.9	21 02.7	175 52.8
18	88 26.8	21 01.8	205 57.7
20	118 26.7	21 00.9	236 02.6
22	148 26.6	N21 00.0	266 07.6
Wednesday, 18th July			
00	178 26.5	N20 59.1	296 12.5
02	208 26.4	20 58.3	326 17.4
04	238 26.4	20 57.4	356 22.4
06	268 26.3	20 56.5	26 27.3
08	298 26.2	20 55.6	56 32.2
10	328 26.1	20 54.7	86 37.1
12	358 26.0	20 53.8	116 42.1
14	28 25.9	20 52.9	146 47.0
16	58 25.8	20 52.0	176 51.9
18	88 25.7	20 51.1	206 56.8
20	118 25.6	20 50.2	237 01.8
22	148 25.6	N20 49.3	267 06.7
Thursday, 19th July			
00	178 25.5	N20 48.4	297 11.6
02	208 25.4	20 47.4	327 16.6
04	238 25.3	20 46.5	357 21.5
06	268 25.2	20 45.6	27 26.4
08	298 25.1	20 44.7	57 31.3
10	328 25.1	20 43.8	87 36.3
12	358 25.0	20 42.8	117 41.2
14	28 24.9	20 41.9	147 46.1
16	58 24.8	20 41.0	177 51.1
18	88 24.8	20 40.0	207 56.0
20	118 24.7	20 39.1	238 00.9
22	148 24.6	N20 38.2	268 05.8
Friday, 20th July			
00	178 24.5	N20 37.2	298 10.8
02	208 24.5	20 36.3	328 15.7
04	238 24.4	20 35.3	358 20.6
06	268 24.3	20 34.4	28 25.6
08	298 24.3	20 33.4	58 30.5
10	328 24.2	20 32.5	88 35.4
12	358 24.1	20 31.5	118 40.3
14	28 24.1	20 30.6	148 45.3
16	58 24.0	20 29.6	178 50.2
18	88 23.9	20 28.6	208 55.1
20	118 23.9	20 27.7	239 00.1
22	148 23.8	N20 26.7	269 05.0
Saturday, 21st July			
00	178 23.8	N20 25.7	299 09.9
02	208 23.7	20 24.8	329 14.8
04	238 23.6	20 23.8	359 19.8
06	268 23.6	20 22.8	29 24.7
08	298 23.5	20 21.8	59 29.6
10	328 23.5	20 20.8	89 34.6
12	358 23.4	20 19.9	119 39.5
14	28 23.4	20 18.9	149 44.4
16	58 23.3	20 17.9	179 49.3
18	88 23.3	20 16.9	209 54.3
20	118 23.2	20 15.9	239 59.2
22	148 23.2	N20 14.9	270 04.1
Sunday, 22nd July			
00	178 23.1	N20 13.9	300 09.1
02	208 23.1	20 12.9	330 14.0
04	238 23.0	20 11.9	0 18.9
06	268 23.0	20 10.9	30 23.8
08	298 22.9	20 09.9	60 28.8
10	328 22.9	20 08.9	90 33.7
12	358 22.8	20 07.9	120 38.6
14	28 22.8	20 06.8	150 43.5
16	58 22.8	20 05.8	180 48.5
18	88 22.7	20 04.8	210 53.4
20	118 22.7	20 03.8	240 58.3
22	148 22.6	N20 02.8	271 03.3
Monday, 23rd July			
00	178 22.6	N20 01.7	301 08.2
02	208 22.6	20 00.7	331 13.1
04	238 22.5	19 59.7	1 18.0
06	268 22.5	19 58.6	31 23.0
08	298 22.5	19 57.6	61 27.9
10	328 22.4	19 56.6	91 32.8
12	358 22.4	19 55.5	121 37.8
14	28 22.4	19 54.5	151 42.7
16	58 22.4	19 53.4	181 47.6
18	88 22.3	19 52.4	211 52.5
20	118 22.3	19 51.3	241 57.5
22	148 22.3	N19 50.3	272 02.4
Tuesday, 24th July			
00	178 22.3	N19 49.2	302 07.3
02	208 22.2	19 48.2	332 12.3
04	238 22.2	19 47.1	2 17.2
06	268 22.2	19 46.1	32 22.1
08	298 22.2	19 45.0	62 27.0
10	328 22.2	19 43.9	92 32.0
12	358 22.1	19 42.9	122 36.9
14	28 22.1	19 41.8	152 41.8
16	58 22.1	19 40.7	182 46.8
18	88 22.1	19 39.6	212 51.7
20	118 22.1	19 38.6	242 56.6
22	148 22.1	N19 37.5	273 01.5
Wednesday, 25th July			
00	178 22.1	N19 36.4	303 06.5
02	208 22.0	19 35.3	333 11.4
04	238 22.0	19 34.2	3 16.3
06	268 22.0	19 33.1	33 21.2
08	298 22.0	19 32.1	63 26.2
10	328 22.0	19 31.0	93 31.1
12	358 22.0	19 29.9	123 36.0
14	28 22.0	19 28.8	153 41.0
16	58 22.0	19 27.7	183 45.9
18	88 22.0	19 26.6	213 50.8
20	118 22.0	19 25.5	243 55.7
22	148 22.0	N19 24.4	274 00.7
Thursday, 26th July			
00	178 22.0	N19 23.3	304 05.6
02	208 22.0	19 22.1	334 10.5
04	238 22.0	19 21.0	4 15.5
06	268 22.0	19 19.9	34 20.4
08	298 22.0	19 18.8	64 25.3
10	328 22.0	19 17.7	94 30.2
12	358 22.0	19 16.6	124 35.2
14	28 22.1	19 15.4	154 40.1
16	58 22.1	19 14.3	184 45.0
18	88 22.1	19 13.2	214 50.0
20	118 22.1	19 12.1	244 54.9
22	148 22.1	N19 10.9	274 59.8
Friday, 27th July			
00	178 22.1	N19 09.8	305 04.7
02	208 22.1	19 08.6	335 09.7
04	238 22.2	19 07.5	5 14.6
06	268 22.2	19 06.4	35 19.5
08	298 22.2	19 05.2	65 24.5
10	328 22.2	19 04.1	95 29.4
12	358 22.2	19 02.9	125 34.3
14	28 22.3	19 01.8	155 39.2
16	58 22.3	19 00.6	185 44.2
18	88 22.3	18 59.5	215 49.1
20	118 22.3	18 58.3	245 54.0
22	148 22.4	N18 57.2	275 59.0
Saturday, 28th July			
00	178 22.4	N18 56.0	306 03.9
02	208 22.4	18 54.8	336 08.8
04	238 22.4	18 53.7	6 13.7
06	268 22.5	18 52.5	36 18.7
08	298 22.5	18 51.3	66 23.6
10	328 22.5	18 50.2	96 28.5
12	358 22.6	18 49.0	126 33.4
14	28 22.6	18 47.8	156 38.4
16	58 22.6	18 46.6	186 43.3
18	88 22.7	18 45.5	216 48.2
20	118 22.7	18 44.3	246 53.2
22	148 22.8	N18 43.1	276 58.1
Sunday, 29th July			
00	178 22.8	N18 41.9	307 03.0
02	208 22.8	18 40.7	337 07.9
04	238 22.9	18 39.5	7 12.9
06	268 22.9	18 38.3	37 17.8
08	298 23.0	18 37.1	67 22.7
10	328 23.0	18 35.9	97 27.7
12	358 23.1	18 34.7	127 32.6
14	28 23.1	18 33.5	157 37.5
16	58 23.2	18 32.3	187 42.4
18	88 23.2	18 31.1	217 47.4
20	118 23.3	18 29.9	247 52.3
22	148 23.3	N18 28.7	277 57.2
Monday, 30th July			
00	178 23.4	N18 27.5	308 02.2
02	208 23.4	18 26.3	338 07.1
04	238 23.5	18 25.1	8 12.0
06	268 23.5	18 23.9	38 16.9
08	298 23.6	18 22.6	68 21.9
10	328 23.7	18 21.4	98 26.8
12	358 23.7	18 20.2	128 31.7
14	28 23.8	18 19.0	158 36.7
16	58 23.9	18 17.7	188 41.6
18	88 23.9	18 16.5	218 46.5
20	118 24.0	18 15.3	248 51.4
22	148 24.0	N18 14.0	278 56.4
Tuesday, 31st July			
00	178 24.1	N18 12.8	309 01.3
02	208 24.2	18 11.6	339 06.2
04	238 24.2	18 10.3	9 11.2
06	268 24.3	N18 09.1	39 16.1
08	298 24.4	N18 07.8	69 21.0
10	328 24.5	18 06.6	99 25.9
12	358 24.5	18 05.3	129 30.9
14	28 24.6	N18 04.1	159 35.8
16	58 24.7	N18 02.8	189 40.7
18	88 24.8	18 01.6	219 45.7
20	118 24.8	18 00.3	249 50.6
22	148 24.9	N17 59.1	279 55.5

MOON

Day	GMT hr	GHA ° '	Mean Var/hr 14°+	Dec ° '	Mean Var/hr '
1 Sun	0	40 27.7	23.0	S20 42.7	3.7
	6	126 46.2	22.8	S21 05.6	2.8
	12	213 03.3	22.7	S21 23.0	1.9
	18	299 19.4	22.6	S21 34.9	1.0
2 Mon	0	25 34.9	22.5	S21 41.1	0.0
	6	111 50.3	22.6	S21 41.6	0.9
	12	198 05.9	22.8	S21 36.4	1.9
	18	284 22.4	23.0	S21 25.5	2.8
3 Tu	0	10 40.0	23.2	S21 09.1	3.7
	6	96 59.2	23.5	S20 47.2	4.6
	12	183 20.5	23.9	S20 20.2	5.5
	18	269 44.0	24.4	S19 48.1	6.2
4 Wed	0	356 10.1	24.8	S19 11.3	7.0
	6	82 39.0	25.4	S18 30.0	7.7
	12	169 11.0	25.9	S17 44.5	8.3
	18	255 46.1	26.5	S16 55.2	8.9
5 Th	0	342 24.5	27.0	S16 02.3	9.4
	6	69 06.3	27.5	S15 06.3	9.8
	12	155 51.3	28.1	S14 07.3	10.3
	18	242 39.7	28.7	S13 05.8	10.6
6 Fri	0	329 31.2	29.1	S12 02.1	11.0
	6	56 25.9	29.6	S10 56.4	11.2
	12	143 23.6	30.2	S 9 49.2	11.4
	18	230 24.2	30.5	S 8 40.6	11.7
7 Sat	0	317 27.4	31.0	S 7 30.9	11.8
	6	44 33.2	31.4	S 6 20.5	11.8
	12	131 41.4	31.7	S 5 09.5	11.9
	18	218 51.7	32.0	S 3 58.1	11.9
8 Sun	0	306 03.9	32.3	S 2 46.7	11.9
	6	33 18.0	32.6	S 1 35.3	11.8
	12	120 33.6	32.9	S 0 24.2	11.8
	18	207 50.6	33.1	N 0 46.4	11.7
9 Mon	0	295 08.7	33.2	N 1 56.3	11.6
	6	22 27.9	33.3	N 3 05.5	11.3
	12	109 47.8	33.4	N 4 13.8	11.2
	18	197 08.3	33.5	N 5 20.9	10.9
10 Tu	0	284 29.2	33.5	N 6 26.9	10.8
	6	11 50.4	33.5	N 7 31.5	10.5
	12	99 11.6	33.5	N 8 34.6	10.2
	18	186 32.7	33.5	N 9 36.1	10.0
11 Wed	0	273 53.6	33.4	N10 35.9	9.7
	6	1 14.1	33.3	N11 33.9	9.3
	12	88 34.0	33.2	N12 29.9	9.0
	18	175 53.3	33.0	N13 23.8	8.6
12 Th	0	263 11.7	32.9	N14 15.6	8.2
	6	350 29.3	32.7	N15 05.1	7.8
	12	77 45.8	32.6	N15 52.2	7.4
	18	165 01.3	32.4	N16 36.8	6.9
13 Fri	0	252 15.6	32.2	N17 18.7	6.5
	6	339 28.7	32.0	N17 57.9	6.0
	12	66 40.6	31.8	N18 34.3	5.5
	18	153 51.2	31.5	N19 07.8	5.1
14 Sat	0	241 00.5	31.3	N19 38.1	4.5
	6	328 08.5	31.1	N20 05.4	4.0
	12	55 15.3	30.9	N20 29.3	3.4
	18	142 20.9	30.8	N20 49.9	2.8
15 Sun	0	229 25.3	30.6	N21 07.1	2.2
	6	316 28.7	30.4	N21 20.8	1.7
	12	43 31.0	30.3	N21 30.8	1.0
	18	130 32.4	30.1	N21 37.2	0.4
16 Mon	0	217 33.1	30.0	N21 39.9	0.2
	6	304 33.1	29.9	N21 38.8	0.9
	12	31 32.5	29.8	N21 34.0	1.5
	18	118 31.6	29.8	N21 25.3	2.2
17 Tu	0	205 30.3	29.8	N21 12.7	2.8
	6	292 28.9	29.8	N20 56.4	3.4
	12	19 27.5	29.8	N20 36.3	4.0
	18	106 26.3	29.8	N20 12.4	4.6
18 Wed	0	193 25.2	29.9	N19 44.8	5.3
	6	280 24.5	29.9	N19 13.5	5.8
	12	7 24.3	30.0	N18 38.7	6.4
	18	94 24.6	30.1	N18 00.3	7.0
19 Th	0	181 25.5	30.2	N17 18.6	7.6
	6	268 27.0	30.4	N16 33.6	8.1
	12	355 29.2	30.5	N15 45.4	8.6
	18	82 32.2	30.6	N14 54.2	9.0
20 Fri	0	169 35.8	30.8	N14 00.2	9.5
	6	256 40.2	30.9	N13 03.4	10.0
	12	343 45.2	31.0	N12 04.1	10.3
	18	70 50.8	31.0	N11 02.4	10.7
21 Sat	0	157 57.0	31.1	N 9 58.5	11.0
	6	245 03.7	31.2	N 8 52.6	11.4
	12	332 10.7	31.2	N 7 44.9	11.6
	18	59 18.0	31.2	N 6 35.5	11.8
22 Sun	0	146 25.4	31.3	N 5 24.7	12.0
	6	233 32.9	31.3	N 4 12.7	12.2
	12	320 40.2	31.2	N 2 59.7	12.3
	18	47 47.2	31.1	N 1 45.8	12.4
23 Mon	0	134 53.8	31.0	N 0 31.4	12.5
	6	221 59.8	30.9	S 0 43.3	12.5
	12	309 05.0	30.7	S 1 58.2	12.4
	18	36 09.3	30.5	S 3 12.9	12.4
24 Tu	0	123 12.5	30.3	S 4 27.4	12.3
	6	210 14.4	30.0	S 5 41.2	12.1
	12	297 14.8	29.8	S 6 54.2	11.9
	18	24 13.6	29.5	S 8 06.1	11.7
25 Wed	0	111 10.6	29.2	S 9 16.6	11.4
	6	198 05.6	28.8	S10 25.5	11.1
	12	284 58.6	28.4	S11 32.6	10.8
	18	11 49.4	28.0	S12 37.5	10.4
26 Th	0	98 37.8	27.6	S13 39.9	9.9
	6	185 23.9	27.2	S14 39.7	9.4
	12	272 07.6	26.8	S15 36.5	8.8
	18	358 48.7	26.4	S16 30.0	8.3
27 Fri	0	85 27.4	26.0	S17 20.0	7.7
	6	172 03.7	25.7	S18 06.3	7.0
	12	258 37.6	25.3	S18 48.5	6.2
	18	345 09.3	24.9	S19 26.4	5.5
28 Sat	0	71 38.9	24.5	S19 59.7	4.7
	6	158 06.6	24.3	S20 28.4	3.9
	12	244 32.6	24.1	S20 52.2	3.0
	18	330 57.3	23.9	S21 10.9	2.2
29 Sun	0	57 20.9	23.8	S21 24.4	1.3
	6	143 43.7	23.7	S21 32.7	0.4
	12	230 06.1	23.7	S21 35.6	0.5
	18	316 28.4	23.8	S21 33.1	1.3
30 Mon	0	42 51.0	23.9	S21 25.3	2.3
	6	129 14.3	24.0	S21 12.1	3.1
	12	215 38.6	24.3	S20 53.8	4.0
	18	302 04.2	24.5	S20 30.4	4.8
31 Tu	0	28 31.4	24.9	S20 02.0	5.6
	6	115 00.6	25.3	S19 28.9	6.3
	12	201 31.9	25.7	S18 51.4	7.0
	18	288 05.5	26.1	S18 09.5	7.7

PLANETS

VENUS Mer Pass h m	VENUS GHA ° '	VENUS Mean Var/hr 15°+	VENUS Dec ° '	VENUS Mean Var/hr '	Day	JUPITER GHA ° '	JUPITER Mean Var/hr 15°+	JUPITER Dec ° '	JUPITER Mean Var/hr '	JUPITER Mer Pass h m
09 47	212 49.8	2.1	N17 31.3	0.1	1 SUN	217 01.2	1.9	N20 14.3	0.1	09 31
09 44	213 39.3	2.0	N17 28.3	0.1	2 Mon	217 47.0	1.9	N20 16.7	0.1	09 28
09 41	214 26.7	1.9	N17 26.0	0.1	3 Tu	218 32.8	1.9	N20 19.0	0.1	09 25
09 38	215 12.0	1.8	N17 24.4	0.0	4 Wed	219 18.8	1.9	N20 21.3	0.1	09 22
09 35	215 55.3	1.7	N17 23.4	0.0	5 Th	220 04.7	1.9	N20 23.5	0.1	09 18
09 32	216 36.7	1.6	N17 23.1	0.0	6 Fri	220 50.8	1.9	N20 25.7	0.1	09 15
09 30	217 16.2	1.6	N17 23.4	0.0	7 Sat	221 36.8	1.9	N20 27.9	0.1	09 12
09 27	217 53.9	1.5	N17 24.3	0.1	8 SUN	222 23.0	1.9	N20 30.1	0.1	09 09
09 25	218 29.8	1.4	N17 25.7	0.1	9 Mon	223 09.2	1.9	N20 32.2	0.1	09 06
09 23	219 03.9	1.4	N17 27.5	0.1	10 Tu	223 55.5	1.9	N20 34.3	0.1	09 03
09 21	219 36.4	1.3	N17 29.9	0.1	11 Wed	224 41.9	1.9	N20 36.4	0.1	09 00
09 19	220 07.3	1.2	N17 32.7	0.1	12 Th	225 28.3	1.9	N20 38.4	0.1	08 57
09 17	220 36.6	1.2	N17 35.8	0.2	13 Fri	226 14.8	1.9	N20 40.5	0.1	08 54
09 15	221 04.5	1.1	N17 39.4	0.2	14 Sat	227 01.3	1.9	N20 42.4	0.1	08 51
09 13	221 30.8	1.0	N17 43.2	0.2	15 SUN	227 48.0	1.9	N20 44.4	0.1	08 48
09 12	221 55.8	1.0	N17 47.4	0.2	16 Mon	228 34.7	1.9	N20 46.3	0.1	08 45
09 10	222 19.4	0.9	N17 51.8	0.2	17 Tu	229 21.5	2.0	N20 48.2	0.1	08 41
09 09	222 41.6	0.9	N17 56.4	0.2	18 Wed	230 08.4	2.0	N20 50.1	0.1	08 38
09 07	223 02.7	0.8	N18 01.3	0.2	19 Th	230 55.4	2.0	N20 51.9	0.1	08 35
09 06	223 22.4	0.8	N18 06.3	0.2	20 Fri	231 42.4	2.0	N20 53.7	0.1	08 32
09 05	223 41.0	0.7	N18 11.5	0.2	21 Sat	232 29.5	2.0	N20 55.5	0.1	08 29
09 04	223 58.5	0.7	N18 16.7	0.2	22 SUN	233 16.8	2.0	N20 57.3	0.1	08 26
09 03	224 14.8	0.6	N18 22.1	0.2	23 Mon	234 04.1	2.0	N20 59.0	0.1	08 23
09 02	224 30.1	0.6	N18 27.6	0.2	24 Tu	234 51.5	2.0	N21 00.7	0.1	08 19
09 01	224 44.3	0.6	N18 33.0	0.2	25 Wed	235 39.0	2.0	N21 02.4	0.1	08 16
09 00	224 57.5	0.5	N18 38.5	0.2	26 Th	236 26.6	2.0	N21 04.0	0.1	08 13
08 59	225 09.8	0.5	N18 44.0	0.2	27 Fri	237 14.3	2.0	N21 05.6	0.1	08 10
08 58	225 21.1	0.4	N18 49.4	0.2	28 Sat	238 02.1	2.0	N21 07.2	0.1	08 07
08 58	225 31.4	0.4	N18 54.8	0.2	29 SUN	238 50.0	2.0	N21 08.7	0.1	08 04
08 57	225 40.9	0.4	N19 00.1	0.2	30 Mon	239 37.9	2.0	N21 10.2	0.1	08 00
08 56	225 49.5	0.3	N19 05.3	0.2	31 Tu	240 26.0	2.0	N21 11.7	0.1	07 57

VENUS, Av. Mag. –4.7
SHA July 5 293; 10 291; 15 288; 20 285; 25 282; 30 278

JUPITER, Av. Mag. –2.1
SHA July 5 297; 10 296; 15 295; 20 294; 25 293; 30 292

MARS Mer Pass h m	MARS GHA ° '	MARS Mean Var/hr 15°+	MARS Dec ° '	MARS Mean Var/hr '	Day	SATURN GHA ° '	SATURN Mean Var/hr 15°+	SATURN Dec ° '	SATURN Mean Var/hr '	SATURN Mer Pass h m
17 17	100 29.5	1.3	N 0 54.7	0.6	1 SUN	77 25.1	2.4	S 6 29.5	0.0	18 47
17 14	101 00.5	1.3	N 0 41.2	0.6	2 Mon	78 23.7	2.4	S 6 30.0	0.0	18 43
17 12	101 31.3	1.3	N 0 27.6	0.6	3 Tu	79 22.3	2.4	S 6 30.5	0.0	18 39
17 10	102 02.0	1.3	N 0 14.1	0.6	4 Wed	80 20.8	2.4	S 6 31.0	0.0	18 36
17 08	102 32.5	1.3	N 0 00.4	0.6	5 Th	81 19.2	2.4	S 6 31.6	0.0	18 32
17 06	103 02.9	1.3	S 0 13.3	0.6	6 Fri	82 17.4	2.4	S 6 32.2	0.0	18 28
17 04	103 33.0	1.3	S 0 27.1	0.6	7 Sat	83 15.7	2.4	S 6 32.9	0.0	18 24
17 02	104 03.1	1.2	S 0 40.9	0.6	8 SUN	84 13.8	2.4	S 6 33.6	0.0	18 20
17 00	104 32.9	1.2	S 0 54.7	0.6	9 Mon	85 11.8	2.4	S 6 34.3	0.0	18 16
16 58	105 02.7	1.2	S 1 08.7	0.6	10 Tu	86 09.7	2.4	S 6 35.0	0.0	18 12
16 56	105 32.2	1.2	S 1 22.6	0.6	11 Wed	87 07.6	2.4	S 6 35.8	0.0	18 09
16 55	106 01.6	1.2	S 1 36.6	0.6	12 Th	88 05.3	2.4	S 6 36.7	0.0	18 05
16 53	106 30.9	1.2	S 1 50.7	0.6	13 Fri	89 03.0	2.4	S 6 37.5	0.0	18 01
16 51	107 00.0	1.2	S 2 04.8	0.6	14 Sat	90 00.5	2.4	S 6 38.4	0.0	17 57
16 49	107 28.9	1.2	S 2 18.9	0.6	15 SUN	90 58.0	2.4	S 6 39.4	0.0	17 53
16 47	107 57.7	1.2	S 2 33.1	0.6	16 Mon	91 55.4	2.4	S 6 40.3	0.0	17 49
16 45	108 26.3	1.2	S 2 47.4	0.6	17 Tu	92 52.7	2.4	S 6 41.3	0.0	17 46
16 43	108 54.8	1.2	S 3 01.6	0.6	18 Wed	93 49.9	2.4	S 6 42.4	0.0	17 42
16 41	109 23.1	1.2	S 3 15.9	0.6	19 Th	94 47.0	2.4	S 6 43.4	0.0	17 38
16 39	109 51.3	1.2	S 3 30.3	0.6	20 Fri	95 44.1	2.4	S 6 44.5	0.0	17 34
16 37	110 19.3	1.2	S 3 44.6	0.6	21 Sat	96 41.0	2.4	S 6 45.7	0.0	17 30
16 36	110 47.2	1.2	S 3 59.0	0.6	22 SUN	97 37.9	2.4	S 6 46.9	0.0	17 27
16 34	111 14.9	1.2	S 4 13.4	0.6	23 Mon	98 34.6	2.4	S 6 48.0	0.1	17 23
16 32	111 42.5	1.1	S 4 27.9	0.6	24 Tu	99 31.3	2.4	S 6 49.3	0.0	17 19
16 30	112 09.9	1.1	S 4 42.3	0.6	25 Wed	100 27.9	2.4	S 6 50.5	0.1	17 15
16 28	112 37.1	1.1	S 4 56.8	0.6	26 Th	101 24.5	2.3	S 6 51.8	0.1	17 12
16 26	113 04.3	1.1	S 5 11.3	0.6	27 Fri	102 20.9	2.3	S 6 53.2	0.1	17 08
16 25	113 31.2	1.1	S 5 25.9	[illegible]	28 Sat	103 17.2	2.3	S 6 54.5	0.1	17 04
16 23	113 58.1	1.1	S 5 40.4	0.6	29 SUN	104 13.5	2.3	S 6 55.9	0.1	17 00
16 21	114 24.7	1.1	S 5 55.0	0.6	30 Mon	105 09.7	2.3	S 6 57.3	0.1	16 57
16 19	114 51.3	1.1	S 6 09.5	0.6	31 Tu	106 05.8	2.3	S 6 58.8	0.1	16 53

MARS, Av. Mag. +1.0
SHA July 5 179; 10 177; 15 174; 20 172; 25 169; 30 166

SATURN, Av. Mag. +0.7
SHA July 5 158; 10 158; 15 158; 20 158; 25 157; 30 157

27

No.	Name	Mag	Transit h m	Dec ° ′	SHA ° ′
	Oh GMT July 1				
♈	ARIES	–	5 21	–	–
1	Alpheratz	2.1	5 30	N29 09.6	357 44.0
2	Ankaa	2.4	5 48	S42 13.9	353 16.2
3	Schedar	2.2	6 02	N56 36.2	349 41.1
4	Diphda	2.0	6 05	S17 54.8	348 56.4
5	Achernar	0.5	6 59	S57 10.0	335 27.3
6	POLARIS	2.0	8 08	N89 18.8	318 13.2
7	Hamal	2.0	7 29	N23 31.2	328 01.5
8	Acamar	3.2	8 20	S40 15.0	315 19.0
9	Menkar	2.5	8 24	N 4 08.3	314 15.8
10	Mirfak	1.8	8 46	N49 54.1	308 41.5
11	Aldebaran	0.9	9 57	N16 32.0	290 50.4
12	Rigel	0.1	10 36	S 8 11.3	281 12.9
13	Capella	0.1	10 38	N46 00.4	280 35.8
14	Bellatrix	1.6	10 46	N 6 21.6	278 33.0
15	Elnath	1.7	10 48	N28 36.9	278 13.7
16	Alnilam	1.7	10 57	S 1 11.7	275 47.3
17	Betelgeuse	0.1–1.2	11 16	N 7 24.5	271 02.3
18	Canopus	–0.7	11 44	S52 42.2	263 57.0
19	Sirius	–1.5	12 06	S16 44.1	258 34.6
20	Adhara	1.5	12 19	S28 59.4	255 13.4
21	Castor	1.6	12 55	N31 51.5	246 09.2
22	Procyon	0.4	13 00	N 5 11.5	245 00.7
23	Pollux	1.1	13 06	N27 59.6	243 28.9
24	Avior	1.9	13 43	S59 33.2	234 18.9
25	Suhail	2.2	14 28	S43 29.2	222 53.3
26	Miaplacidus	1.7	14 33	S69 46.4	221 40.7
27	Alphard	2.0	14 48	S 8 42.9	217 57.0
28	Regulus	1.4	15 29	N11 54.3	207 44.4
29	Dubhe	1.8	16 24	N61 41.1	193 52.9
30	Denebola	2.1	17 09	N14 30.1	182 34.4
31	Gienah	2.6	17 36	S17 36.8	175 53.0
32	Acrux	1.3	17 47	S63 10.5	173 10.1
33	Gacrux	1.6	17 51	S57 11.3	172 01.7
34	Mimosa	1.3	18 08	S59 45.8	167 52.8
35	Alioth	1.8	18 14	N55 53.7	166 21.3
36	Spica	1.0	18 45	S11 13.7	158 31.9
37	Alkaid	1.9	19 07	N49 15.3	152 59.3
38	Hadar	0.6	19 24	S60 26.3	148 48.6
39	Menkent	2.1	19 26	S36 26.1	148 08.2
40	Arcturus	0.0	19 35	N19 07.2	145 56.2
41	Rigil Kent	–0.3	19 59	S60 53.4	139 52.4
42	Zuben'ubi	2.8	20 10	S16 05.7	137 06.0
43	Kochab	2.1	20 10	N74 06.5	137 19.6
44	Alphecca	2.2	20 54	N26 40.6	126 11.3
45	Antares	1.0	21 49	S26 27.6	112 26.7
46	Atria	1.9	22 09	S69 03.1	107 28.7
47	Sabik	2.4	22 30	S15 44.3	102 13.0
48	Shaula	1.6	22 53	S37 06.7	96 22.4
49	Rasalhague	2.1	22 54	N12 33.3	96 06.7
50	Eltanin	2.2	23 15	N51 29.5	90 45.9
51	Kaus Aust.	1.9	23 43	S34 22.6	83 44.2
52	Vega	0.0	23 56	N38 47.9	80 39.0
53	Nunki	2.0	0 18	S26 16.7	75 58.7
54	Altair	0.8	1 13	N 8 54.3	62 08.5
55	Peacock	1.9	1 49	S56 41.4	53 19.6
56	Deneb	1.3	2 04	N45 19.6	49 31.4
57	Enif	2.4	3 07	N 9 56.1	33 47.4
58	Al Na'ir	1.7	3 31	S46 53.7	27 44.1
59	Fomalhaut	1.2	4 20	S29 33.0	15 24.4
60	Markab	2.5	4 27	N15 16.5	13 38.7

SUN

Yr	Day of Mth	Day of Week	Transit h m	Semi-Diam	Lat 52°N Twilight h m	Lat 52°N Sunrise h m	Lat 52°N Sunset h m	Lat 52°N Twilight h m
183	1	Sun	12 04	15.8	02 56	03 45	20 23	21 11
184	2	Mon	12 04	15.8	02 57	03 45	20 23	21 11
185	3	Tu	12 04	15.8	02 58	03 46	20 22	21 10
186	4	Wed	12 04	15.8	02 59	03 47	20 22	21 09
187	5	Th	12 05	15.8	03 00	03 48	20 21	21 09
188	6	Fri	12 05	15.8	03 01	03 49	20 20	21 08
189	7	Sat	12 05	15.8	03 02	03 50	20 20	21 07
190	8	Sun	12 05	15.8	03 03	03 51	20 19	21 06
191	9	Mon	12 05	15.8	03 05	03 52	20 18	21 05
192	10	Tu	12 05	15.8	03 06	03 53	20 18	21 04
193	11	Wed	12 06	15.8	03 07	03 54	20 17	21 03
194	12	Th	12 06	15.8	03 09	03 55	20 16	21 02
195	13	Fri	12 06	15.8	03 10	03 56	20 15	21 01
196	14	Sat	12 06	15.8	03 11	03 57	20 14	20 59
197	15	Sun	12 06	15.8	03 13	03 58	20 13	20 58
198	16	Mon	12 06	15.8	03 14	04 00	20 12	20 57
199	17	Tu	12 06	15.8	03 16	04 01	20 11	20 56
200	18	Wed	12 06	15.8	03 17	04 02	20 10	20 54
201	19	Th	12 06	15.8	03 19	04 03	20 08	20 53
202	20	Fri	12 06	15.8	03 20	04 05	20 07	20 51
203	21	Sat	12 06	15.8	03 22	04 06	20 06	20 50
204	22	Sun	12 06	15.8	03 24	04 08	20 05	20 48
205	23	Mon	12 07	15.8	03 25	04 09	20 03	20 46
206	24	Tu	12 07	15.8	03 27	04 10	20 02	20 45
207	25	Wed	12 07	15.8	03 29	04 12	20 00	20 43
208	26	Th	12 07	15.8	03 30	04 13	19 59	20 41
209	27	Fri	12 07	15.8	03 32	04 15	19 57	20 40
210	28	Sat	12 06	15.8	03 34	04 16	19 56	20 38
211	29	Sun	12 06	15.8	03 36	04 18	19 54	20 36
212	30	Mon	12 06	15.8	03 37	04 19	19 53	20 34
213	31	Tu	12 06	15.8	03 39	04 21	19 51	20 32

Lat Corr to Sunrise, Sunset etc.

Lat °	Twilight h m	Sunrise h m	Sunset h m	Twilight h m
N70	SAH	SAH	SAH	SAH
68	SAH	SAH	SAH	SAH
66	SAH	–2 21	+2 19	TAN
64	TAN	–1 42	+1 40	TAN
62	–2 21	–1 14	+1 14	+2 21
N60	–1 27	–0 54	+0 54	+1 27
58	–0 55	–0 37	+0 37	+0 56
56	–0 33	–0 22	+0 23	+0 34
54	–0 15	–0 10	+0 11	+0 15
50	+0 13	+0 10	–0 10	–0 13
N45	+0 39	+0 30	–0 30	–0 39
40	+0 59	+0 46	–0 46	–0 59
35	+1 16	+1 00	–0 59	–1 16
30	+1 30	+1 12	–1 11	–1 30
20	+ 1 53	+1 32	–1 31	–1 52
N10	+2 12	+1 46	–1 48	–2 11
0	+2 28	+2 05	–2 04	–2 27
S10	+2 43	+2 21	–2 19	–2 43
20	+2 59	+2 38	–2 36	–2 59
30	+3 15	+2 57	–2 54	–3 15
S35	+3 25	+3 08	–3 06	–3 25
40	+3 35	+3 21	–3 18	–3 35
45	+3 28	+3 36	–3 33	–3 47
S50	+4 02	+3 55	–3 51	–4 02

NOTES
The corrections to sunrise etc. are for middle of July. SAH means Sun above Horizon. TAN means Twilight all night.

MOON

Yr	Day of Mth	Day of Week	Age days	Transit (Upper) h m	Diff m	Semi-diam ′	Hor Par ′	Lat 52°N Moonrise h m	Lat 52°N Moonset h m
183	1	Sun	12	22 13	63	16.5	60.5	18 06	01 24
184	2	Mon	13	23 16	60	16.5	60.4	19 07	02 20
185	3	Tu	14	24 16	–	16.4	60.1	19 55	03 27
186	4	Wed	15	00 16	57	16.2	59.6	20 33	04 43
187	5	Th	16	01 13	53	16.0	58.9	21 03	06 02
188	6	Fri	17	02 06	50	15.8	58.0	21 28	07 20
189	7	Sat	18	02 56	47	15.6	57.2	21 49	08 36
190	8	Sun	19	03 43	44	15.4	56.3	22 09	09 49
191	9	Mon	20	04 27	44	15.2	55.6	22 28	10 59
192	10	Tu	21	05 11	44	15.0	55.0	22 49	12 07
193	11	Wed	22	05 55	44	14.9	54.6	23 11	13 14
194	12	Th	23	06 39	46	14.8	54.3	23 37	14 20
195	13	Fri	24	07 25	47	14.8	54.2	– –	15 23
196	14	Sat	25	08 12	48	14.8	54.2	00 08	16 23
197	15	Sun	26	09 00	49	14.8	54.4	00 46	17 18
198	16	Mon	27	09 49	50	14.9	54.7	01 31	18 06
199	17	Tu	28	10 39	50	15.0	55.1	02 25	18 47
200	18	Wed	29	11 29	50	15.1	55.6	03 27	19 22
201	19	Th	00	12 19	48	15.3	56.1	04 34	19 51
202	20	Fri	01	13 07	48	15.4	56.6	05 44	20 16
203	21	Sat	02	13 55	48	15.5	57.1	06 57	20 38
204	22	Sun	03	14 43	48	15.7	57.6	08 12	21 00
205	23	Mon	04	15 31	49	15.8	58.0	09 27	21 21
206	24	Tu	05	16 20	51	15.9	58.4	10 44	21 44
207	25	Wed	06	17 11	54	16.0	58.8	12 02	22 10
208	26	Th	07	18 05	57	16.1	59.2	13 21	22 41
209	27	Fri	08	19 02	59	16.2	59.5	14 38	23 20
210	28	Sat	09	20 01	60	16.3	59.6	15 50	– –
211	29	Sun	10	21 01	60	16.3	59.7	16 54	00 09
212	30	Mon	11	22 01	58	16.2	59.6	17 47	01 10
213	31	Tu	12	22 59	55	16.2	59.3	18 29	02 20

Phases of the Moon

	Phase	d	h	m
○	Full Moon	3	18	52
☽	Last Quarter	11	01	48
●	New Moon	19	04	24
☾	First Quarter	26	08	56

	d	h
Perigee	1	18
Apogee	13	17
Perigee	29	08

SUN AND ARIES

GMT	SUN GHA ° ′	SUN Dec ° ′	ARIES GHA ° ′
Wednesday, 1st August			
00	178 25.0	**N17 57.8**	310 00.4
02	208 25.1	**17 56.5**	340 05.4
04	238 25.2	**17 55.3**	10 10.3
06	268 25.2	**17 54.0**	40 15.2
08	298 25.3	**17 52.7**	70 20.2
10	328 25.4	**17 51.5**	100 25.1
12	358 25.5	**17 50.2**	130 30.0
14	28 25.6	**17 48.9**	160 34.9
16	58 25.7	**17 47.6**	190 39.9
18	88 25.8	**17 46.4**	220 44.8
20	118 25.9	**17 45.1**	250 49.7
22	148 25.9	**N17 43.8**	280 54.7
Thursday, 2nd August			
00	178 26.0	**N17 42.5**	310 59.6
02	208 26.1	**17 41.2**	341 04.5
04	238 26.2	**17 39.9**	11 09.4
06	268 26.3	**17 38.6**	41 14.4
08	298 26.4	**17 37.3**	71 19.3
10	328 26.5	**17 36.0**	101 24.2
12	358 26.6	**17 34.7**	131 29.2
14	28 26.7	**17 33.4**	161 34.1
16	58 26.8	**17 32.1**	191 39.0
18	88 26.9	**17 30.8**	221 43.9
20	118 27.0	**17 29.5**	251 48.9
22	148 27.1	**N17 28.2**	281 53.8
Friday, 3rd August			
00	178 27.2	**N17 26.9**	311 58.7
02	208 27.3	**17 25.6**	342 03.6
04	238 27.4	**17 24.3**	12 08.6
06	268 27.6	**17 23.0**	42 13.5
08	298 27.7	**17 21.7**	72 18.4
10	328 27.8	**17 20.3**	102 23.4
12	358 27.9	**17 19.0**	132 28.3
14	28 28.0	**17 17.7**	162 33.2
16	58 28.1	**17 16.4**	192 38.1
18	88 28.2	**17 15.1**	222 43.1
20	118 28.3	**17 13.7**	252 48.0
22	148 28.5	**N17 12.4**	282 52.9
Saturday, 4th August			
00	178 28.6	**N17 11.1**	312 57.9
02	208 28.7	**17 09.7**	343 02.8
04	238 28.8	**17 08.4**	13 07.7
06	268 28.9	**17 07.0**	43 12.6
08	298 29.1	**17 05.7**	73 17.6
10	328 29.2	**17 04.4**	103 22.5
12	358 29.3	**17 03.0**	133 27.4
14	28 29.4	**17 01.7**	163 32.4
16	58 29.5	**17 00.3**	193 37.3
18	88 29.7	**16 59.0**	223 42.2
20	118 29.8	**16 57.6**	253 47.1
22	148 29.9	**N16 56.3**	283 52.1
Sunday, 5th August			
00	178 30.1	**N16 54.9**	313 57.0
02	208 30.2	**16 53.6**	344 01.9
04	238 30.3	**16 52.2**	14 06.9
06	268 30.5	**16 50.8**	44 11.8
08	298 30.6	**16 49.5**	74 16.7
10	328 30.7	**16 48.1**	104 21.6
12	358 30.9	**16 46.7**	134 26.6
14	28 31.0	**16 45.4**	164 31.5
16	58 31.1	**16 44.0**	194 36.4
18	88 31.3	**16 42.6**	224 41.3
20	118 31.4	**16 41.3**	254 46.3
22	148 31.6	**N16 39.9**	284 51.2

GMT	SUN GHA ° ′	SUN Dec ° ′	ARIES GHA ° ′
Monday, 6th August			
00	178 31.7	**N16 38.5**	314 56.1
02	208 31.8	**16 37.1**	345 01.1
04	238 32.0	**16 35.7**	15 06.0
06	268 32.1	**16 34.4**	45 10.9
08	298 32.3	**16 33.0**	75 15.8
10	328 32.4	**16 31.6**	105 20.8
12	358 32.6	**16 30.2**	135 25.7
14	28 32.7	**16 28.8**	165 30.6
16	58 32.9	**16 27.4**	195 35.6
18	88 33.0	**16 26.0**	225 40.5
20	118 33.2	**16 24.6**	255 45.4
22	148 33.3	**N16 23.2**	285 50.3
Tuesday, 7th August			
00	178 33.5	**N16 21.8**	315 55.3
02	208 33.6	**16 20.4**	346 00.2
04	238 33.8	**16 19.0**	16 05.1
06	268 34.0	**16 17.6**	46 10.1
08	298 34.1	**16 16.2**	76 15.0
10	328 34.3	**16 14.8**	106 19.9
12	358 34.4	**16 13.4**	136 24.8
14	28 34.6	**16 12.0**	166 29.8
16	58 34.8	**16 10.5**	196 34.7
18	88 34.9	**16 09.1**	226 39.6
20	118 35.1	**16 07.7**	256 44.6
22	148 35.2	**N16 06.3**	286 49.5
Wednesday, 8th August			
00	178 35.4	**N16 04.9**	316 54.4
02	208 35.6	**16 03.4**	346 59.3
04	238 35.7	**16 02.0**	17 04.3
06	268 35.9	**16 00.6**	47 09.2
08	298 36.1	**15 59.2**	77 14.1
10	328 36.3	**15 57.7**	107 19.0
12	358 36.4	**15 56.3**	137 24.0
14	28 36.6	**15 54.8**	167 28.9
16	58 36.8	**15 53.4**	197 33.8
18	88 36.9	**15 52.0**	227 38.8
20	118 37.1	**15 50.5**	257 43.7
22	148 37.3	**N15 49.1**	287 48.6
Thursday, 9th August			
00	178 37.5	**N15 47.6**	317 53.5
02	208 37.7	**15 46.2**	347 58.5
04	238 37.8	**15 44.8**	18 03.4
06	268 38.0	**15 43.3**	48 08.3
08	298 38.2	**15 41.9**	78 13.3
10	328 38.4	**15 40.4**	108 18.2
12	358 38.6	**15 38.9**	138 23.1
14	28 38.7	**15 37.5**	168 28.0
16	58 38.9	**15 36.0**	198 33.0
18	88 39.1	**15 34.6**	228 37.9
20	118 39.3	**15 33.1**	258 42.8
22	148 39.5	**N15 31.6**	288 47.8
Friday, 10th August			
00	178 39.7	**N15 30.2**	318 52.7
02	208 39.9	**15 28.7**	348 57.6
04	238 40.1	**15 27.2**	19 02.5
06	268 40.3	**15 25.8**	49 07.5
08	298 40.5	**15 24.3**	79 12.4
10	328 40.6	**15 22.8**	109 17.3
12	358 40.8	**15 21.4**	139 22.3
14	28 41.0	**15 19.9**	169 27.2
16	58 41.2	**15 18.4**	199 32.1
18	88 41.4	**15 16.9**	229 37.0
20	118 41.6	**15 15.4**	259 42.0
22	148 41.8	**N15 14.0**	289 46.9

GMT	SUN GHA ° ′	SUN Dec ° ′	ARIES GHA ° ′
Saturday, 11th August			
00	178 42.0	**N15 12.5**	319 51.8
02	208 42.2	**15 11.0**	349 56.8
04	238 42.4	**15 09.5**	20 01.7
06	268 42.6	**15 08.0**	50 06.6
08	298 42.8	**15 06.5**	80 11.5
10	328 43.1	**15 05.0**	110 16.5
12	358 43.3	**15 03.5**	140 21.4
14	28 43.5	**15 02.0**	170 26.3
16	58 43.7	**15 00.5**	200 31.2
18	88 43.9	**14 59.0**	230 36.2
20	118 44.1	**14 57.5**	260 41.1
22	148 44.3	**N14 56.0**	290 46.0
Sunday, 12th August			
00	178 44.5	**N14 54.5**	320 51.0
02	208 44.7	**14 53.0**	350 55.9
04	238 44.9	**14 51.5**	21 00.8
06	268 45.2	**14 50.0**	51 05.7
08	298 45.4	**14 48.5**	81 10.7
10	328 45.6	**14 47.0**	111 15.6
12	358 45.8	**14 45.4**	141 20.5
14	28 46.0	**14 43.9**	171 25.5
16	58 46.2	**14 42.4**	201 30.4
18	88 46.5	**14 40.9**	231 35.3
20	118 46.7	**14 39.4**	261 40.2
22	148 46.9	**N14 37.8**	291 45.2
Monday, 13th August			
00	178 47.1	**N14 36.3**	321 50.1
02	208 47.4	**14 34.8**	351 55.0
04	238 47.6	**14 33.3**	22 00.0
06	268 47.8	**14 31.7**	52 04.9
08	298 48.0	**14 30.2**	82 09.8
10	328 48.3	**14 28.7**	112 14.7
12	358 48.5	**14 27.1**	142 19.7
14	28 48.7	**14 25.6**	172 24.6
16	58 49.0	**14 24.0**	202 29.5
18	88 49.2	**14 22.5**	232 34.5
20	118 49.4	**14 21.0**	262 39.4
22	148 49.7	**N14 19.4**	292 44.3
Tuesday, 14th August			
00	178 49.9	**N14 17.9**	322 49.2
02	208 50.1	**14 16.3**	352 54.2
04	238 50.4	**14 14.8**	22 59.1
06	268 50.6	**14 13.2**	53 04.0
08	298 50.8	**14 11.7**	83 09.0
10	328 51.1	**14 10.1**	113 13.9
12	358 51.3	**14 08.6**	143 18.8
14	28 51.6	**14 07.0**	173 23.7
16	58 51.8	**14 05.5**	203 28.7
18	88 52.0	**14 03.9**	233 33.6
20	118 52.3	**14 02.3**	263 38.5
22	148 52.5	**N14 00.8**	293 43.5
Wednesday, 15th August			
00	178 52.8	**N13 59.2**	323 48.4
02	208 53.0	**13 57.6**	353 53.3
04	238 53.3	**13 56.1**	23 58.2
06	268 53.5	**13 54.5**	54 03.2
08	298 53.8	**13 52.9**	84 08.1
10	328 54.0	**13 51.4**	114 13.0
12	358 54.3	**13 49.8**	144 18.0
14	28 54.5	**13 48.2**	174 22.9
16	58 54.8	**13 46.7**	204 27.8
18	88 55.0	**13 45.1**	234 32.7
20	118 55.3	**13 43.5**	264 37.7
22	148 55.5	**N13 41.9**	294 42.6

GMT	SUN GHA ° ′	SUN Dec ° ′	ARIES GHA ° ′
Thursday, 16th August			
00	178 55.8	**N13 40.3**	324 47.5
02	208 56.0	**13 38.7**	354 52.4
04	238 56.3	**13 37.2**	24 57.4
06	268 56.6	**13 35.6**	55 02.3
08	298 56.8	**13 34.0**	85 07.2
10	328 57.1	**13 32.4**	115 12.2
12	358 57.3	**13 30.8**	145 17.1
14	28 57.6	**13 29.2**	175 22.0
16	58 57.9	**13 27.6**	205 26.9
18	88 58.1	**13 26.0**	235 31.9
20	118 58.4	**13 24.4**	265 36.8
22	148 58.7	**N13 22.8**	295 41.7
Friday, 17th August			
00	178 58.9	**N13 21.2**	325 46.7
02	208 59.2	**13 19.6**	355 51.6
04	238 59.5	**13 18.0**	25 56.5
06	268 59.7	**13 16.4**	56 01.4
08	299 00.0	**13 14.8**	86 06.4
10	329 00.3	**13 13.2**	116 11.3
12	359 00.6	**13 11.6**	146 16.2
14	29 00.8	**13 10.0**	176 21.2
16	59 01.1	**13 08.4**	206 26.1
18	89 01.4	**13 06.8**	236 31.0
20	119 01.7	**13 05.2**	266 35.9
22	149 01.9	**N13 03.5**	296 40.9
Saturday, 18th August			
00	179 02.2	**N13 01.9**	326 45.8
02	209 02.5	**13 00.3**	356 50.7
04	239 02.8	**12 58.7**	26 55.7
06	269 03.0	**12 57.1**	57 00.6
08	299 03.3	**12 55.4**	87 05.5
10	329 03.6	**12 53.8**	117 10.4
12	359 03.9	**12 52.2**	147 15.4
14	29 04.2	**12 50.6**	177 20.3
16	59 04.5	**12 48.9**	207 25.2
18	89 04.7	**12 47.3**	237 30.1
20	119 05.0	**12 45.7**	267 35.1
22	149 05.3	**N12 44.0**	297 40.0
Sunday, 19th August			
00	179 05.6	**N12 42.4**	327 44.9
02	209 05.9	**12 40.8**	357 49.9
04	239 06.2	**12 39.1**	27 54.8
06	269 06.5	**12 37.5**	57 59.7
08	299 06.8	**12 35.9**	88 04.6
10	329 07.1	**12 34.2**	118 09.6
12	359 07.3	**12 32.6**	148 14.5
14	29 07.6	**12 30.9**	178 19.4
16	59 07.9	**12 29.3**	208 24.4
18	89 08.2	**12 27.6**	238 29.3
20	119 08.5	**12 26.0**	268 34.2
22	149 08.8	**N12 24.3**	298 39.1
Monday, 20th August			
00	179 09.1	**N12 22.7**	328 44.1
02	209 09.4	**12 21.0**	358 49.0
04	239 09.7	**12 19.4**	28 53.9
06	269 10.0	**12 17.7**	58 58.9
08	299 10.3	**12 16.1**	89 03.8
10	329 10.6	**12 14.4**	119 08.7
12	359 10.9	**12 12.8**	149 13.6
14	29 11.2	**12 11.1**	179 18.6
16	59 11.5	**12 09.4**	209 23.5
18	89 11.8	**12 07.8**	239 28.4
20	119 12.1	**12 06.1**	269 33.4
22	149 12.4	**N12 04.4**	299 38.3
00	179 55.1	**N 8 34.1**	339 34.6
02	209 55.5	**8 32.3**	9 39.5
04	239 55.9	**8 30.5**	39 44.5
06	269 56.3	**N 8 28.7**	69 49.4

GMT	SUN GHA ° ′	SUN Dec ° ′	ARIES GHA ° ′
Tuesday, 21st August			
00	179 12.8	**N12 02.8**	329 43.2
02	209 13.1	**12 01.1**	359 48.1
04	239 13.4	**11 59.4**	29 53.1
06	269 13.7	**11 57.8**	59 58.0
08	299 14.0	**11 56.1**	90 02.9
10	329 14.3	**11 54.4**	120 07.8
12	359 14.6	**11 52.8**	150 12.8
14	29 14.9	**11 51.1**	180 17.7
16	59 15.2	**11 49.4**	210 22.6
18	89 15.6	**11 47.7**	240 27.6
20	119 15.9	**11 46.0**	270 32.5
22	149 16.2	**N11 44.4**	300 37.4
Wednesday, 22nd August			
00	179 16.5	**N11 42.7**	330 42.3
02	209 16.8	**11 41.0**	0 47.3
04	239 17.2	**11 39.3**	30 52.2
06	269 17.5	**11 37.6**	60 57.1
08	299 17.8	**11 35.9**	91 02.1
10	329 18.1	**11 34.2**	121 07.0
12	359 18.4	**11 32.6**	151 11.9
14	29 18.8	**11 30.9**	181 16.8
16	59 19.1	**11 29.2**	211 21.8
18	89 19.4	**11 27.5**	241 26.7
20	119 19.7	**11 25.8**	271 31.6
22	149 20.1	**N11 24.1**	301 36.6
Thursday, 23rd August			
00	179 20.4	**N11 22.4**	331 41.5
02	209 20.7	**11 20.7**	1 46.4
04	239 21.0	**11 19.0**	31 51.3
06	269 21.4	**11 17.3**	61 56.3
08	299 21.7	**11 15.6**	92 01.2
10	329 22.0	**11 13.9**	122 06.1
12	359 22.4	**11 12.2**	152 11.1
14	29 22.7	**11 10.5**	182 16.0
16	59 23.0	**11 08.8**	212 20.9
18	89 23.4	**11 07.1**	242 25.8
20	119 23.7	**11 05.4**	272 30.8
22	149 24.0	**N11 03.6**	302 35.7
Friday, 24th August			
00	179 24.4	**N11 01.9**	332 40.6
02	209 24.7	**11 00.2**	2 45.6
04	239 25.0	**10 58.5**	32 50.5
06	269 25.4	**10 56.8**	62 55.4
08	299 25.7	**10 55.1**	93 00.3
10	329 26.1	**10 53.4**	123 05.3
12	359 26.4	**10 51.6**	153 10.2
14	29 26.7	**10 49.9**	183 15.1
16	59 27.1	**10 48.2**	213 20.0
18	89 27.4	**10 46.5**	243 25.0
20	119 27.8	**10 44.7**	273 29.9
22	149 28.1	**N10 43.0**	303 34.8
Saturday, 25th August			
00	179 28.5	**N10 41.3**	333 39.8
02	209 28.8	**10 39.6**	3 44.7
04	239 29.2	**10 37.8**	33 49.6
06	269 29.5	**10 36.1**	63 54.5
08	299 29.9	**10 34.4**	93 59.5
10	329 30.2	**10 32.6**	124 04.4
12	359 30.6	**10 30.9**	154 09.3
14	29 30.9	**10 29.2**	184 14.3
16	59 31.3	**10 27.4**	214 19.2
18	89 31.6	**10 25.7**	244 24.1
20	119 32.0	**10 24.0**	274 29.0
22	149 32.3	**N10 22.2**	304 34.0
Friday, 31st August			
08	299 56.7	**N 8 26.9**	99 54.3
10	329 57.1	**8 25.1**	129 59.2
12	359 57.5	**8 23.3**	160 04.2
14	29 57.9	**N 8 21.5**	190 09.1

GMT	SUN GHA ° ′	SUN Dec ° ′	ARIES GHA ° ′
Sunday, 26th August			
00	179 32.7	**N10 20.5**	334 38.9
02	209 33.0	**10 18.8**	4 43.8
04	239 33.4	**10 17.0**	34 48.8
06	269 33.7	**10 15.3**	64 53.7
08	299 34.1	**10 13.5**	94 58.6
10	329 34.5	**10 11.8**	125 03.5
12	359 34.8	**10 10.0**	155 08.5
14	29 35.2	**10 08.3**	185 13.4
16	59 35.5	**10 06.5**	215 18.3
18	89 35.9	**10 04.8**	245 23.3
20	119 36.3	**10 03.0**	275 28.2
22	149 36.6	**N10 01.3**	305 33.1
Monday, 27th August			
00	179 37.0	**N 9 59.5**	335 38.0
02	209 37.3	**9 57.8**	5 43.0
04	239 37.7	**9 56.0**	35 47.9
06	269 38.1	**9 54.3**	65 52.8
08	299 38.4	**9 52.5**	95 57.8
10	329 38.8	**9 50.7**	126 02.7
12	359 39.2	**9 49.0**	156 07.6
14	29 39.5	**9 47.2**	186 12.5
16	59 39.9	**9 45.5**	216 17.5
18	89 40.3	**9 43.7**	246 22.4
20	119 40.6	**9 41.9**	276 27.3
22	149 41.0	**N 9 40.2**	306 32.3
Tuesday, 28th August			
00	179 41.4	**N 9 38.4**	336 37.2
02	209 41.7	**9 36.6**	6 42.1
04	239 42.1	**9 34.9**	36 47.0
06	269 42.5	**9 33.1**	66 52.0
08	299 42.9	**9 31.3**	96 56.9
10	329 43.2	**9 29.6**	127 01.8
12	359 43.6	**9 27.8**	157 06.8
14	29 44.0	**9 26.0**	187 11.7
16	59 44.4	**9 24.2**	217 16.6
18	89 44.7	**9 22.5**	247 21.5
20	119 45.1	**9 20.7**	277 26.5
22	149 45.5	**N 9 18.9**	307 31.4
Wednesday, 29th August			
00	179 45.9	**N 9 17.1**	337 36.3
02	209 46.3	**9 15.4**	7 41.2
04	239 46.6	**9 13.6**	37 46.2
06	269 47.0	**9 11.8**	67 51.1
08	299 47.4	**9 10.0**	97 56.0
10	329 47.8	**9 08.2**	128 01.0
12	359 48.2	**9 06.4**	158 05.9
14	29 48.5	**9 04.7**	188 10.8
16	59 48.9	**9 02.9**	218 15.7
18	89 49.3	**9 01.1**	248 20.7
20	119 49.7	**8 59.3**	278 25.6
22	149 50.1	**N 8 57.5**	308 30.5
Thursday, 30th August			
00	179 50.5	**N 8 55.7**	338 35.5
02	209 50.8	**8 53.9**	8 40.4
04	239 51.2	**8 52.1**	38 45.3
06	269 51.6	**8 50.3**	68 50.2
08	299 52.0	**8 48.5**	98 55.2
10	329 52.4	**8 46.7**	129 00.1
12	359 52.8	**8 44.9**	159 05.0
14	29 53.2	**8 43.1**	189 10.0
16	59 53.6	**8 41.3**	219 14.9
18	89 54.0	**8 39.6**	249 19.8
20	119 54.3	**8 37.7**	279 24.7
22	149 54.7	**N 8 35.9**	309 29.7
16	59 58.3	**N 8 19.7**	220 14.0
18	89 58.7	**8 17.9**	250 18.9
20	119 59.1	**8 16.1**	280 23.9
22	149 59.5	**N 8 14.3**	310 28.8

AUGUST 2012

MOON

Day	GMT hr	GHA ° ′	Mean Var/hr 14°+	Dec ° ′	Mean Var/hr ′
1 Wed	0	14 41.7	26.5	S17 23.6	8.3
	6	101 20.5	26.9	S16 34.0	8.9
	12	188 02.1	27.4	S15 40.9	9.4
	18	274 46.4	27.9	S14 44.7	9.9
2 Th	0	1 33.5	28.3	S13 45.6	10.3
	6	88 23.5	28.9	S12 44.0	10.7
	12	175 16.1	29.3	S11 40.1	11.0
	18	262 11.4	29.7	S10 34.3	11.2
3 Fri	0	349 09.3	30.1	S 9 26.8	11.5
	6	76 09.6	30.5	S 8 18.0	11.7
	12	163 12.3	30.8	S 7 08.0	11.8
	18	250 17.2	31.2	S 5 57.2	11.9
4 Sat	0	337 24.1	31.5	S 4 45.8	11.9
	6	64 33.0	31.8	S 3 34.0	12.0
	12	151 43.5	32.0	S 2 22.1	11.9
	18	238 55.6	32.3	S 1 10.4	11.9
5 Sun	0	326 09.1	32.5	N 0 01.1	11.8
	6	53 23.8	32.6	N 1 12.1	11.7
	12	140 39.6	32.8	N 2 22.4	11.5
	18	227 56.2	32.9	N 3 31.9	11.4
6 Mon	0	315 13.6	33.0	N 4 40.3	11.2
	6	42 31.4	33.0	N 5 47.5	10.9
	12	129 49.7	33.1	N 6 53.4	10.7
	18	217 08.2	33.1	N 7 57.9	10.5
7 Tu	0	304 26.8	33.1	N 9 00.7	10.2
	6	31 45.4	33.1	N10 01.8	9.8
	12	119 03.7	33.0	N11 01.0	9.5
	18	206 21.8	32.9	N11 58.3	9.2
8 Wed	0	293 39.3	32.9	N12 53.4	8.8
	6	20 56.4	32.7	N13 46.4	8.4
	12	108 12.7	32.6	N14 37.0	8.0
	18	195 28.3	32.5	N15 25.1	7.5
9 Th	0	282 43.1	32.3	N16 10.8	7.1
	6	9 56.9	32.1	N16 53.7	6.7
	12	97 09.8	32.0	N17 34.0	6.2
	18	184 21.7	31.8	N18 11.3	5.7
10 Fri	0	271 32.5	31.6	N18 45.8	5.2
	6	358 42.3	31.4	N19 17.2	4.7
	12	85 50.9	31.3	N19 45.5	4.1
	18	172 58.5	31.1	N20 10.5	3.6
11 Sat	0	260 05.0	30.9	N20 32.3	3.0
	6	347 10.5	30.7	N20 50.7	2.4
	12	74 15.0	30.6	N21 05.6	1.9
	18	161 18.6	30.5	N21 17.0	1.3
12 Sun	0	248 21.3	30.3	N21 24.8	0.7
	6	335 23.2	30.2	N21 29.0	0.0
	12	62 24.5	30.1	N21 29.5	0.6
	18	149 25.1	30.0	N21 26.2	1.2
13 Mon	0	236 25.2	29.9	N21 19.2	1.8
	6	323 24.9	29.9	N21 08.5	2.5
	12	50 24.3	29.9	N20 54.0	3.1
	18	137 23.5	29.8	N20 35.7	3.7
14 Tu	0	224 22.6	29.8	N20 13.6	4.4
	6	311 21.8	29.9	N19 47.8	5.0
	12	38 21.0	29.9	N19 18.4	5.6
	18	125 20.3	30.0	N18 45.3	6.1
15 Wed	0	212 20.0	30.0	N18 08.7	6.7
	6	299 19.9	30.1	N17 28.5	7.3
	12	26 20.2	30.1	N16 45.0	7.8
	18	113 20.9	30.2	N15 58.3	8.4
16 Th	0	200 22.0	30.3	N15 08.3	8.9
	6	287 23.6	30.4	N14 15.3	9.4
	12	14 25.6	30.4	N13 19.5	9.8
	18	101 27.9	30.4	N12 20.9	10.2

Day	GMT hr	GHA ° ′	Mean Var/hr 14°+	Dec ° ′	Mean Var/hr ′
17 Fri	0	188 30.7	30.5	N11 19.7	10.6
	6	275 33.8	30.6	N10 16.2	10.9
	12	2 37.1	30.6	N 9 10.4	11.4
	18	89 40.7	30.6	N 8 02.6	11.6
18 Sat	0	176 44.3	30.6	N 6 53.0	11.8
	6	263 47.9	30.6	N 5 41.9	12.1
	12	350 51.5	30.5	N 4 29.3	12.3
	18	77 54.8	30.4	N 3 15.6	12.5
19 Sun	0	164 57.8	30.4	N 2 01.1	12.5
	6	252 00.3	30.4	N 0 45.8	12.6
	12	339 02.2	30.2	S 0 29.8	12.7
	18	66 03.4	30.1	S 1 45.6	12.6
20 Mon	0	153 03.6	29.9	S 3 01.3	12.5
	6	240 02.9	29.6	S 4 16.7	12.4
	12	327 01.0	29.5	S 5 31.4	12.2
	18	53 57.7	29.2	S 6 45.1	12.0
21 Tu	0	140 53.1	28.9	S 7 57.7	11.9
	6	227 46.8	28.7	S 9 08.8	11.5
	12	314 38.9	28.4	S10 18.1	11.1
	18	41 29.3	28.1	S11 25.4	10.9
22 Wed	0	128 17.7	27.7	S12 30.4	10.4
	6	215 04.3	27.4	S13 32.9	9.9
	12	301 48.9	27.1	S14 32.4	9.3
	18	28 31.5	26.8	S15 28.9	8.7
23 Th	0	115 12.1	26.5	S16 21.9	8.2
	6	201 50.7	26.1	S17 11.4	7.5
	12	288 27.5	25.8	S17 57.0	6.9
	18	15 02.4	25.5	S18 38.4	6.1
24 Fri	0	101 35.7	25.3	S19 15.6	5.4
	6	188 07.5	25.0	S19 48.3	4.6
	12	274 37.9	24.9	S20 16.4	3.8
	18	1 07.2	24.7	S20 39.6	3.0
25 Sat	0	87 35.6	24.6	S20 58.0	2.1
	6	174 03.4	24.6	S21 11.3	1.3
	12	260 30.8	24.6	S21 19.6	0.4
	18	346 58.1	24.5	S21 22.7	0.4
26 Sun	0	73 25.7	24.7	S21 20.8	1.2
	6	159 53.7	24.9	S21 13.8	2.1
	12	246 22.6	25.0	S21 01.8	2.9
	18	332 52.5	25.2	S20 44.8	3.7
27 Mon	0	59 23.7	25.5	S20 23.1	4.5
	6	145 56.5	25.8	S19 56.7	5.2
	12	232 31.0	26.1	S19 25.8	5.9
	18	319 07.5	26.4	S18 50.6	6.6
28 Tu	0	45 46.1	26.9	S18 11.3	7.3
	6	132 26.9	27.2	S17 28.1	7.9
	12	219 10.0	27.6	S16 41.3	8.4
	18	305 55.5	28.0	S15 51.1	9.0
29 Wed	0	32 43.4	28.5	S14 57.7	9.4
	6	119 33.7	28.8	S14 01.5	9.9
	12	206 26.4	29.2	S13 02.6	10.2
	18	293 21.4	29.6	S12 01.4	10.6
30 Th	0	20 18.7	29.9	S10 58.1	10.9
	6	107 18.2	30.3	S 9 53.0	11.1
	12	194 19.8	30.6	S 8 46.3	11.3
	18	281 23.4	31.0	S 7 38.3	11.5
31 Fri	0	8 28.9	31.2	S 6 29.2	11.7
	6	95 36.1	31.5	S 5 19.3	11.8
	12	182 45.0	31.8	S 4 08.9	11.8
	18	269 55.3	32.0	S 2 58.1	11.8

PLANETS

VENUS Mer Pass h m	VENUS GHA ° ′	VENUS Mean Var/hr 15°+	VENUS Dec ° ′	VENUS Mean Var/hr ′	Day	JUPITER GHA ° ′	JUPITER Mean Var/hr 15°+	JUPITER Dec ° ′	JUPITER Mean Var/hr ′	JUPITER Mer Pass h m
08 56	225 57.3	0.3	N19 10.4	0.2	1 Wed	241 14.2	2.0	N21 13.2	0.1	07 54
08 56	226 04.2	0.3	N19 15.4	0.2	2 Th	242 02.6	2.0	N21 14.6	0.1	07 51
08 55	226 10.3	0.2	N19 20.1	0.2	3 Fri	242 51.0	2.0	N21 16.0	0.1	07 48
08 55	226 15.7	0.2	N19 24.7	0.2	4 Sat	243 39.5	2.0	N21 17.4	0.1	07 44
08 55	226 20.3	0.2	N19 29.1	0.2	5 SUN	244 28.1	2.0	N21 18.8	0.1	07 41
08 54	226 24.2	0.1	N19 33.3	0.2	6 Mon	245 16.9	2.0	N21 20.1	0.1	07 38
08 54	226 27.4	0.1	N19 37.2	0.2	7 Tu	246 05.8	2.0	N21 21.4	0.1	07 35
08 54	226 30.0	0.1	N19 40.9	0.1	8 Wed	246 54.7	2.0	N21 22.7	0.1	07 31
08 54	226 31.8	0.1	N19 44.3	0.1	9 Th	247 43.9	2.1	N21 23.9	0.0	07 28
08 54	226 33.1	0.0	N19 47.4	0.1	10 Fri	248 33.1	2.1	N21 25.1	0.1	07 25
08 54	226 33.7	0.0	N19 50.2	0.1	11 Sat	249 22.4	2.1	N21 26.3	0.1	07 21
08 54	226 33.7	0.0	N19 52.7	0.1	12 SUN	250 11.9	2.1	N21 27.5	0.0	07 18
08 54	226 33.2	0.0	N19 54.9	0.1	13 Mon	251 01.5	2.1	N21 28.6	0.0	07 15
08 54	226 32.1	14	N19 56.7	0.1	14 Tu	251 51.3	2.1	N21 29.7	0.0	07 12
08 54	226 30.6	59.9	N19 58.2	0.0	15 Wed	252 41.1	2.1	N21 30.8	0.0	07 08
08 54	226 28.5	59.9	N19 59.3	0.0	16 Th	253 31.1	2.1	N21 31.9	0.0	07 05
08 54	226 25.9	59.9	N20 00.0	0.0	17 Fri	254 21.3	2.1	N21 32.9	0.0	07 02
08 55	226 22.9	59.9	N20 00.4	0.0	18 Sat	255 11.6	2.1	N21 33.9	0.0	06 58
08 55	226 19.4	59.8	N20 00.3	0.0	19 SUN	256 02.0	2.1	N21 34.9	0.0	06 55
08 55	226 15.5	59.8	N19 59.9	0.0	20 Mon	256 52.6	2.1	N21 35.8	0.0	06 52
08 55	226 11.2	59.8	N19 59.0	0.1	21 Tu	257 43.3	2.1	N21 36.8	0.0	06 48
08 56	226 06.6	59.8	N19 57.7	0.1	22 Wed	258 34.2	2.1	N21 37.7	0.0	06 45
08 56	226 01.5	59.8	N19 55.9	0.1	23 Th	259 25.2	2.1	N21 38.6	0.0	06 41
08 56	225 56.1	59.8	N19 53.7	0.1	24 Fri	260 16.3	2.1	N21 39.4	0.0	06 38
08 57	225 50.4	59.7	N19 51.1	0.1	25 Sat	261 07.7	2.1	N21 40.2	0.0	06 35
08 57	225 44.4	59.7	N19 48.0	0.1	26 SUN	261 59.1	2.2	N21 41.0	0.0	06 31
08 58	225 38.0	59.7	N19 44.4	0.2	27 Mon	262 50.8	2.2	N21 41.8	0.0	06 28
08 58	225 31.4	59.7	N19 40.4	0.2	28 Tu	263 42.5	2.2	N21 42.6	0.0	06 24
08 59	225 24.5	59.7	N19 35.9	0.2	29 Wed	264 34.5	2.2	N21 43.3	0.0	06 21
08 59	225 17.4	59.7	N19 30.9	0.2	30 Th	265 26.6	2.2	N21 44.0	0.0	06 17
09 00	225 10.0	59.7	N19 25.5	0.3	31 Fri	266 18.9	2.2	N21 44.7	0.0	06 14

VENUS, Av. Mag. –4.4
SHA August 5 272; 10 268; 15 263; 20 258; 25 252; 30 247

JUPITER, Av. Mag. –2.2
SHA August 5 291; 10 290; 15 289; 20 288; 25 287; 30 287

MARS Mer Pass h m	MARS GHA ° ′	MARS Mean Var/hr 15°+	MARS Dec ° ′	MARS Mean Var/hr ′	Day	SATURN GHA ° ′	SATURN Mean Var/hr 15°+	SATURN Dec ° ′	SATURN Mean Var/hr ′	SATURN Mer Pass h m
16 18	115 17.6	1.1	S 6 24.1	0.6	1 Wed	107 01.8	2.3	S 7 00.2	0.1	16 49
16 16	115 43.9	1.1	S 6 38.7	0.6	2 Th	107 57.8	2.3	S 7 01.7	0.1	16 46
16 14	116 10.0	1.1	S 6 53.3	0.6	3 Fri	108 53.6	2.3	S 7 03.3	0.1	16 42
16 12	116 35.9	1.1	S 7 07.8	0.6	4 Sat	109 49.4	2.3	S 7 04.8	0.1	16 38
16 11	117 01.7	1.1	S 7 22.4	0.6	5 SUN	110 45.1	2.3	S 7 06.4	0.1	16 34
16 09	117 27.3	1.1	S 7 37.0	0.6	6 Mon	111 40.8	2.3	S 7 08.0	0.1	16 31
16 07	117 52.8	1.1	S 7 51.6	0.6	7 Tu	112 36.3	2.3	S 7 09.7	0.1	16 27
16 06	118 18.1	1.0	S 8 06.2	0.6	8 Wed	113 31.8	2.3	S 7 11.3	0.1	16 23
16 04	118 43.3	1.0	S 8 20.7	0.6	9 Th	114 27.2	2.3	S 7 13.0	0.1	16 20
16 02	119 08.4	1.0	S 8 35.3	0.6	10 Fri	115 22.6	2.3	S 7 14.7	0.1	16 16
16 01	119 33.2	1.0	S 8 49.8	0.6	11 Sat	116 17.8	2.3	S 7 16.5	0.1	16 12
15 59	119 57.9	1.0	S 9 04.4	0.6	12 SUN	117 13.0	2.3	S 7 18.3	0.1	16 09
15 57	120 22.5	1.0	S 9 18.9	0.6	13 Mon	118 08.1	2.3	S 7 20.1	0.1	16 05
15 56	120 46.9	1.0	S 9 33.4	0.6	14 Tu	119 03.2	2.3	S 7 21.9	0.1	16 01
15 54	121 11.1	1.0	S 9 47.9	0.6	15 Wed	119 58.1	2.3	S 7 23.7	0.1	15 58
15 53	121 35.2	1.0	S10 02.3	0.6	16 Th	120 53.0	2.3	S 7 25.6	0.1	15 54
15 51	121 59.1	1.0	S10 16.8	0.6	17 Fri	121 47.9	2.3	S 7 27.5	0.1	15 50
15 49	122 22.9	1.0	S10 31.2	0.6	18 Sat	122 42.6	2.3	S 7 29.4	0.1	15 47
15 48	122 46.5	1.0	S10 45.6	0.6	19 SUN	123 37.3	2.3	S 7 31.3	0.1	15 43
15 46	123 09.9	1.0	S10 59.9	0.6	20 Mon	124 31.9	2.3	S 7 33.3	0.1	15 39
15 45	123 33.2	1.0	S11 14.2	0.6	21 Tu	125 26.5	2.3	S 7 35.3	0.1	15 36
15 43	123 56.3	1.0	S11 28.5	0.6	22 Wed	126 21.0	2.3	S 7 37.3	0.1	15 32
15 42	124 19.2	1.0	S11 42.7	0.6	23 Th	127 15.4	2.3	S 7 39.3	0.1	15 29
15 40	124 42.0	0.9	S11 56.9	0.6	24 Fri	128 09.8	2.3	S 7 41.3	0.1	15 25
15 39	125 04.6	0.9	S12 11.1	0.6	25 Sat	129 04.1	2.3	S 7 43.4	0.1	15 21
15 37	125 27.0	0.9	S12 25.2	0.6	26 SUN	129 58.3	2.3	S 7 45.5	0.1	15 18
15 36	125 49.3	0.9	S12 39.2	0.6	27 Mon	130 52.5	2.3	S 7 47.6	0.1	15 14
15 34	126 11.4	0.9	S12 53.3	0.6	28 Tu	131 46.6	2.3	S 7 49.7	0.1	15 11
15 33	126 33.4	0.9	S13 07.2	0.6	29 Wed	132 40.7	2.3	S 7 51.8	0.1	15 07
15 31	126 55.2	0.9	S13 21.1	0.6	30 Th	133 34.7	2.3	S 7 54.0	0.1	15 03
15 30	127 16.8	0.9	S13 34.9	0.6	31 Fri	134 28.7	2.2	S 7 56.2	0.1	15 00

MARS, Av. Mag. +1.1
SHA August 5 163; 10 160; 15 157; 20 154; 25 151; 30 148

SATURN, Av. Mag. +0.8
SHA August 5 157; 10 156; 15 156; 20 156; 25 155; 30 155

STARS

No.	Name	Mag	Transit h m	Dec ° ′	SHA ° ′
	Oh GMT August 1				
Ψ	ARIES	–	3 19	–	–
1	Alpheratz	2.1	3 28	N29 09.7	357 43.8
2	Ankaa	2.4	3 46	S42 13.9	353 15.9
3	Schedar	2.2	4 01	N56 36.3	349 40.8
4	Diphda	2.0	4 04	S17 54.8	348 56.2
5	Achernar	0.5	4 57	S57 10.0	335 26.9
6	POLARIS	2.0	6 07	N89 18.8	317 58.4
7	Hamal	2.0	5 27	N23 31.3	328 01.3
8	Acamar	3.2	6 18	S40 15.0	315 18.7
9	Menkar	2.5	6 22	N 4 08.4	314 15.6
10	Mirfak	1.8	6 44	N49 54.1	308 41.1
11	Aldebaran	0.9	7 55	N16 32.0	290 50.1
12	Rigel	0.1	8 34	S 8 11.2	281 12.7
13	Capella	0.1	8 36	N46 00.4	280 35.5
14	Bellatrix	1.6	8 44	N 6 21.6	278 32.8
15	Elnath	1.7	8 46	N28 36.9	278 13.5
16	Alnilam	1.7	8 55	S 1 11.6	275 47.1
17	Betelgeuse	0.1–1.2	9 14	N 7 24.5	271 02.1
18	Canopus	–0.7	9 43	S52 42.0	263 56.8
19	Sirius	–1.5	10 04	S16 44.0	258 34.5
20	Adhara	1.5	10 17	S28 59.3	255 13.3
21	Castor	1.6	10 54	N31 51.5	246 09.0
22	Procyon	0.4	10 58	N 5 11.5	245 00.6
23	Pollux	1.1	11 04	N27 59.6	243 28.7
24	Avior	1.9	11 41	S59 33.0	234 18.9
25	Suhail	2.2	12 26	S43 29.1	222 53.3
26	Miaplacidus	1.7	12 31	S69 46.2	221 40.7
27	Alphard	2.0	12 46	S 8 42.9	217 57.0
28	Regulus	1.4	13 27	N11 54.3	207 44.4
29	Dubhe	1.8	14 22	N61 41.0	193 53.0
30	Denebola	2.1	15 07	N14 30.1	182 34.5
31	Gienah	2.6	15 34	S17 36.8	175 53.1
32	Acrux	1.3	15 45	S63 10.4	173 10.4
33	Gacrux	1.6	15 49	S57 11.3	172 01.9
34	Mimosa	1.3	16 06	S59 45.7	167 53.0
35	Alioth	1.8	16 12	N55 53.7	166 21.5
36	Spica	1.0	16 43	S11 13.6	158 32.0
37	Alkaid	1.9	17 05	N49 15.3	152 59.5
38	Hadar	0.6	17 22	S60 26.3	148 48.9
39	Menkent	2.1	17 25	S36 26.0	148 08.4
40	Arcturus	0.0	17 33	N19 07.2	145 56.3
41	Rigil Kent	–0.3	17 58	S60 53.4	139 52.6
42	Zuben'ubi	2.8	18 09	S16 05.6	137 06.1
43	Kochab	2.1	18 08	N74 06.5	137 20.2
44	Alphecca	2.2	18 52	N26 40.6	126 11.5
45	Antares	1.0	19 47	S26 27.6	112 26.8
46	Atria	1.9	20 07	S69 03.2	107 28.9
47	Sabik	2.4	20 28	S15 44.3	102 13.0
48	Shaula	1.6	20 51	S37 06.7	96 22.4
49	Rasalhague	2.1	20 52	N12 33.3	96 06.8
50	Eltanin	2.2	21 13	N51 29.6	90 46.1
51	Kaus Aust.	1.9	21 41	S34 22.6	83 44.3
52	Vega	0.0	21 54	N38 48.1	80 39.0
53	Nunki	2.0	22 12	S26 16.7	75 58.7
54	Altair	0.8	23 08	N 8 54.4	62 08.5
55	Peacock	1.9	23 43	S56 41.5	53 19.5
56	Deneb	1.3	0 02	N45 19.8	49 31.4
57	Enif	2.4	1 05	N 9 56.2	33 47.3
58	Al Na'ir	1.7	1 29	S46 53.7	27 44.0
59	Fomalhaut	1.2	2 18	S29 33.0	15 24.3
60	Markab	2.5	2 25	N15 16.6	13 38.6

SUN

Yr	Day of Mth	Week	Transit h m	Semi-Diam	Lat 52°N Twilight h m	Sunrise h m	Sunset h m	Twilight h m
214	1	Wed	12 06	15.8	03 41	04 22	19 49	20 30
215	2	Th	12 06	15.8	03 43	04 24	19 48	20 28
216	3	Fri	12 06	15.8	03 45	04 25	19 46	20 26
217	4	Sat	12 06	15.8	03 46	04 27	19 44	20 24
218	5	Sun	12 06	15.8	03 48	04 28	19 42	20 22
219	6	Mon	12 06	15.8	03 50	04 30	19 41	20 20
220	7	Tu	12 06	15.8	03 52	04 32	19 39	20 18
221	8	Wed	12 06	15.8	03 54	04 33	19 37	20 16
222	9	Th	12 05	15.8	03 55	04 35	19 35	20 14
223	10	Fri	12 05	15.8	03 57	04 36	19 33	20 12
224	11	Sat	12 05	15.8	03 59	04 38	19 31	20 10
225	12	Sun	12 05	15.8	04 01	04 40	19 29	20 08
226	13	Mon	12 05	15.8	04 03	04 41	19 27	20 05
227	14	Tu	12 05	15.8	04 05	04 43	19 25	20 03
228	15	Wed	12 04	15.8	04 06	04 45	19 23	20 01
229	16	Th	12 04	15.8	04 08	04 46	19 21	19 59
230	17	Fri	12 04	15.8	04 10	04 48	19 19	19 57
231	18	Sat	12 04	15.8	04 12	04 49	19 17	19 54
232	19	Sun	12 04	15.8	04 14	04 51	19 15	19 52
233	20	Mon	12 03	15.8	04 16	04 53	19 13	19 50
234	21	Tu	12 03	15.8	04 17	04 54	19 11	19 47
235	22	Wed	12 03	15.8	04 19	04 56	19 08	19 45
236	23	Th	12 03	15.8	04 21	04 58	19 06	19 43
237	24	Fri	12 02	15.8	04 23	04 59	19 04	19 40
238	25	Sat	12 02	15.8	04 25	05 01	19 02	19 38
239	26	Sun	12 02	15.9	04 26	05 02	19 00	19 36
240	27	Mon	12 01	15.9	04 28	05 04	18 58	19 33
241	28	Tu	12 01	15.9	04 30	05 06	18 55	19 31
242	29	Wed	12 01	15.9	04 32	05 07	18 53	19 29
243	30	Th	12 00	15.9	04 33	05 09	18 51	19 26
244	31	Fri	12 00	15.9	04 35	05 11	18 49	19 24

Lat Corr to Sunrise, Sunset etc.

Lat °	Twilight h m	Sunrise h m	Sunset h m	Twilight h m
N70	TAN	–1 50	+1 46	TAN
68	–2 28	–1 26	+1 23	+2 26
66	–1 48	–1 08	+1 06	+1 47
64	–1 21	–0 53	+0 51	+1 20
62	–1 01	–0 41	+0 39	+1 00
N60	–0 45	–0 30	+0 29	+0 44
58	–0 32	–0 21	+0 21	+0 30
56	–0 20	–0 13	+0 13	+0 18
54	–0 10	–0 06	+0 06	+0 09
50	+0 07	+0 06	–0 06	–0 04
N45	+0 17	+0 18	–0 18	–0 27
40	+0 23	+0 28	–0 28	–0 38
35	+0 36	+0 37	–0 37	–0 48
30	+0 47	+0 44	–0 45	–0 58
20	+1 12	+0 57	–0 57	–1 13
N10	+1 24	+1 09	–1 08	–1 25
0	+1 35	+1 19	–1 18	–1 35
S10	+1 44	+1 29	–1 28	–1 46
20	+1 54	+1 40	–1 39	–1 56
30	+2 05	+1 52	–1 51	–2 06
S35	+2 10	+2 00	–1 57	–2 11
40	+2 16	+2 07	–2 05	–2 17
45	+2 23	+2 17	–2 15	–2 24
S50	+2 30	+2 27	–2 26	–2 32

NOTES
The corrections to sunrise, etc. are for middle of August. TAN means Twilight all night.

MOON

Yr	Day of Mth	Week	Age days	Transit (Upper) h m	Diff m	Semi-diam ′	Hor Par ′	Lat 52°N Moonrise h m	Moonset h m
214	1	Wed	13	23 54	51	16.0	58.9	19 02	03 37
215	2	Th	14	24 45	–	15.9	58.3	19 29	04 55
216	3	Fri	15	00 45	48	15.7	57.6	19 52	06 12
217	4	Sat	16	01 33	47	15.5	56.9	20 13	07 27
218	5	Sun	17	02 20	45	15.3	56.2	20 33	08 39
219	6	Mon	18	03 05	44	15.1	55.5	20 53	09 50
220	7	Tu	19	03 49	45	15.0	55.0	21 15	10 58
221	8	Wed	20	04 34	45	14.9	54.6	21 40	12 05
222	9	Th	21	05 19	46	14.8	54.3	22 09	13 09
223	10	Fri	22	06 05	48	14.8	54.3	22 44	14 11
224	11	Sat	23	06 53	49	14.8	54.4	23 26	15 07
225	12	Sun	24	07 42	49	14.9	54.6	– –	15 58
226	13	Mon	25	08 31	50	15.0	55.0	00 16	16 42
227	14	Tu	26	09 21	50	15.1	55.5	01 14	17 20
228	15	Wed	27	10 11	49	15.3	56.1	02 18	17 52
229	16	Th	28	11 00	49	15.5	56.7	03 28	18 19
230	17	Fri	29	11 49	49	15.6	57.3	04 41	18 43
231	18	Sat	01	12 38	49	15.8	57.9	05 56	19 05
232	19	Sun	02	13 27	50	15.9	58.4	07 13	19 27
233	20	Mon	03	14 17	51	16.0	58.8	08 31	19 50
234	21	Tu	04	15 08	54	16.1	59.1	09 49	20 16
235	22	Wed	05	16 02	55	16.1	59.2	11 09	20 45
236	23	Th	06	16 57	58	16.2	59.3	12 27	21 22
237	24	Fri	07	17 55	59	16.2	59.3	13 40	22 07
238	25	Sat	08	18 54	59	16.1	59.2	14 46	23 03
239	26	Sun	09	19 53	57	16.1	59.0	15 41	– –
240	27	Mon	10	20 50	54	16.0	58.8	16 26	00 08
241	28	Tu	11	21 44	52	15.9	58.4	17 01	01 20
242	29	Wed	12	22 36	49	15.8	58.0	17 30	02 36
243	30	Th	13	23 25	47	15.7	57.5	17 55	03 52
244	31	Fri	14	24 12	–	15.5	57.0	18 17	05 07

Phases of the Moon

		d	h	m
○	Full Moon	2	03	27
☽	Last Quarter	9	18	55
●	New Moon	17	15	54
☾	First Quarter	24	13	54
○	Full Moon	31	13	58

	d	h
Apogee	10	11
Perigee	23	19

31

SEPTEMBER 2012

Saturday, 1st September

GMT	SUN GHA ° ′	SUN Dec ° ′	ARIES GHA ° ′
00	179 59.9	N 8 12.4	340 33.7
02	210 00.3	8 10.6	10 38.7
04	240 00.7	8 08.8	40 43.6
06	270 01.1	8 07.0	70 48.5
08	300 01.5	8 05.2	100 53.4
10	330 01.9	8 03.4	130 58.4
12	0 02.3	8 01.5	161 03.3
14	30 02.7	7 59.7	191 08.2
16	60 03.1	7 57.9	221 13.2
18	90 03.5	7 56.1	251 18.1
20	120 03.9	7 54.3	281 23.0
22	150 04.3	N 7 52.4	311 27.9

Sunday, 2nd September

GMT	SUN GHA ° ′	SUN Dec ° ′	ARIES GHA ° ′
00	180 04.7	N 7 50.6	341 32.9
02	210 05.1	7 48.8	11 37.8
04	240 05.5	7 47.0	41 42.7
06	270 05.9	7 45.1	71 47.7
08	300 06.3	7 43.3	101 52.6
10	330 06.7	7 41.5	131 57.5
12	0 07.1	7 39.6	162 02.4
14	30 07.5	7 37.8	192 07.4
16	60 07.9	7 36.0	222 12.3
18	90 08.4	7 34.2	252 17.2
20	120 08.8	7 32.3	282 22.2
22	150 09.2	N 7 30.5	312 27.1

Monday, 3rd September

GMT	SUN GHA ° ′	SUN Dec ° ′	ARIES GHA ° ′
00	180 09.6	N 7 28.7	342 32.0
02	210 10.0	7 26.8	12 36.9
04	240 10.4	7 25.0	42 41.9
06	270 10.8	7 23.1	72 46.8
08	300 11.2	7 21.3	102 51.7
10	330 11.6	7 19.5	132 56.6
12	0 12.1	7 17.6	163 01.6
14	30 12.5	7 15.8	193 06.5
16	60 12.9	7 13.9	223 11.4
18	90 13.3	7 12.1	253 16.4
20	120 13.7	7 10.3	283 21.3
22	150 14.1	N 7 08.4	313 26.2

Tuesday, 4th September

GMT	SUN GHA ° ′	SUN Dec ° ′	ARIES GHA ° ′
00	180 14.5	N 7 06.6	343 31.1
02	210 15.0	7 04.7	13 36.1
04	240 15.4	7 02.9	43 41.0
06	270 15.8	7 01.0	73 45.9
08	300 16.2	6 59.2	103 50.9
10	330 16.6	6 57.3	133 55.8
12	0 17.0	6 55.5	164 00.7
14	30 17.5	6 53.6	194 05.6
16	60 17.9	6 51.8	224 10.6
18	90 18.3	6 49.9	254 15.5
20	120 18.7	6 48.1	284 20.4
22	150 19.1	N 6 46.2	314 25.4

Wednesday, 5th September

GMT	SUN GHA ° ′	SUN Dec ° ′	ARIES GHA ° ′
00	180 19.5	N 6 44.4	344 30.3
02	210 20.0	6 42.5	14 35.2
04	240 20.4	6 40.7	44 40.1
06	270 20.8	6 38.8	74 45.1
08	300 21.2	6 36.9	104 50.0
10	330 21.6	6 35.1	134 54.9
12	0 22.1	6 33.2	164 59.9
14	30 22.5	6 31.4	195 04.8
16	60 22.9	6 29.5	225 09.7
18	90 23.3	6 27.6	255 14.6
20	120 23.8	6 25.8	285 19.6
22	150 24.2	N 6 23.9	315 24.5

Thursday, 6th September

GMT	SUN GHA ° ′	SUN Dec ° ′	ARIES GHA ° ′
00	180 24.6	N 6 22.1	345 29.4
02	210 25.0	6 20.2	15 34.3
04	240 25.5	6 18.3	45 39.3
06	270 25.9	6 16.5	75 44.2
08	300 26.3	6 14.6	105 49.1
10	330 26.7	6 12.7	135 54.1
12	0 27.2	6 10.9	165 59.0
14	30 27.6	6 09.0	196 03.9
16	60 28.0	6 07.1	226 08.8
18	90 28.4	6 05.3	256 13.8
20	120 28.9	6 03.4	286 18.7
22	150 29.3	N 6 01.5	316 23.6

Friday, 7th September

GMT	SUN GHA ° ′	SUN Dec ° ′	ARIES GHA ° ′
00	180 29.7	N 5 59.6	346 28.6
02	210 30.1	5 57.8	16 33.5
04	240 30.6	5 55.9	46 38.4
06	270 31.0	5 54.0	76 43.3
08	300 31.4	5 52.1	106 48.3
10	330 31.9	5 50.3	136 53.2
12	0 32.3	5 48.4	166 58.1
14	30 32.7	5 46.5	197 03.1
16	60 33.1	5 44.6	227 08.0
18	90 33.6	5 42.8	257 12.9
20	120 34.0	5 40.9	287 17.8
22	150 34.4	N 5 39.0	317 22.8

Saturday, 8th September

GMT	SUN GHA ° ′	SUN Dec ° ′	ARIES GHA ° ′
00	180 34.9	N 5 37.1	347 27.7
02	210 35.3	5 35.2	17 32.6
04	240 35.7	5 33.4	47 37.6
06	270 36.2	5 31.5	77 42.5
08	300 36.6	5 29.6	107 47.4
10	330 37.0	5 27.7	137 52.3
12	0 37.5	5 25.8	167 57.3
14	30 37.9	5 23.9	198 02.2
16	60 38.3	5 22.1	228 07.1
18	90 38.8	5 20.2	258 12.1
20	120 39.2	5 18.3	288 17.0
22	150 39.6	N 5 16.4	318 21.9

Sunday, 9th September

GMT	SUN GHA ° ′	SUN Dec ° ′	ARIES GHA ° ′
00	180 40.1	N 5 14.5	348 26.8
02	210 40.5	5 12.6	18 31.8
04	240 40.9	5 10.7	48 36.7
06	270 41.4	5 08.8	78 41.6
08	300 41.8	5 07.0	108 46.5
10	330 42.2	5 05.1	138 51.5
12	0 42.7	5 03.2	168 56.4
14	30 43.1	5 01.3	199 01.3
16	60 43.5	4 59.4	229 06.3
18	90 44.0	4 57.5	259 11.2
20	120 44.4	4 55.6	289 16.1
22	150 44.9	N 4 53.7	319 21.0

Monday, 10th September

GMT	SUN GHA ° ′	SUN Dec ° ′	ARIES GHA ° ′
00	180 45.3	N 4 51.8	349 26.0
02	210 45.7	4 49.9	19 30.9
04	240 46.2	4 48.0	49 35.8
06	270 46.6	4 46.1	79 40.8
08	300 47.0	4 44.2	109 45.7
10	330 47.5	4 42.3	139 50.6
12	0 47.9	4 40.4	169 55.5
14	30 48.4	4 38.5	200 00.5
16	60 48.8	4 36.6	230 05.4
18	90 49.2	4 34.7	260 10.3
20	120 49.7	4 32.8	290 15.3
22	150 50.1	N 4 30.9	320 20.2

Tuesday, 11th September

GMT	SUN GHA ° ′	SUN Dec ° ′	ARIES GHA ° ′
00	180 50.5	N 4 29.0	350 25.1
02	210 51.0	4 27.1	20 30.0
04	240 51.4	4 25.2	50 35.0
06	270 51.9	4 23.3	80 39.9
08	300 52.3	4 21.4	110 44.8
10	330 52.7	4 19.5	140 49.8
12	0 53.2	4 17.6	170 54.7
14	30 53.6	4 15.7	200 59.6
16	60 54.1	4 13.8	231 04.5
18	90 54.5	4 11.9	261 09.5
20	120 54.9	4 10.0	291 14.4
22	150 55.4	N 4 08.1	321 19.3

Wednesday, 12th September

GMT	SUN GHA ° ′	SUN Dec ° ′	ARIES GHA ° ′
00	180 55.8	N 4 06.1	351 24.3
02	210 56.3	4 04.2	21 29.2
04	240 56.7	4 02.3	51 34.1
06	270 57.1	4 00.4	81 39.0
08	300 57.6	3 58.5	111 44.0
10	330 58.0	3 56.6	141 48.9
12	0 58.5	3 54.7	171 53.8
14	30 58.9	3 52.8	201 58.8
16	60 59.3	3 50.9	232 03.7
18	90 59.8	3 49.0	262 08.6
20	121 00.2	3 47.0	292 13.5
22	151 00.7	N 3 45.1	322 18.5

Thursday, 13th September

GMT	SUN GHA ° ′	SUN Dec ° ′	ARIES GHA ° ′
00	181 01.1	N 3 43.2	352 23.4
02	211 01.6	3 41.3	22 28.3
04	241 02.0	3 39.4	52 33.2
06	271 02.4	3 37.5	82 38.2
08	301 02.9	3 35.5	112 43.1
10	331 03.3	3 33.6	142 48.0
12	1 03.8	3 31.7	172 53.0
14	31 04.2	3 29.8	202 57.9
16	61 04.7	3 27.9	233 02.8
18	91 05.1	3 26.0	263 07.7
20	121 05.5	3 24.0	293 12.7
22	151 06.0	N 3 22.1	323 17.6

Friday, 14th September

GMT	SUN GHA ° ′	SUN Dec ° ′	ARIES GHA ° ′
00	181 06.4	N 3 20.2	353 22.5
02	211 06.9	3 18.3	23 27.5
04	241 07.3	3 16.4	53 32.4
06	271 07.8	3 14.4	83 37.3
08	301 08.2	3 12.5	113 42.2
10	331 08.6	3 10.6	143 47.2
12	1 09.1	3 08.7	173 52.1
14	31 09.5	3 06.8	203 57.0
16	61 10.0	3 04.8	234 02.0
18	91 10.4	3 02.9	264 06.9
20	121 10.9	3 01.0	294 11.8
22	151 11.3	N 2 59.1	324 16.7

Saturday, 15th September

GMT	SUN GHA ° ′	SUN Dec ° ′	ARIES GHA ° ′
00	181 11.8	N 2 57.1	354 21.7
02	211 12.2	2 55.2	24 26.6
04	241 12.6	2 53.3	54 31.5
06	271 13.1	2 51.4	84 36.5
08	301 13.5	2 49.4	114 41.4
10	331 14.0	2 47.5	144 46.3
12	1 14.4	2 45.6	174 51.2
14	31 14.9	2 43.6	204 56.2
16	61 15.3	2 41.7	235 01.1
18	91 15.8	2 39.8	265 06.0
20	121 16.2	2 37.9	295 10.9
22	151 16.6	N 2 35.9	325 15.9

Sunday, 16th September

GMT	SUN GHA ° ′	SUN Dec ° ′	ARIES GHA ° ′
00	181 17.1	N 2 34.0	355 20.8
02	211 17.5	2 32.1	25 25.7
04	241 18.0	2 30.1	55 30.7
06	271 18.4	2 28.2	85 35.6
08	301 18.9	2 26.3	115 40.5
10	331 19.3	2 24.4	145 45.4
12	1 19.8	2 22.4	175 50.4
14	31 20.2	2 20.5	205 55.3
16	61 20.6	2 18.6	236 00.2
18	91 21.1	2 16.6	266 05.2
20	121 21.5	2 14.7	296 10.1
22	151 22.0	N 2 12.8	326 15.0

Monday, 17th September

GMT	SUN GHA ° ′	SUN Dec ° ′	ARIES GHA ° ′
00	181 22.4	N 2 10.8	356 19.9
02	211 22.9	2 08.9	26 24.9
04	241 23.3	2 07.0	56 29.8
06	271 23.8	2 05.0	86 34.7
08	301 24.2	2 03.1	116 39.7
10	331 24.6	2 01.2	146 44.6
12	1 25.1	1 59.2	176 49.5
14	31 25.5	1 57.3	206 54.4
16	61 26.0	1 55.4	236 59.4
18	91 26.4	1 53.4	267 04.3
20	121 26.9	1 51.5	297 09.2
22	151 27.3	N 1 49.5	327 14.1

Tuesday, 18th September

GMT	SUN GHA ° ′	SUN Dec ° ′	ARIES GHA ° ′
00	181 27.8	N 1 47.6	357 19.1
02	211 28.2	1 45.7	27 24.0
04	241 28.6	1 43.7	57 28.9
06	271 29.1	1 41.8	87 33.9
08	301 29.5	1 39.9	117 38.8
10	331 30.0	1 37.9	147 43.7
12	1 30.4	1 36.0	177 48.6
14	31 30.9	1 34.1	207 53.6
16	61 31.3	1 32.1	237 58.5
18	91 31.7	1 30.2	268 03.4
20	121 32.2	1 28.2	298 08.4
22	151 32.6	N 1 26.3	328 13.3

Wednesday, 19th September

GMT	SUN GHA ° ′	SUN Dec ° ′	ARIES GHA ° ′
00	181 33.1	N 1 24.4	358 18.2
02	211 33.5	1 22.4	28 23.1
04	241 34.0	1 20.5	58 28.1
06	271 34.4	1 18.5	88 33.0
08	301 34.9	1 16.6	118 37.9
10	331 35.3	1 14.7	148 42.9
12	1 35.7	1 12.7	178 47.8
14	31 36.2	1 10.8	208 52.7
16	61 36.6	1 08.8	238 57.6
18	91 37.1	1 06.9	269 02.6
20	121 37.5	1 05.0	299 07.5
22	151 38.0	N 1 03.0	329 12.4

Thursday, 20th September

GMT	SUN GHA ° ′	SUN Dec ° ′	ARIES GHA ° ′
00	181 38.4	N 1 01.1	359 17.4
02	211 38.8	0 59.1	29 22.3
04	241 39.3	0 57.2	59 27.2
06	271 39.7	0 55.2	89 32.1
08	301 40.2	0 53.3	119 37.1
10	331 40.6	0 51.4	149 42.0
12	1 41.1	0 49.4	179 46.9
14	31 41.5	0 47.5	209 51.8
16	61 41.9	0 45.5	239 56.8
18	91 42.4	0 43.6	270 01.7
20	121 42.8	0 41.6	300 06.6
22	151 43.3	N 0 39.7	330 11.6

Friday, 21st September

GMT	SUN GHA ° ′	SUN Dec ° ′	ARIES GHA ° ′
00	181 43.7	N 0 37.8	0 16.5
02	211 44.1	0 35.8	30 21.4
04	241 44.6	0 33.9	60 26.3
06	271 45.0	0 31.9	90 31.3
08	301 45.5	0 30.0	120 36.2
10	331 45.9	0 28.0	150 41.1
12	1 46.4	0 26.1	180 46.1
14	31 46.8	0 24.1	210 51.0
16	61 47.2	0 22.2	240 55.9
18	91 47.7	0 20.3	271 00.8
20	121 48.1	0 18.3	301 05.8
22	151 48.6	N 0 16.4	331 10.7

Saturday, 22nd September

GMT	SUN GHA ° ′	SUN Dec ° ′	ARIES GHA ° ′
00	181 49.0	N 0 14.4	1 15.6
02	211 49.4	0 12.5	31 20.6
04	241 49.9	0 10.5	61 25.5
06	271 50.3	0 08.6	91 30.4
08	301 50.8	0 06.6	121 35.3
10	331 51.2	0 04.7	151 40.3
12	1 51.6	0 02.7	181 45.2
14	31 52.1	0 00.8	211 50.1
16	61 52.5	S 0 01.2	241 55.1
18	91 52.9	0 03.1	272 00.0
20	121 53.4	0 05.0	302 04.9
22	151 53.8	S 0 07.0	332 09.8

Sunday, 23rd September

GMT	SUN GHA ° ′	SUN Dec ° ′	ARIES GHA ° ′
00	181 54.3	S 0 08.9	2 14.8
02	211 54.7	0 10.9	32 19.7
04	241 55.1	0 12.8	62 24.6
06	271 55.6	0 14.8	92 29.6
08	301 56.0	0 16.7	122 34.5
10	331 56.4	0 18.7	152 39.4
12	1 56.9	0 20.6	182 44.3
14	31 57.3	0 22.6	212 49.3
16	61 57.8	0 24.5	242 54.2
18	91 58.2	0 26.5	272 59.1
20	121 58.6	0 28.4	303 04.1
22	151 59.1	S 0 30.3	333 09.0

Monday, 24th September

GMT	SUN GHA ° ′	SUN Dec ° ′	ARIES GHA ° ′
00	181 59.5	S 0 32.3	3 13.9
02	211 59.9	0 34.2	33 18.8
04	242 00.4	0 36.2	63 23.8
06	272 00.8	0 38.1	93 28.7
08	302 01.2	0 40.1	123 33.6
10	332 01.7	0 42.0	153 38.6
12	2 02.1	0 44.0	183 43.5
14	32 02.5	0 45.9	213 48.4
16	62 03.0	0 47.9	243 53.3
18	92 03.4	0 49.8	273 58.3
20	122 03.8	0 51.8	304 03.2
22	152 04.3	S 0 53.7	334 08.1

Tuesday, 25th September

GMT	SUN GHA ° ′	SUN Dec ° ′	ARIES GHA ° ′
00	182 04.7	S 0 55.7	4 13.0
02	212 05.1	0 57.6	34 18.0
04	242 05.6	0 59.6	64 22.9
06	272 06.0	1 01.5	94 27.8
08	302 06.4	1 03.4	124 32.8
10	332 06.9	1 05.4	154 37.7
12	2 07.3	1 07.3	184 42.6
14	32 07.7	1 09.3	214 47.5
16	62 08.2	1 11.2	244 52.5
18	92 08.6	1 13.2	274 57.4
20	122 09.0	1 15.1	305 02.3
22	152 09.4	S 1 17.1	335 07.3

Wednesday, 26th September

GMT	SUN GHA ° ′	SUN Dec ° ′	ARIES GHA ° ′
00	182 09.9	S 1 19.0	5 12.2
02	212 10.3	1 21.0	35 17.1
04	242 10.7	1 22.9	65 22.0
06	272 11.2	1 24.9	95 27.0
08	302 11.6	1 26.8	125 31.9
10	332 12.0	1 28.8	155 36.8
12	2 12.4	1 30.7	185 41.8
14	32 12.9	1 32.6	215 46.7
16	62 13.3	1 34.6	245 51.6
18	92 13.7	1 36.5	275 56.5
20	122 14.2	1 38.5	306 01.5
22	152 14.6	S 1 40.4	336 06.4

Thursday, 27th September

GMT	SUN GHA ° ′	SUN Dec ° ′	ARIES GHA ° ′
00	182 15.0	S 1 42.4	6 11.3
02	212 15.4	1 44.3	36 16.3
04	242 15.9	1 46.3	66 21.2
06	272 16.3	1 48.2	96 26.1
08	302 16.7	1 50.2	126 31.0
10	332 17.1	1 52.1	156 36.0
12	2 17.5	1 54.0	186 40.9
14	32 18.0	1 56.0	216 45.8
16	62 18.4	1 57.9	246 50.7
18	92 18.8	1 59.9	276 55.7
20	122 19.2	2 01.8	307 00.6
22	152 19.7	S 2 03.8	337 05.5

Friday, 28th September

GMT	SUN GHA ° ′	SUN Dec ° ′	ARIES GHA ° ′
00	182 20.1	S 2 05.7	7 10.5
02	212 20.5	2 07.7	37 15.4
04	242 20.9	2 09.6	67 20.3
06	272 21.3	2 11.5	97 25.2
08	302 21.8	2 13.5	127 30.2
10	332 22.2	2 15.4	157 35.1
12	2 22.6	2 17.4	187 40.0
14	32 23.0	2 19.3	217 45.0
16	62 23.4	2 21.3	247 49.9
18	92 23.8	2 23.2	277 54.8
20	122 24.3	2 25.1	307 59.7
22	152 24.7	S 2 27.1	338 04.7

Saturday, 29th September

GMT	SUN GHA ° ′	SUN Dec ° ′	ARIES GHA ° ′
00	182 25.1	S 2 29.0	8 09.6
02	212 25.5	2 31.0	38 14.5
04	242 25.9	2 32.9	68 19.5
06	272 26.3	2 34.9	98 24.4
08	302 26.8	2 36.8	128 29.3
10	332 27.2	2 38.7	158 34.2
12	2 27.6	2 40.7	188 39.2
14	32 28.0	2 42.6	218 44.1
16	62 28.4	2 44.6	248 49.0
18	92 28.8	2 46.5	278 54.0
20	122 29.2	2 48.5	308 58.9
22	152 29.6	S 2 50.4	339 03.8

Sunday, 30th September

GMT	SUN GHA ° ′	SUN Dec ° ′	ARIES GHA ° ′
00	182 30.1	S 2 52.3	9 08.7
02	212 30.5	2 54.3	39 13.7
04	242 30.9	2 56.2	69 18.6
06	272 31.3	2 58.2	99 23.5
08	302 31.7	3 00.1	129 28.4
10	332 32.1	3 02.0	159 33.4
12	2 32.5	3 04.0	189 38.3
14	32 32.9	3 05.9	219 43.2
16	62 33.3	3 07.8	249 48.2
18	92 33.7	3 09.8	279 53.1
20	122 34.1	3 11.7	309 58.0
22	152 34.5	S 3 13.7	340 02.9

MOON

Day	GMT hr	GHA ° ′	Mean Var/hr 14°+	Dec ° ′	Mean Var/hr ′
1 Sat	0	357 06.9	32.2	S 1 47.1	11.8
	6	84 19.7	32.3	S 0 36.3	11.7
	12	171 33.5	32.4	N 0 34.3	11.7
	18	258 48.2	32.6	N 1 44.4	11.6
2 Sun	0	346 03.7	32.6	N 2 53.8	11.5
	6	73 19.7	32.7	N 4 02.4	11.3
	12	160 36.3	32.9	N 5 09.9	11.0
	18	247 53.1	32.8	N 6 16.3	10.8
3 Mon	0	335 10.1	32.8	N 7 21.3	10.5
	6	62 27.2	32.9	N 8 24.7	10.3
	12	149 44.2	32.8	N 9 26.5	9.9
	18	237 01.1	32.8	N10 26.5	9.7
4 Tu	0	324 17.7	32.7	N11 24.6	9.4
	6	51 33.9	32.7	N12 20.6	8.9
	12	138 49.6	32.5	N13 14.4	8.6
	18	226 04.8	32.4	N14 05.9	8.1
5 Wed	0	313 19.4	32.3	N14 54.9	7.7
	6	40 33.3	32.2	N15 41.4	7.3
	12	127 46.5	32.1	N16 25.3	6.8
	18	214 58.8	31.9	N17 06.4	6.3
6 Th	0	302 10.4	31.8	N17 44.7	5.9
	6	29 21.1	31.6	N18 20.0	5.3
	12	116 30.9	31.5	N18 52.3	4.8
	18	203 39.9	31.4	N19 21.5	4.3
7 Fri	0	290 48.1	31.2	N19 47.5	3.8
	6	17 55.4	31.1	N20 10.3	3.2
	12	105 01.9	31.0	N20 29.7	2.6
	18	192 07.7	30.8	N20 45.7	2.1
8 Sat	0	279 12.7	30.7	N20 58.3	1.5
	6	6 17.1	30.6	N21 07.3	0.9
	12	93 20.9	30.5	N21 12.8	0.2
	18	180 24.2	30.5	N21 14.8	0.3
9 Sun	0	267 26.9	30.4	N21 13.1	0.9
	6	354 29.3	30.3	N21 07.8	1.6
	12	81 31.4	30.3	N20 58.8	2.2
	18	168 33.2	30.3	N20 46.2	2.7
10 Mon	0	255 34.8	30.3	N20 30.0	3.4
	6	342 36.4	30.3	N20 10.1	4.0
	12	69 37.9	30.2	N19 46.5	4.5
	18	156 39.4	30.3	N19 19.4	5.2
11 Tu	0	243 40.9	30.3	N18 48.8	5.8
	6	330 42.6	30.3	N18 14.6	6.3
	12	57 44.4	30.3	N17 37.1	6.9
	18	144 46.4	30.4	N16 56.1	7.4
12 Wed	0	231 48.5	30.4	N16 11.9	8.0
	6	318 50.8	30.4	N15 24.5	8.5
	12	45 53.3	30.4	N14 34.0	9.0
	18	132 55.9	30.4	N13 40.5	9.5
13 Th	0	219 58.6	30.5	N12 44.1	9.8
	6	307 01.4	30.5	N11 45.1	10.3
	12	34 04.2	30.4	N10 43.5	10.7
	18	121 06.9	30.4	N 9 39.5	11.1
14 Fri	0	208 09.5	30.4	N 8 33.2	11.4
	6	295 11.8	30.3	N 7 24.9	11.7
	12	22 13.8	30.2	N 6 14.8	12.0
	18	109 15.4	30.2	N 5 03.0	12.3
15 Sat	0	196 16.4	30.1	N 3 49.8	12.4
	6	283 16.7	29.9	N 2 35.4	12.6
	12	10 16.2	29.7	N 1 20.1	12.7
	18	97 14.9	29.6	N 0 04.1	12.8
16 Sun	0	184 12.4	29.3	S 1 12.3	12.8
	6	271 08.8	29.2	S 2 28.8	12.7
	12	358 03.9	28.9	S 3 45.2	12.6
	18	84 57.5	28.6	S 5 01.1	12.5
17 Mon	0	171 49.6	28.4	S 6 16.2	12.3
	6	258 40.1	28.1	S 7 30.3	12.1
	12	345 28.8	27.8	S 8 43.0	11.8
	18	72 15.7	27.5	S 9 53.9	11.5
18 Tu	0	159 00.7	27.2	S11 02.9	11.1
	6	245 43.7	26.8	S12 09.5	10.7
	12	332 24.8	26.4	S13 13.5	10.1
	18	59 03.9	26.2	S14 14.5	9.5
19 Wed	0	145 41.0	25.9	S15 12.2	9.0
	6	232 16.3	25.5	S16 06.4	8.3
	12	318 49.8	25.3	S16 56.8	7.7
	18	45 21.6	25.0	S17 43.1	7.0
20 Th	0	131 51.8	24.8	S18 25.2	6.2
	6	218 20.8	24.6	S19 02.7	5.5
	12	304 48.5	24.5	S19 35.6	4.6
	18	31 15.4	24.4	S20 03.6	3.8
21 Fri	0	117 41.5	24.3	S20 26.7	2.9
	6	204 07.3	24.2	S20 44.7	2.1
	12	290 33.0	24.3	S20 57.7	1.2
	18	16 58.9	24.4	S21 05.5	0.4
22 Sat	0	103 25.3	24.5	S21 08.1	0.4
	6	189 52.4	24.8	S21 05.7	1.3
	12	276 20.6	24.9	S20 58.3	2.1
	18	2 50.1	25.2	S20 45.9	2.9
23 Sun	0	89 21.2	25.5	S20 28.7	3.7
	6	175 54.0	25.8	S20 06.9	4.5
	12	262 28.9	26.2	S19 40.6	5.2
	18	349 05.8	26.5	S19 09.9	5.9
24 Mon	0	75 45.1	26.9	S18 35.2	6.5
	6	162 26.7	27.4	S17 56.6	7.1
	12	249 10.7	27.7	S17 14.3	7.7
	18	335 57.2	28.2	S16 28.5	8.2
25 Tu	0	62 46.2	28.6	S15 39.6	8.7
	6	149 37.7	29.0	S14 47.7	9.1
	12	236 31.6	29.4	S13 53.1	9.5
	18	323 27.9	29.8	S12 56.0	9.9
26 Wed	0	50 26.5	30.2	S11 56.6	10.2
	6	137 27.3	30.5	S10 55.3	10.6
	12	224 30.2	30.9	S 9 52.2	10.8
	18	311 35.0	31.2	S 8 47.6	11.0
27 Th	0	38 41.7	31.4	S 7 41.7	11.2
	6	125 50.1	31.7	S 6 34.6	11.3
	12	213 00.1	31.9	S 5 26.8	11.5
	18	300 11.5	32.2	S 4 18.2	11.5
28 Fri	0	27 24.1	32.3	S 3 09.2	11.5
	6	114 37.9	32.5	S 2 00.0	11.6
	12	201 52.6	32.6	S 0 50.8	11.5
	18	289 08.2	32.7	N 0 18.4	11.5
29 Sat	0	16 24.4	32.8	N 1 27.1	11.4
	6	103 41.2	32.8	N 2 35.4	11.2
	12	190 58.4	32.9	N 3 42.9	11.1
	18	278 15.8	32.9	N 4 49.6	10.9
30 Sun	0	5 33.4	32.9	N 5 55.2	10.7
	6	92 51.0	32.9	N 6 59.6	10.4
	12	180 08.5	32.9	N 8 02.7	10.2
	18	267 25.8	32.8	N 9 04.2	10.0

PLANETS

VENUS Mer Pass h m	GHA ° ′	Mean Var/hr 14°+	Dec ° ′	Mean Var/hr ′	Day	JUPITER GHA ° ′	Mean Var/hr 15°+	Dec ° ′	Mean Var/hr ′	Mer Pass h m
09 00	225 02.4	59.7	N19 19.5	0.3	1 Sat	267 11.3	2.2	N21 45.4	0.0	06 10
09 01	224 54.5	59.7	N19 13.1	0.3	2 SUN	268 03.9	2.2	N21 46.0	0.0	06 07
09 01	224 46.5	59.7	N19 06.1	0.3	3 Mon	268 56.7	2.2	N21 46.6	0.0	06 03
09 02	224 38.3	59.7	N18 58.7	0.3	4 Tu	269 49.7	2.2	N21 47.2	0.0	06 00
09 02	224 29.9	59.6	N18 50.8	0.4	5 Wed	270 42.8	2.2	N21 47.8	0.0	05 56
09 03	224 21.3	59.6	N18 42.3	0.4	6 Th	271 36.1	2.2	N21 48.4	0.0	05 53
09 03	224 12.7	59.6	N18 33.4	0.4	7 Fri	272 29.6	2.2	N21 48.9	0.0	05 49
09 04	224 03.9	59.6	N18 24.0	0.4	8 Sat	273 23.2	2.2	N21 49.4	0.0	05 46
09 05	223 54.9	59.6	N18 14.0	0.4	9 SUN	274 17.1	2.3	N21 49.9	0.0	05 42
09 05	223 45.9	59.6	N18 03.6	0.5	10 Mon	275 11.1	2.3	N21 50.3	0.0	05 38
09 06	223 36.8	59.6	N17 52.7	0.5	11 Tu	276 05.3	2.3	N21 50.8	0.0	05 35
09 06	223 27.6	59.6	N17 41.3	0.5	12 Wed	276 59.7	2.3	N21 51.2	0.0	05 31
09 07	223 18.3	59.6	N17 29.3	0.5	13 Th	277 54.3	2.3	N21 51.6	0.0	05 28
09 08	223 09.0	59.6	N17 16.9	0.5	14 Fri	278 49.1	2.3	N21 52.0	0.0	05 24
09 08	222 59.7	59.6	N17 04.0	0.6	15 Sat	279 44.1	2.3	N21 52.3	0.0	05 20
09 09	222 50.3	59.6	N16 50.7	0.6	16 SUN	280 39.3	2.3	N21 52.7	0.0	05 17
09 10	222 40.9	59.6	N16 36.8	0.6	17 Mon	281 34.6	2.3	N21 53.0	0.0	05 13
09 10	222 31.4	59.6	N16 22.5	0.6	18 Tu	282 30.2	2.3	N21 53.3	0.0	05 09
09 11	222 22.0	59.6	N16 07.7	0.6	19 Wed	283 26.0	2.3	N21 53.6	0.0	05 05
09 11	222 12.6	59.6	N15 52.4	0.7	20 Th	284 22.0	2.3	N21 53.8	0.0	05 02
09 12	222 03.2	59.6	N15 36.6	0.7	21 Fri	285 18.1	2.4	N21 54.0	0.0	04 58
09 13	221 53.8	59.6	N15 20.4	0.7	22 Sat	286 14.5	2.4	N21 54.3	0.0	04 54
09 13	221 44.5	59.6	N15 03.8	0.7	23 SUN	287 11.1	2.4	N21 54.5	0.0	04 50
09 14	221 35.1	59.6	N14 46.7	0.7	24 Mon	288 07.9	2.4	N21 54.6	0.0	04 47
09 15	221 25.8	59.6	N14 29.2	0.8	25 Tu	289 04.9	2.4	N21 54.8	0.0	04 43
09 15	221 16.6	59.6	N14 11.2	0.8	26 Wed	290 02.1	2.4	N21 54.9	0.0	04 39
09 16	221 07.3	59.6	N13 52.8	0.8	27 Th	290 59.5	2.4	N21 55.0	0.0	04 35
09 16	220 58.2	59.6	N13 34.0	0.8	28 Fri	291 57.2	2.4	N21 55.1	0.0	04 31
09 17	220 49.1	59.6	N13 14.8	0.8	29 Sat	292 55.0	2.4	N21 55.2	0.0	04 28
09 18	220 40.0	59.6	N12 55.2	0.8	30 SUN	293 53.1	2.4	N21 55.3	0.0	04 24

VENUS, Av. Mag. –4.2
SHA September 5 240; 10 234; 15 229; 20 223; 25 217; 30 212

JUPITER, Av. Mag. –2.4
SHA September 5 286; 10 286; 15 285; 20 285; 25 285; 30 285

MARS Mer Pass h m	GHA ° ′	Mean Var/hr 15°+	Dec ° ′	Mean Var/hr ′	Day	SATURN GHA ° ′	Mean Var/hr 15°+	Dec ° ′	Mean Var/hr ′	Mer Pass h m
15 29	127 38.3	0.9	S13 48.7	0.6	1 Sat	135 22.5	2.2	S 7 58.3	0.1	14 56
15 27	127 59.5	0.9	S14 02.4	0.6	2 SUN	136 16.4	2.2	S 8 00.6	0.1	14 53
15 26	128 20.7	0.9	S14 16.0	0.6	3 Mon	137 10.2	2.2	S 8 02.8	0.1	14 49
15 24	128 41.6	0.9	S14 29.6	0.6	4 Tu	138 03.9	2.2	S 8 05.0	0.1	14 46
15 23	129 02.4	0.9	S14 43.1	0.6	5 Wed	138 57.6	2.2	S 8 07.3	0.1	14 42
15 22	129 23.0	0.8	S14 56.5	0.6	6 Th	139 51.2	2.2	S 8 09.5	0.1	14 38
15 20	129 43.4	0.8	S15 09.8	0.6	7 Fri	140 44.8	2.2	S 8 11.8	0.1	14 35
15 19	130 03.6	0.8	S15 23.1	0.5	8 Sat	141 38.3	2.2	S 8 14.1	0.1	14 31
15 18	130 23.7	0.8	S15 36.2	0.5	9 SUN	142 31.7	2.2	S 8 16.4	0.1	14 28
15 16	130 43.6	0.8	S15 49.3	0.5	10 Mon	143 25.2	2.2	S 8 18.7	0.1	14 24
15 15	131 03.3	0.8	S16 02.3	0.5	11 Tu	144 18.5	2.2	S 8 21.1	0.1	14 21
15 14	131 22.8	0.8	S16 15.2	0.5	12 Wed	145 11.9	2.2	S 8 23.4	0.1	14 17
15 12	131 42.1	0.8	S16 28.0	0.5	13 Th	146 05.1	2.2	S 8 25.8	0.1	14 14
15 11	132 01.3	0.8	S16 40.7	0.5	14 Fri	146 58.4	2.2	S 8 28.2	0.1	14 10
15 10	132 20.3	0.8	S16 53.3	0.5	15 Sat	147 51.6	2.2	S 8 30.6	0.1	14 06
15 09	132 39.1	0.8	S17 05.8	0.5	16 SUN	148 44.7	2.2	S 8 33.0	0.1	14 03
15 07	132 57.7	0.8	S17 18.2	0.5	17 Mon	149 37.8	2.2	S 8 35.4	0.1	13 59
15 06	133 16.1	0.8	S17 30.5	0.5	18 Tu	150 30.9	2.2	S 8 37.8	0.1	13 56
15 05	133 34.3	0.8	S17 42.7	0.5	19 Wed	151 23.9	2.2	S 8 40.2	0.1	13 52
15 04	133 52.4	0.7	S17 54.7	0.5	20 Th	152 16.9	2.2	S 8 42.6	0.1	13 49
15 03	134 10.3	0.7	S18 06.7	0.5	21 Fri	153 09.8	2.2	S 8 45.1	0.1	13 45
15 01	134 28.0	0.7	S18 18.5	0.5	22 Sat	154 02.7	2.2	S 8 47.5	0.1	13 42
15 00	134 45.5	0.7	S18 30.2	0.5	23 SUN	154 55.6	2.2	S 8 50.0	0.1	13 38
14 59	135 02.8	0.7	S18 41.8	0.5	24 Mon	155 48.4	2.2	S 8 52.5	0.1	13 35
14 58	135 20.0	0.7	S18 53.2	0.5	25 Tu	156 41.2	2.2	S 8 54.9	0.1	13 31
14 57	135 36.9	0.7	S19 04.5	0.5	26 Wed	157 34.0	2.2	S 8 57.4	0.1	13 28
14 56	135 53.7	0.7	S19 15.6	0.5	27 Th	158 26.7	2.2	S 8 59.9	0.1	13 24
14 55	136 10.3	0.7	S19 26.7	0.4	28 Fri	159 19.4	2.2	S 9 02.4	0.1	13 21
14 54	136 26.7	0.7	S19 37.5	0.4	29 Sat	160 12.1	2.2	S 9 04.9	0.1	13 17
14 52	136 43.0	0.7	S19 48.3	0.4	30 SUN	161 04.7	2.2	S 9 07.4	0.1	13 14

MARS, Av. Mag. +1.2
SHA September 5 145; 10 141; 15 138; 20 135; 25 131; 30 128

SATURN, Av. Mag. +0.8
SHA September 5 154; 10 154; 15 153; 20 153; 25 152; 30 152

SEPTEMBER 2012

STARS

No.	Name	Mag	Transit h m	Dec ° '	SHA ° '
	Oh GMT September 1				
Ψ	**ARIES**	–	1 18	–	–
1	**Alpheratz**	2.1	1 27	N29 09.8	357 43.7
2	**Ankaa**	2.4	1 44	S42 14.0	353 15.8
3	**Schedar**	2.2	1 59	N56 36.5	349 40.6
4	**Diphda**	2.0	2 02	S17 54.8	348 56.1
5	**Achernar**	0.5	2 55	S57 10.1	335 26.7
6	**POLARIS**	2.0	4 06	N89 18.9	317 45.1
7	**Hamal**	2.0	3 25	N23 31.4	328 01.1
8	**Acamar**	3.2	4 16	S40 15.0	315 18.5
9	**Menkar**	2.5	4 20	N 4 08.5	314 15.4
10	**Mirfak**	1.8	4 42	N49 54.2	308 40.8
11	**Aldebaran**	0.9	5 53	N16 32.0	290 49.9
12	**Rigel**	0.1	6 32	S 8 11.1	281 12.5
13	**Capella**	0.1	6 34	N46 00.4	280 35.2
14	**Bellatrix**	1.6	6 42	N 6 21.7	278 32.5
15	**Elnath**	1.7	6 44	N28 36.9	278 13.3
16	**Alnilam**	1.7	6 53	S 1 11.6	275 46.9
17	**Betelgeuse**	0.1–1.2	7 12	N 7 24.5	271 01.9
18	**Canopus**	–0.7	7 41	S52 41.9	263 56.5
19	**Sirius**	–1.5	8 02	S16 43.9	258 34.3
20	**Adhara**	1.5	8 16	S28 59.2	255 13.1
21	**Castor**	1.6	8 52	N31 51.4	246 08.8
22	**Procyon**	0.4	8 56	N 5 11.5	245 00.4
23	**Pollux**	1.1	9 02	N27 59.6	243 28.6
24	**Avior**	1.9	9 39	S59 32.9	234 18.7
25	**Suhail**	2.2	10 24	S43 29.0	222 53.2
26	**Miaplacidus**	1.7	10 29	S69 46.1	221 40.5
27	**Alphard**	2.0	10 44	S 8 42.8	217 56.9
28	**Regulus**	1.4	11 25	N11 54.3	207 44.4
29	**Dubhe**	1.8	12 20	N61 40.9	193 53.0
30	**Denebola**	2.1	13 05	N14 30.1	182 34.5
31	**Gienah**	2.6	13 32	S17 36.7	175 53.1
32	**Acrux**	1.3	13 43	S63 10.3	173 10.5
33	**Gacrux**	1.6	13 47	S57 11.1	172 02.0
34	**Mimosa**	1.3	14 04	S59 45.6	167 53.1
35	**Alioth**	1.8	14 10	N55 53.5	166 21.6
36	**Spica**	1.0	14 41	S11 13.6	158 32.1
37	**Alkaid**	1.9	15 03	N49 15.2	152 59.7
38	**Hadar**	0.6	15 20	S60 26.2	148 49.1
39	**Menkent**	2.1	15 23	S36 26.0	148 08.5
40	**Arcturus**	0.0	15 31	N19 07.2	145 56.4
41	**Rigil Kent**	–0.3	15 56	S60 53.4	139 52.9
42	**Zuben'ubi**	2.8	16 07	S16 05.6	137 06.2
43	**Kochab**	2.1	16 06	N74 06.4	137 20.7
44	**Alphecca**	2.2	16 50	N26 40.6	126 11.6
45	**Antares**	1.0	17 45	S26 27.5	112 27.0
46	**Atria**	1.9	18 05	S69 03.2	107 29.3
47	**Sabik**	2.4	18 26	S15 44.3	102 13.1
48	**Shaula**	1.6	18 49	S37 06.7	96 22.6
49	**Rasalhague**	2.1	18 50	N12 33.4	96 06.9
50	**Eltanin**	2.2	19 12	N51 29.7	90 46.3
51	**Kaus Aust.**	1.9	19 40	S34 22.6	83 44.4
52	**Vega**	0.0	19 52	N38 48.1	80 39.2
53	**Nunki**	2.0	20 11	S26 16.7	75 58.8
54	**Altair**	0.8	21 06	N 8 54.4	62 08.6
55	**Peacock**	1.9	21 41	S56 41.6	53 19.6
56	**Deneb**	1.3	21 56	N45 19.9	49 31.5
57	**Enif**	2.4	22 59	N 9 56.3	33 47.3
58	**Al Na'ir**	1.7	23 23	S46 53.8	27 43.9
59	**Fomalhaut**	1.2	0 16	S29 33.1	15 24.2
60	**Markab**	2.5	0 23	N15 16.7	13 38.5

SUN AND MOON

SUN

Yr	Day of Mth	Week	Transit h m	Semi-Diam	Twilight h m (Lat 52°N)	Sunrise h m	Sunset h m	Twilight h m
245	1	Sat	12 00	15.9	04 37	05 12	18 46	19 22
246	2	Sun	12 00	15.9	04 39	05 14	18 44	19 19
247	3	Mon	11 59	15.9	04 40	05 15	18 42	19 17
248	4	Tu	11 59	15.9	04 42	05 17	18 40	19 14
249	5	Wed	11 59	15.9	04 44	05 19	18 37	19 12
250	6	Th	11 58	15.9	04 46	05 20	18 35	19 10
251	7	Fri	11 58	15.9	04 47	05 22	18 33	19 07
252	8	Sat	11 58	15.9	04 49	05 24	18 30	19 05
253	9	Sun	11 57	15.9	04 51	05 25	18 28	19 02
254	10	Mon	11 57	15.9	04 52	05 27	18 26	19 00
255	11	Tu	11 56	15.9	04 54	05 29	18 23	18 58
256	12	Wed	11 56	15.9	04 56	05 30	18 21	18 55
257	13	Th	11 56	15.9	04 58	05 32	18 19	18 53
258	14	Fri	11 55	15.9	04 59	05 33	18 16	18 50
259	15	Sat	11 55	15.9	05 01	05 35	18 14	18 48
260	16	Sun	11 55	15.9	05 03	05 37	18 12	18 46
261	17	Mon	11 54	15.9	05 04	05 38	18 09	18 43
262	18	Tu	11 54	15.9	05 06	05 40	18 07	18 41
263	19	Wed	11 54	15.9	05 08	05 42	18 05	18 38
264	20	Th	11 53	16.0	05 09	05 43	18 02	18 36
265	21	Fri	11 53	16.0	05 11	05 45	18 00	18 34
266	22	Sat	11 53	16.0	05 13	05 46	17 58	18 31
267	23	Sun	11 52	16.0	05 14	05 48	17 55	18 29
268	24	Mon	11 52	16.0	05 16	05 50	17 53	18 27
269	25	Tu	11 52	16.0	05 18	05 51	17 51	18 24
270	26	Wed	11 51	16.0	05 19	05 53	17 48	18 22
271	27	Th	11 51	16.0	05 21	05 55	17 46	18 19
272	28	Fri	11 51	16.0	05 23	05 56	17 44	18 17
273	29	Sat	11 50	16.0	05 24	05 58	17 41	18 15
274	30	Sun	11 50	16.0	05 26	06 00	17 39	18 13

Lat Corr to Sunrise, Sunset etc.

Lat °	Twilight h m	Sunrise h m	Sunset h m	Twilight h m
N70	–0 53	–0 21	+0 20	+0 51
68	–0 42	–0 17	+0 16	+0 41
66	–0 34	–0 14	+0 13	+0 32
64	–0 27	–0 11	+0 10	+0 25
62	–0 20	–0 08	+0 08	+0 19
N60	–0 15	–0 06	+0 06	+0 15
58	–0 11	–0 04	+0 05	+0 10
56	–0 07	–0 02	+0 03	+0 07
54	–0 03	–0 01	+0 02	+0 03
50	+0 03	+0 02	–0 01	–0 03
N45	+0 09	+0 04	–0 03	–0 09
40	+0 12	+0 06	–0 06	–0 13
35	+0 16	+0 08	–0 08	–0 17
30	+0 20	+0 10	–0 09	–0 20
20	+0 24	+0 13	–0 57	–0 25
N10	+0 28	+0 16	–1 08	–0 28
0	+0 31	+0 18	–1 18	–0 32
S10	+0 33	+0 19	–1 28	–0 32
20	+0 33	+0 21	–1 39	–0 33
30	+0 34	+0 24	–1 51	–0 33
S35	+0 34	+0 26	–1 57	–0 32
40	+0 33	+0 27	–2 05	–0 32
45	+0 32	+0 28	–2 15	–0 31
S50	+0 31	+0 30	–2 26	–0 30

NOTES
The corrections to sunrise etc. are for middle of September.

MOON

Yr	Day of Mth	Week	Age days	Transit (Upper) h m	Diff m	Semi-diam '	Hor Par '	Moonrise h m (Lat 52°N)	Moonset h m
245	1	Sat	15	00 12	46	15.4	56.4	18 37	06 20
246	2	Sun	16	00 58	44	15.2	55.8	18 58	07 31
247	3	Mon	17	01 42	45	15.1	55.3	19 19	08 40
248	4	Tu	18	02 27	46	14.9	54.8	19 43	09 48
249	5	Wed	19	03 13	46	14.8	54.5	20 11	10 54
250	6	Th	20	03 59	47	14.8	54.3	20 43	11 57
251	7	Fri	21	04 46	48	14.8	54.2	21 22	12 56
252	8	Sat	22	05 34	49	14.8	54.4	22 08	13 49
253	9	Sun	23	06 23	49	14.9	54.7	23 02	14 35
254	10	Mon	24	07 12	49	15.0	55.1	– –	15 15
255	11	Tu	25	08 01	49	15.2	55.7	00 03	15 49
256	12	Wed	26	08 50	49	15.4	56.4	01 09	16 18
257	13	Th	27	09 39	49	15.6	57.2	02 20	16 44
258	14	Fri	28	10 28	49	15.8	58.0	03 35	17 07
259	15	Sat	29	11 17	51	16.0	58.7	04 51	17 30
260	16	Sun	00	12 08	52	16.2	59.3	06 10	17 53
261	17	Mon	01	13 00	55	16.3	59.7	07 30	18 19
262	18	Tu	02	13 55	56	16.3	59.9	08 51	18 48
263	19	Wed	03	14 51	59	16.3	59.9	10 12	19 23
264	20	Th	04	15 50	59	16.3	59.8	11 29	20 06
265	21	Fri	05	16 49	59	16.2	59.5	12 38	20 59
266	22	Sat	06	17 48	57	16.1	59.1	13 37	22 02
267	23	Sun	07	18 45	55	16.0	58.7	14 24	23 12
268	24	Mon	08	19 40	51	15.9	58.2	15 02	– –
269	25	Tu	09	20 31	49	15.7	57.7	15 33	00 25
270	26	Wed	10	21 20	47	15.6	57.2	15 58	01 39
271	27	Th	11	22 07	45	15.4	56.7	16 21	02 53
272	28	Fri	12	22 52	45	15.3	56.2	16 42	04 05
273	29	Sat	13	23 37	45	15.2	55.7	17 02	05 16
274	30	Sun	14	24 22	–	15.1	55.3	17 23	06 25

Phases of the Moon

		d	h	m
☽	Last Quarter	8	13	15
●	New Moon	16	02	11
☾	First Quarter	22	19	41
○	Full Moon	30	03	19

	d	h
Apogee	7	06
Perigee	19	03

GMT	SUN GHA ° '	SUN Dec ° '	ARIES GHA ° '
Monday, 1st October			
00	182 34.9	**S 3 15.6**	10 07.9
02	212 35.4	**3 17.5**	40 12.8
04	242 35.8	**3 19.5**	70 17.7
06	272 36.2	**3 21.4**	100 22.7
08	302 36.6	**3 23.3**	130 27.6
10	332 37.0	**3 25.3**	160 32.5
12	2 37.4	**3 27.2**	190 37.4
14	32 37.8	**3 29.2**	220 42.4
16	62 38.2	**3 31.1**	250 47.3
18	92 38.6	**3 33.0**	280 52.2
20	122 39.0	**3 35.0**	310 57.2
22	152 39.4	**S 3 36.9**	341 02.1
Tuesday, 2nd October			
00	182 39.8	**S 3 38.8**	11 07.0
02	212 40.2	**3 40.8**	41 11.9
04	242 40.6	**3 42.7**	71 16.9
06	272 41.0	**3 44.6**	101 21.8
08	302 41.4	**3 46.6**	131 26.7
10	332 41.7	**3 48.5**	161 31.6
12	2 42.1	**3 50.4**	191 36.6
14	32 42.5	**3 52.4**	221 41.5
16	62 42.9	**3 54.3**	251 46.4
18	92 43.3	**3 56.2**	281 51.4
20	122 43.7	**3 58.2**	311 56.3
22	152 44.1	**S 4 00.1**	342 01.2
Wednesday, 3rd October			
00	182 44.5	**S 4 02.0**	12 06.1
02	212 44.9	**4 04.0**	42 11.1
04	242 45.3	**4 05.9**	72 16.0
06	272 45.7	**4 07.8**	102 20.9
08	302 46.1	**4 09.8**	132 25.9
10	332 46.5	**4 11.7**	162 30.8
12	2 46.8	**4 13.6**	192 35.7
14	32 47.2	**4 15.5**	222 40.6
16	62 47.6	**4 17.5**	252 45.6
18	92 48.0	**4 19.4**	282 50.5
20	122 48.4	**4 21.3**	312 55.4
22	152 48.8	**S 4 23.2**	343 00.4
Thursday, 4th October			
00	182 49.2	**S 4 25.2**	13 05.3
02	212 49.5	**4 27.1**	43 10.2
04	242 49.9	**4 29.0**	73 15.1
06	272 50.3	**4 31.0**	103 20.1
08	302 50.7	**4 32.9**	133 25.0
10	332 51.1	**4 34.8**	163 29.9
12	2 51.4	**4 36.7**	193 34.9
14	32 51.8	**4 38.7**	223 39.8
16	62 52.2	**4 40.6**	253 44.7
18	92 52.6	**4 42.5**	283 49.6
20	122 53.0	**4 44.4**	313 54.6
22	152 53.3	**S 4 46.4**	343 59.5
Friday, 5th October			
00	182 53.7	**S 4 48.3**	14 04.4
02	212 54.1	**4 50.2**	44 09.4
04	242 54.5	**4 52.1**	74 14.3
06	272 54.8	**4 54.0**	104 19.2
08	302 55.2	**4 56.0**	134 24.1
10	332 55.6	**4 57.9**	164 29.1
12	2 56.0	**4 59.8**	194 34.0
14	32 56.3	**5 01.7**	224 38.9
16	62 56.7	**5 03.6**	254 43.8
18	92 57.1	**5 05.6**	284 48.8
20	122 57.4	**5 07.5**	314 53.7
22	152 57.8	**S 5 09.4**	344 58.6

GMT	SUN GHA ° '	SUN Dec ° '	ARIES GHA ° '
Saturday, 6th October			
00	182 58.2	**S 5 11.3**	15 03.6
02	212 58.5	**5 13.2**	45 08.5
04	242 58.9	**5 15.1**	75 13.4
06	272 59.3	**5 17.1**	105 18.3
08	302 59.6	**5 19.0**	135 23.3
10	333 00.0	**5 20.9**	165 28.2
12	3 00.4	**5 22.8**	195 33.1
14	33 00.7	**5 24.7**	225 38.1
16	63 01.1	**5 26.6**	255 43.0
18	93 01.5	**5 28.5**	285 47.9
20	123 01.8	**5 30.5**	315 52.8
22	153 02.2	**S 5 32.4**	345 57.8
Sunday, 7th October			
00	183 02.5	**S 5 34.3**	16 02.7
02	213 02.9	**5 36.2**	46 07.6
04	243 03.3	**5 38.1**	76 12.6
06	273 03.6	**5 40.0**	106 17.5
08	303 04.0	**5 41.9**	136 22.4
10	333 04.3	**5 43.8**	166 27.3
12	3 04.7	**5 45.7**	196 32.3
14	33 05.0	**5 47.7**	226 37.2
16	63 05.4	**5 49.6**	256 42.1
18	93 05.7	**5 51.5**	286 47.1
20	123 06.1	**5 53.4**	316 52.0
22	153 06.4	**S 5 55.3**	346 56.9
Monday, 8th October			
00	183 06.8	**S 5 57.2**	17 01.8
02	213 07.1	**5 59.1**	47 06.8
04	243 07.5	**6 01.0**	77 11.7
06	273 07.8	**6 02.9**	107 16.6
08	303 08.2	**6 04.8**	137 21.6
10	333 08.5	**6 06.7**	167 26.5
12	3 08.9	**6 08.6**	197 31.4
14	33 09.2	**6 10.5**	227 36.3
16	63 09.6	**6 12.4**	257 41.3
18	93 09.9	**6 14.3**	287 46.2
20	123 10.3	**6 16.2**	317 51.1
22	153 10.6	**S 6 18.1**	347 56.1
Tuesday, 9th October			
00	183 10.9	**S 6 20.0**	18 01.0
02	213 11.3	**6 21.9**	48 05.9
04	243 11.6	**6 23.8**	78 10.8
06	273 12.0	**6 25.7**	108 15.8
08	303 12.3	**6 27.6**	138 20.7
10	333 12.6	**6 29.5**	168 25.6
12	3 13.0	**6 31.4**	198 30.5
14	33 13.3	**6 33.3**	228 35.5
16	63 13.6	**6 35.2**	258 40.4
18	93 14.0	**6 37.1**	288 45.3
20	123 14.3	**6 39.0**	318 50.3
22	153 14.6	**S 6 40.9**	348 55.2
Wednesday, 10th October			
00	183 15.0	**S 6 42.8**	19 00.1
02	213 15.3	**6 44.7**	49 05.0
04	243 15.6	**6 46.6**	79 10.0
06	273 16.0	**6 48.4**	109 14.9
08	303 16.3	**6 50.3**	139 19.8
10	333 16.6	**6 52.2**	169 24.8
12	3 16.9	**6 54.1**	199 29.7
14	33 17.3	**6 56.0**	229 34.6
16	63 17.6	**6 57.9**	259 39.5
18	93 17.9	**6 59.8**	289 44.5
20	123 18.2	**7 01.7**	319 49.4
22	153 18.6	**S 7 03.5**	349 54.3

GMT	SUN GHA ° '	SUN Dec ° '	ARIES GHA ° '
Thursday, 11th October			
00	183 18.9	**S 7 05.4**	19 59.3
02	213 19.2	**7 07.3**	50 04.2
04	243 19.5	**7 09.2**	80 09.1
06	273 19.8	**7 11.1**	110 14.0
08	303 20.1	**7 13.0**	140 19.0
10	333 20.5	**7 14.8**	170 23.9
12	3 20.8	**7 16.7**	200 28.8
14	33 21.1	**7 18.6**	230 33.8
16	63 21.4	**7 20.5**	260 38.7
18	93 21.7	**7 22.4**	290 43.6
20	123 22.0	**7 24.2**	320 48.5
22	153 22.3	**S 7 26.1**	350 53.5
Friday, 12th October			
00	183 22.7	**S 7 28.0**	20 58.4
02	213 23.0	**7 29.9**	51 03.3
04	243 23.3	**7 31.7**	81 08.3
06	273 23.6	**7 33.6**	111 13.2
08	303 23.9	**7 35.5**	141 18.1
10	333 24.2	**7 37.4**	171 23.0
12	3 24.5	**7 39.2**	201 28.0
14	33 24.8	**7 41.1**	231 32.9
16	63 25.1	**7 43.0**	261 37.8
18	93 25.4	**7 44.8**	291 42.7
20	123 25.7	**7 46.7**	321 47.7
22	153 26.0	**S 7 48.6**	351 52.6
Saturday, 13th October			
00	183 26.3	**S 7 50.5**	21 57.5
02	213 26.6	**7 52.3**	52 02.5
04	243 26.9	**7 54.2**	82 07.4
06	273 27.2	**7 56.0**	112 12.3
08	303 27.5	**7 57.9**	142 17.2
10	333 27.8	**7 59.8**	172 22.2
12	3 28.1	**8 01.6**	202 27.1
14	33 28.4	**8 03.5**	232 32.0
16	63 28.7	**8 05.4**	262 37.0
18	93 28.9	**8 07.2**	292 41.9
20	123 29.2	**8 09.1**	322 46.8
22	153 29.5	**S 8 10.9**	352 51.7
Sunday, 14th October			
00	183 29.8	**S 8 12.8**	22 56.7
02	213 30.1	**8 14.7**	53 01.6
04	243 30.4	**8 16.5**	83 06.5
06	273 30.7	**8 18.4**	113 11.5
08	303 31.0	**8 20.2**	143 16.4
10	333 31.2	**8 22.1**	173 21.3
12	3 31.5	**8 23.9**	203 26.2
14	33 31.8	**8 25.8**	233 31.2
16	63 32.1	**8 27.6**	263 36.1
18	93 32.4	**8 29.5**	293 41.0
20	123 32.6	**8 31.3**	323 45.9
22	153 32.9	**S 8 33.2**	353 50.9
Monday, 15th October			
00	183 33.2	**S 8 35.0**	23 55.8
02	213 33.5	**8 36.9**	54 00.7
04	243 33.7	**8 38.7**	84 05.7
06	273 34.0	**8 40.6**	114 10.6
08	303 34.3	**8 42.4**	144 15.5
10	333 34.5	**8 44.3**	174 20.4
12	3 34.8	**8 46.1**	204 25.4
14	33 35.1	**8 48.0**	234 30.3
16	63 35.4	**8 49.8**	264 35.2
18	93 35.6	**8 51.6**	294 40.2
20	123 35.9	**8 53.5**	324 45.1
22	153 36.2	**S 8 55.3**	354 50.0

GMT	SUN GHA ° '	SUN Dec ° '	ARIES GHA ° '
Tuesday, 16th October			
00	183 36.4	**S 8 57.2**	24 54.9
02	213 36.7	**8 59.0**	54 59.9
04	243 36.9	**9 00.8**	85 04.8
06	273 37.2	**9 02.7**	115 09.7
08	303 37.5	**9 04.5**	145 14.7
10	333 37.7	**9 06.3**	175 19.6
12	3 38.0	**9 08.2**	205 24.5
14	33 38.2	**9 10.0**	235 29.4
16	63 38.5	**9 11.8**	265 34.4
18	93 38.7	**9 13.7**	295 39.3
20	123 39.0	**9 15.5**	325 44.2
22	153 39.3	**S 9 17.3**	355 49.2
Wednesday, 17th October			
00	183 39.5	**S 9 19.1**	25 54.1
02	213 39.8	**9 21.0**	55 59.0
04	243 40.0	**9 22.8**	86 03.9
06	273 40.3	**9 24.6**	116 08.9
08	303 40.5	**9 26.5**	146 13.8
10	333 40.8	**9 28.3**	176 18.7
12	3 41.0	**9 30.1**	206 23.6
14	33 41.2	**9 31.9**	236 28.6
16	63 41.5	**9 33.7**	266 33.5
18	93 41.7	**9 35.6**	296 38.4
20	123 42.0	**9 37.4**	326 43.4
22	153 42.2	**S 9 39.2**	356 48.3
Thursday, 18th October			
00	183 42.4	**S 9 41.0**	26 53.2
02	213 42.7	**9 42.8**	56 58.1
04	243 42.9	**9 44.6**	87 03.1
06	273 43.2	**9 46.4**	117 08.0
08	303 43.4	**9 48.3**	147 12.9
10	333 43.6	**9 50.1**	177 17.9
12	3 43.9	**9 51.9**	207 22.8
14	33 44.1	**9 53.7**	237 27.7
16	63 44.3	**9 55.5**	267 32.6
18	93 44.6	**9 57.3**	297 37.6
20	123 44.8	**9 59.1**	327 42.5
22	153 45.0	**S10 00.9**	357 47.4
Friday, 19th October			
00	183 45.2	**S10 02.7**	27 52.4
02	213 45.5	**10 04.5**	57 57.3
04	243 45.7	**10 06.3**	88 02.2
06	273 45.9	**10 08.1**	118 07.1
08	303 46.1	**10 09.9**	148 12.1
10	333 46.4	**10 11.7**	178 17.0
12	3 46.6	**10 13.5**	208 21.9
14	33 46.8	**10 15.3**	238 26.9
16	63 47.0	**10 17.1**	268 31.8
18	93 47.2	**10 18.9**	298 36.7
20	123 47.4	**10 20.7**	328 41.6
22	153 47.7	**S10 22.5**	358 46.6
Saturday, 20th October			
00	183 47.9	**S10 24.3**	28 51.5
02	213 48.1	**10 26.1**	58 56.4
04	243 48.3	**10 27.9**	89 01.4
06	273 48.5	**10 29.7**	119 06.3
08	303 48.7	**10 31.4**	149 11.2
10	333 48.9	**10 33.2**	179 16.1
12	3 49.1	**10 35.0**	209 21.1
14	33 49.3	**10 36.8**	239 26.0
16	63 49.5	**10 38.6**	269 30.9
18	93 49.8	**10 40.4**	299 35.9
20	123 50.0	**10 42.1**	329 40.8
22	153 50.2	**S10 43.9**	359 45.7
00	184 05.7	**S14 09.9**	39 42.0
02	214 05.7	**14 11.5**	69 46.9
04	244 05.8	**14 13.1**	99 51.9
06	274 05.8	**S14 14.7**	129 56.8

M GMT	SUN GHA ° '	SUN Dec ° '	ARIES GHA ° '
Sunday, 21st October			
00	183 50.4	**S10 45.7**	29 50.6
02	213 50.6	**10 47.5**	59 55.6
04	243 50.8	**10 49.3**	90 00.5
06	273 51.0	**10 51.0**	120 05.4
08	303 51.1	**10 52.8**	150 10.4
10	333 51.3	**10 54.6**	180 15.3
12	3 51.5	**10 56.3**	210 20.2
14	33 51.7	**10 58.1**	240 25.1
16	63 51.9	**10 59.9**	270 30.1
18	93 52.1	**11 01.7**	300 35.0
20	123 52.3	**11 03.4**	330 39.9
22	153 52.5	**S11 05.2**	0 44.9
Monday, 22nd October			
00	183 52.7	**S11 06.9**	30 49.8
02	213 52.9	**11 08.7**	60 54.7
04	243 53.0	**11 10.5**	90 59.6
06	273 53.2	**11 12.2**	121 04.6
08	303 53.4	**11 14.0**	151 09.5
10	333 53.6	**11 15.8**	181 14.4
12	3 53.8	**11 17.5**	211 19.3
14	33 54.0	**11 19.3**	241 24.3
16	63 54.1	**11 21.0**	271 29.2
18	93 54.3	**11 22.8**	301 34.1
20	123 54.5	**11 24.5**	331 39.1
22	153 54.7	**S11 26.3**	1 44.0
Tuesday, 23rd October			
00	183 54.8	**S11 28.0**	31 48.9
02	213 55.0	**11 29.8**	61 53.8
04	243 55.2	**11 31.5**	91 58.8
06	273 55.3	**11 33.3**	122 03.7
08	303 55.5	**11 35.0**	152 08.6
10	333 55.7	**11 36.8**	182 13.6
12	3 55.8	**11 38.5**	212 18.5
14	33 56.0	**11 40.2**	242 23.4
16	63 56.2	**11 42.0**	272 28.3
18	93 56.3	**11 43.7**	302 33.3
20	123 56.5	**11 45.5**	332 38.2
22	153 56.7	**S11 47.2**	2 43.1
Wednesday, 24th October			
00	183 56.8	**S11 48.9**	32 48.1
02	213 57.0	**11 50.7**	62 53.0
04	243 57.1	**11 52.4**	92 57.9
06	273 57.3	**11 54.1**	123 02.8
08	303 57.4	**11 55.9**	153 07.8
10	333 57.6	**11 57.6**	183 12.7
12	3 57.7	**11 59.3**	213 17.6
14	33 57.9	**12 01.0**	243 22.6
16	63 58.0	**12 02.8**	273 27.5
18	93 58.2	**12 04.5**	303 32.4
20	123 58.3	**12 06.2**	333 37.3
22	153 58.5	**S12 07.9**	3 42.3
Thursday, 25th October			
00	183 58.6	**S12 09.7**	33 47.2
02	213 58.8	**12 11.4**	63 52.1
04	243 58.9	**12 13.1**	93 57.0
06	273 59.1	**12 14.8**	124 02.0
08	303 59.2	**12 16.5**	154 06.9
10	333 59.3	**12 18.2**	184 11.8
12	3 59.5	**12 20.0**	214 16.8
14	33 59.6	**12 21.7**	244 21.7
16	63 59.7	**12 23.4**	274 26.6
18	93 59.9	**12 25.1**	304 31.5
20	124 00.0	**12 26.8**	334 36.5
22	154 00.1	**S12 28.5**	4 41.4
Wednesday, 31st October			
08	304 05.9	**S14 16.3**	160 01.7
10	334 05.9	**14 17.9**	190 06.7
12	4 06.0	**14 19.5**	220 11.6
14	34 06.0	**S14 21.1**	250 16.5

GMT	SUN GHA ° '	SUN Dec ° '	ARIES GHA ° '
Friday, 26th October			
00	184 00.3	**S12 30.2**	34 46.3
02	214 00.4	**12 31.9**	64 51.3
04	244 00.5	**12 33.6**	94 56.2
06	274 00.7	**12 35.3**	125 01.1
08	304 00.8	**12 37.0**	155 06.0
10	334 00.9	**12 38.7**	185 11.0
12	4 01.0	**12 40.4**	215 15.9
14	34 01.1	**12 42.1**	245 20.8
16	64 01.3	**12 43.8**	275 25.8
18	94 01.4	**12 45.5**	305 30.7
20	124 01.5	**12 47.2**	335 35.6
22	154 01.6	**S12 48.9**	5 40.5
Saturday, 27th October			
00	184 01.7	**S12 50.5**	35 45.5
02	214 01.8	**12 52.2**	65 50.4
04	244 02.0	**12 53.9**	95 55.3
06	274 02.1	**12 55.6**	126 00.3
08	304 02.2	**12 57.3**	156 05.2
10	334 02.3	**12 59.0**	186 10.1
12	4 02.4	**13 00.6**	216 15.0
14	34 02.5	**13 02.3**	246 20.0
16	64 02.6	**13 04.0**	276 24.9
18	94 02.7	**13 05.7**	306 29.8
20	124 02.8	**13 07.4**	336 34.7
22	154 02.9	**S13 09.0**	6 39.7
Sunday, 28th October			
00	184 03.0	**S13 10.7**	36 44.6
02	214 03.1	**13 12.4**	66 49.5
04	244 03.2	**13 14.0**	96 54.5
06	274 03.3	**13 15.7**	126 59.4
08	304 03.4	**13 17.4**	157 04.3
10	334 03.5	**13 19.0**	187 09.2
12	4 03.6	**13 20.7**	217 14.2
14	34 03.7	**13 22.4**	247 19.1
16	64 03.8	**13 24.0**	277 24.0
18	94 03.8	**13 25.7**	307 29.0
20	124 03.9	**13 27.3**	337 33.9
22	154 04.0	**S13 29.0**	7 38.8
Monday, 29th October			
00	184 04.1	**S13 30.6**	37 43.7
02	214 04.2	**13 32.3**	67 48.7
04	244 04.3	**13 33.9**	97 53.6
06	274 04.3	**13 35.6**	127 58.5
08	304 04.4	**13 37.2**	158 03.5
10	334 04.5	**13 38.9**	188 08.4
12	4 04.6	**13 40.5**	218 13.3
14	34 04.6	**13 42.2**	248 18.2
16	64 04.7	**13 43.8**	278 23.2
18	94 04.8	**13 45.5**	308 28.1
20	124 04.9	**13 47.1**	338 33.0
22	154 04.9	**S13 48.7**	8 38.0
Tuesday, 30th October			
00	184 05.0	**S13 50.4**	38 42.9
02	214 05.1	**13 52.0**	68 47.8
04	244 05.1	**13 53.6**	98 52.7
06	274 05.2	**13 55.3**	128 57.7
08	304 05.2	**13 56.9**	159 02.6
10	334 05.3	**13 58.5**	189 07.5
12	4 05.4	**14 00.1**	219 12.4
14	34 05.4	**14 01.8**	249 17.4
16	64 05.5	**14 03.4**	279 22.3
18	94 05.5	**14 05.0**	309 27.2
20	124 05.6	**14 06.6**	339 32.2
22	154 05.6	**S14 08.3**	9 37.1
16	64 06.0	**S14 22.7**	280 21.4
18	94 06.1	**14 24.4**	310 26.4
20	124 06.1	**14 26.0**	340 31.3
22	154 06.2	**S14 27.6**	10 36.2

OCTOBER 2012

MOON

Day	GMT hr	GHA ° '	Mean Var/hr 14°+	Dec ° '	Mean Var/hr '
1 Mon	0	354 42.8	32.8	N10 04.0	9.6
	6	81 59.3	32.6	N11 02.1	9.4
	12	169 15.4	32.6	N11 58.1	8.9
	18	256 31.0	32.5	N12 52.1	8.6
2 Tu	0	343 45.9	32.4	N13 43.8	8.2
	6	71 00.2	32.3	N14 33.1	7.8
	12	158 13.7	32.1	N15 20.0	7.4
	18	245 26.5	32.0	N16 04.2	6.9
3 Wed	0	332 38.5	31.8	N16 45.7	6.4
	6	59 49.8	31.8	N17 24.5	6.0
	12	147 00.3	31.6	N18 00.3	5.4
	18	234 10.0	31.4	N18 33.1	4.9
4 Th	0	321 18.9	31.3	N19 02.8	4.4
	6	48 27.2	31.2	N19 29.3	3.8
	12	135 34.8	31.2	N19 52.6	3.2
	18	222 41.7	31.0	N20 12.6	2.7
5 Fri	0	309 48.1	30.9	N20 29.3	2.2
	6	36 54.0	30.9	N20 42.5	1.6
	12	123 59.4	30.9	N20 52.3	1.0
	18	211 04.4	30.8	N20 58.6	0.4
6 Sat	0	298 09.2	30.7	N21 01.4	0.1
	6	25 13.7	30.7	N21 00.6	0.7
	12	112 18.1	30.7	N20 56.3	1.4
	18	199 22.4	30.7	N20 48.5	1.9
7 Sun	0	286 26.6	30.7	N20 37.2	2.5
	6	13 31.0	30.7	N20 22.3	3.1
	12	100 35.4	30.7	N20 04.0	3.7
	18	187 40.0	30.8	N19 42.1	4.3
8 Mon	0	274 44.7	30.8	N19 16.9	4.8
	6	1 49.7	30.9	N18 48.3	5.4
	12	88 54.9	30.9	N18 16.3	6.0
	18	176 00.4	31.0	N17 41.0	6.5
9 Tu	0	263 06.1	31.0	N17 02.6	6.9
	6	350 12.0	31.0	N16 20.9	7.5
	12	77 18.2	31.1	N15 36.3	8.0
	18	164 24.5	31.0	N14 48.6	8.5
10 Wed	0	251 30.9	31.1	N13 58.0	8.9
	6	338 37.4	31.1	N13 04.7	9.4
	12	65 43.8	31.0	N12 08.7	9.8
	18	152 50.1	31.0	N11 10.1	10.2
11 Th	0	239 56.2	31.0	N10 09.1	10.6
	6	327 02.0	30.9	N 9 05.7	11.0
	12	54 07.4	30.8	N 8 00.3	11.3
	18	141 12.3	30.7	N 6 52.8	11.6
12 Fri	0	228 16.4	30.5	N 5 43.5	11.9
	6	315 19.7	30.4	N 4 32.6	12.1
	12	42 22.1	30.2	N 3 20.3	12.3
	18	129 23.3	30.0	N 2 06.7	12.4
13 Sat	0	216 23.3	29.7	N 0 52.1	12.6
	6	303 21.9	29.5	S 0 23.2	12.7
	12	30 18.9	29.1	S 1 39.0	12.7
	18	117 14.2	28.9	S 2 55.0	12.7
14 Sun	0	204 07.6	28.5	S 4 11.0	12.6
	6	290 59.0	28.2	S 5 26.5	12.5
	12	17 48.3	27.8	S 6 41.4	12.3
	18	104 35.4	27.4	S 7 55.3	12.1
15 Mon	0	191 20.1	27.1	S 9 07.8	11.8
	6	278 02.4	26.6	S10 18.6	11.4
	12	4 42.2	26.2	S11 27.4	11.0
	18	91 19.6	25.8	S12 33.7	10.5
16 Tu	0	177 54.4	25.4	S13 37.3	10.0
	6	264 26.8	25.0	S14 37.7	9.4
	12	350 56.9	24.6	S15 34.7	8.8
	18	77 24.6	24.3	S16 27.9	8.1
17 Wed	0	163 50.3	23.9	S17 17.0	7.4
	6	250 14.1	23.6	S18 01.7	6.6
	12	336 36.1	23.4	S18 41.8	5.8
	18	62 56.8	23.2	S19 17.0	4.9
18 Th	0	149 16.4	23.2	S19 47.1	4.1
	6	235 35.2	23.1	S20 12.0	3.2
	12	321 53.6	23.1	S20 31.6	2.2
	18	48 11.9	23.1	S20 45.8	1.4
19 Fri	0	134 30.6	23.2	S20 54.5	0.4
	6	220 50.0	23.4	S20 57.8	0.4
	12	307 10.5	23.7	S20 55.8	1.3
	18	33 32.3	24.0	S20 48.5	2.2
20 Sat	0	119 56.0	24.3	S20 36.0	3.0
	6	206 21.6	24.7	S20 18.5	3.8
	12	292 49.6	25.1	S19 56.2	4.6
	18	19 20.1	25.6	S19 29.4	5.3
21 Sun	0	105 53.2	26.0	S18 58.1	5.9
	6	192 29.2	26.5	S18 22.7	6.6
	12	279 08.1	27.0	S17 43.5	7.2
	18	5 50.0	27.5	S17 00.6	7.8
22 Mon	0	92 34.8	28.0	S16 14.3	8.2
	6	179 22.7	28.5	S15 25.0	8.7
	12	266 13.4	29.0	S14 32.8	9.1
	18	353 07.1	29.5	S13 38.0	9.6
23 Tu	0	80 03.5	29.9	S12 41.0	9.9
	6	167 02.5	30.3	S11 41.8	10.2
	12	254 04.1	30.7	S10 40.8	10.5
	18	341 08.0	31.0	S 9 38.2	10.7
24 Wed	0	68 14.2	31.4	S 8 34.2	10.9
	6	155 22.4	31.7	S 7 29.1	11.0
	12	242 32.5	32.0	S 6 22.9	11.2
	18	329 44.3	32.3	S 5 16.1	11.2
25 Th	0	56 57.7	32.4	S 4 08.7	11.3
	6	144 12.4	32.6	S 3 00.9	11.3
	12	231 28.2	32.8	S 1 52.9	11.3
	18	318 45.1	32.9	S 0 45.0	11.3
26 Fri	0	46 02.8	33.1	N 0 22.8	11.3
	6	133 21.2	33.1	N 1 30.3	11.1
	12	220 40.1	33.2	N 2 37.3	11.0
	18	307 59.3	33.2	N 3 43.6	10.9
27 Sat	0	35 18.7	33.2	N 4 49.0	10.7
	6	122 38.2	33.3	N 5 53.5	10.6
	12	209 57.6	33.1	N 6 56.9	10.3
	18	297 16.8	33.1	N 7 58.9	10.1
28 Sun	0	24 35.7	33.0	N 8 59.6	9.8
	6	111 54.1	33.0	N 9 58.6	9.5
	12	199 12.0	32.8	N10 55.9	9.2
	18	286 29.3	32.8	N11 51.4	8.8
29 Mon	0	13 45.9	32.6	N12 44.8	8.5
	6	101 01.7	32.5	N13 36.1	8.1
	12	188 16.8	32.4	N14 25.2	7.8
	18	275 30.9	32.2	N15 11.8	7.3
30 Tu	0	2 44.2	32.1	N15 55.9	6.9
	6	89 56.6	31.9	N16 37.4	6.4
	12	177 08.1	31.8	N17 16.1	6.0
	18	264 18.8	31.6	N17 52.0	5.5
31 Wed	0	351 28.5	31.5	N18 24.9	4.9
	6	78 37.5	31.4	N18 54.7	4.4
	12	165 45.7	31.2	N19 21.4	3.8
	18	252 53.2	31.2	N19 44.9	3.3

PLANETS

VENUS Mer Pass h m	VENUS GHA ° '	VENUS Mean Var/hr 14°+	VENUS Dec ° '	VENUS Mean Var/hr '	Day	JUPITER GHA ° '	JUPITER Mean Var/hr 15°+	JUPITER Dec ° '	JUPITER Mean Var/hr '	JUPITER Mer Pass h m
09 18	220 31.0	59.6	N12 35.1	0.8	1 Mon	294 51.3	2.4	N21 55.3	0.0	04 20
09 19	220 22.0	59.6	N12 14.7	0.9	2 Tu	295 49.8	2.4	N21 55.3	0.0	04 16
09 19	220 13.1	59.6	N11 53.9	0.9	3 Wed	296 48.5	2.5	N21 55.3	0.0	04 12
09 20	220 04.2	59.6	N11 32.7	0.9	4 Th	297 47.4	2.5	N21 55.3	0.0	04 08
09 21	219 55.4	59.6	N11 11.2	0.9	5 Fri	298 46.5	2.5	N21 55.2	0.0	04 04
09 21	219 46.7	59.6	N10 49.3	0.9	6 Sat	299 45.9	2.5	N21 55.2	0.0	04 00
09 22	219 38.0	59.6	N10 27.0	0.9	7 SUN	300 45.4	2.5	N21 55.1	0.0	03 56
09 22	219 29.3	59.6	N10 04.5	1.0	8 Mon	301 45.2	2.5	N21 55.0	0.0	03 52
09 23	219 20.8	59.6	N 9 41.5	1.0	9 Tu	302 45.2	2.5	N21 54.9	0.0	03 48
09 23	219 12.2	59.6	N 9 18.3	1.0	10 Wed	303 45.4	2.5	N21 54.7	0.0	03 44
09 24	219 03.8	59.6	N 8 54.8	1.0	11 Th	304 45.8	2.5	N21 54.6	0.0	03 40
09 25	218 55.3	59.7	N 8 30.9	1.0	12 Fri	305 46.5	2.5	N21 54.4	0.0	03 36
09 25	218 47.0	59.7	N 8 06.8	1.0	13 Sat	306 47.3	2.5	N21 54.2	0.0	03 32
09 26	218 38.7	59.7	N 7 42.4	1.0	14 SUN	307 48.4	2.6	N21 54.0	0.0	03 28
09 26	218 30.4	59.7	N 7 17.8	1.0	15 Mon	308 49.7	2.6	N21 53.8	0.0	03 24
09 27	218 22.2	59.7	N 6 52.8	1.0	16 Tu	309 51.2	2.6	N21 53.5	0.0	03 20
09 27	218 14.0	59.7	N 6 27.7	1.1	17 Wed	310 52.9	2.6	N21 53.2	0.0	03 16
09 28	218 05.8	59.7	N 6 02.3	1.1	18 Th	311 54.8	2.6	N21 53.0	0.0	03 12
09 28	217 57.7	59.7	N 5 36.6	1.1	19 Fri	312 57.0	2.6	N21 52.6	0.0	03 08
09 29	217 49.6	59.7	N 5 10.8	1.1	20 Sat	313 59.3	2.6	N21 52.3	0.0	03 04
09 29	217 41.5	59.7	N 4 44.7	1.1	21 SUN	315 01.9	2.6	N21 52.0	0.0	02 59
09 30	217 33.5	59.7	N 4 18.5	1.1	22 Mon	316 04.7	2.6	N21 51.6	0.0	02 55
09 31	217 25.5	59.7	N 3 52.1	1.1	23 Tu	317 07.6	2.6	N21 51.2	0.0	02 51
09 31	217 17.4	59.7	N 3 25.5	1.1	24 Wed	318 10.8	2.6	N21 50.8	0.0	02 47
09 32	217 09.4	59.7	N 2 58.8	1.1	25 Th	319 14.2	2.6	N21 50.4	0.0	02 43
09 32	217 01.4	59.7	N 2 31.9	1.1	26 Fri	320 17.7	2.7	N21 49.9	0.0	02 38
09 33	216 53.3	59.7	N 2 04.9	1.1	27 Sat	321 21.5	2.7	N21 49.5	0.0	02 34
09 33	216 45.3	59.7	N 1 37.7	1.1	28 SUN	322 25.4	2.7	N21 49.0	0.0	02 30
09 34	216 37.2	59.7	N 1 10.4	1.1	29 Mon	323 29.6	2.7	N21 48.5	0.0	02 26
09 34	216 29.0	59.7	N 0 43.1	1.1	30 Tu	324 33.9	2.7	N21 48.0	0.0	02 21
09 35	216 20.8	59.7	N 0 15.6	1.1	31 Wed	325 38.4	2.7	N21 47.4	0.0	02 17

VENUS, Av. Mag. –4.0
SHA October 5 206; 10 200; 15 195; 20 189; 25 183; 30 178

JUPITER, Av. Mag. –2.6
SHA October 5 285; 10 285; 15 285; 20 285; 25 285; 30 286

MARS Mer Pass h m	MARS GHA ° '	MARS Mean Var/hr 15°+	MARS Dec ° '	MARS Mean Var/hr '	Day	SATURN GHA ° '	SATURN Mean Var/hr 15°+	SATURN Dec ° '	SATURN Mean Var/hr '	SATURN Mer Pass h m
14 51	136 59.1	0.7	S19 58.8	0.4	1 Mon	161 57.3	2.2	S 9 09.9	0.1	13 10
14 50	137 14.9	0.7	S20 09.2	0.4	2 Tu	162 49.9	2.2	S 9 12.4	0.1	13 07
14 49	137 30.6	0.7	S20 19.5	0.4	3 Wed	163 42.5	2.2	S 9 14.9	0.1	13 03
14 48	137 46.2	0.6	S20 29.6	0.4	4 Th	164 35.0	2.2	S 9 17.4	0.1	13 00
14 47	138 01.5	0.6	S20 39.6	0.4	5 Fri	165 27.5	2.2	S 9 19.9	0.1	12 56
14 46	138 16.7	0.6	S20 49.3	0.4	6 Sat	166 20.0	2.2	S 9 22.5	0.1	12 53
14 45	138 31.7	0.6	S20 59.0	0.4	7 SUN	167 12.4	2.2	S 9 25.0	0.1	12 49
14 44	138 46.5	0.6	S21 08.4	0.4	8 Mon	168 04.9	2.2	S 9 27.5	0.1	12 46
14 43	139 01.1	0.6	S21 17.7	0.4	9 Tu	168 57.3	2.2	S 9 30.0	0.1	12 42
14 42	139 15.5	0.6	S21 26.8	0.4	10 Wed	169 49.7	2.2	S 9 32.6	0.1	12 39
14 41	139 29.8	0.6	S21 35.7	0.4	11 Th	170 42.1	2.2	S 9 35.1	0.1	12 35
14 41	139 43.9	0.6	S21 44.4	0.4	12 Fri	171 34.4	2.2	S 9 37.6	0.1	12 32
14 40	139 57.8	0.6	S21 52.9	0.3	13 Sat	172 26.8	2.2	S 9 40.2	0.1	12 28
14 39	140 11.5	0.6	S22 01.3	0.3	14 SUN	173 19.1	2.2	S 9 42.7	0.1	12 25
14 38	140 25.1	0.6	S22 09.4	0.3	15 Mon	174 11.4	2.2	S 9 45.2	0.1	12 21
14 37	140 38.5	0.6	S22 17.4	0.3	16 Tu	175 03.7	2.2	S 9 47.7	0.1	12 18
14 36	140 51.7	0.5	S22 25.2	0.3	17 Wed	175 56.0	2.2	S 9 50.3	0.1	12 14
14 35	141 04.8	0.5	S22 32.7	0.3	18 Th	176 48.2	2.2	S 9 52.8	0.1	12 11
14 34	141 17.7	0.5	S22 40.1	0.3	19 Fri	177 40.5	2.2	S 9 55.3	0.1	12 08
14 33	141 30.4	0.5	S22 47.2	0.3	20 Sat	178 32.7	2.2	S 9 57.8	0.1	12 04
14 33	141 43.0	0.5	S22 54.2	0.3	21 SUN	179 25.0	2.2	S10 00.3	0.1	12 01
14 32	141 55.4	0.5	S23 00.9	0.3	22 Mon	180 17.2	2.2	S10 02.8	0.1	11 57
14 31	142 07.7	0.5	S23 07.4	0.3	23 Tu	181 09.4	2.2	S10 05.3	0.1	11 54
14 30	142 19.8	0.5	S23 13.7	0.3	24 Wed	182 01.6	2.2	S10 07.8	0.1	11 50
14 29	142 31.8	0.5	S23 19.8	0.2	25 Th	182 53.9	2.2	S10 10.3	0.1	11 47
14 29	142 43.6	0.5	S23 25.6	0.2	26 Fri	183 46.1	2.2	S10 12.8	0.1	11 43
14 28	142 55.3	0.5	S23 31.2	0.2	27 Sat	184 38.3	2.2	S10 15.3	0.1	11 40
14 27	143 06.9	0.5	S23 36.6	0.2	28 SUN	185 30.5	2.2	S10 17.7	0.1	11 36
14 26	143 18.3	0.5	S23 41.8	0.2	29 Mon	186 22.7	2.2	S10 20.2	0.1	11 33
14 26	143 29.6	0.5	S23 46.7	0.2	30 Tu	187 14.9	2.2	S10 22.6	0.1	11 29
14 25	143 40.7	0.5	S23 51.4	0.2	31 Wed	188 07.1	2.2	S10 25.1	0.1	11 26

MARS, Av. Mag. +1.2
SHA October 5 124; 10 120; 15 116; 20 113; 25 109; 30 105

SATURN, Av. Mag. +0.6
SHA October 5 151; 10 151; 15 150; 20 150; 25 149; 30 149

OCTOBER 2012

STARS

No.	Name	Mag	Transit h m	Dec ° ′	SHA ° ′
	0h GMT October 1				
♈	ARIES	–	23 16	–	–
1	Alpheratz	2.1	23 25	N29 09.9	357 43.6
2	Ankaa	2.4	23 43	S42 14.1	353 15.8
3	Schedar	2.2	0 01	N56 36.6	349 40.5
4	Diphda	2.0	0 04	S17 54.8	348 56.0
5	Achernar	0.5	0 58	S57 10.2	335 26.5
6	POLARIS	2.0	2 09	N89 19.1	317 35.6
7	Hamal	2.0	1 27	N23 31.5	328 00.9
8	Acamar	3.2	2 18	S40 15.1	315 18.3
9	Menkar	2.5	2 22	N 4 08.5	314 15.2
10	Mirfak	1.8	2 44	N49 54.3	308 40.6
11	Aldebaran	0.9	3 56	N16 32.1	290 49.7
12	Rigel	0.1	4 34	S 8 11.2	281 12.3
13	Capella	0.1	4 36	N46 00.4	280 34.9
14	Bellatrix	1.6	4 45	N 6 21.7	278 32.3
15	Elnath	1.7	4 46	N28 37.0	278 13.0
16	Alnilam	1.7	4 56	S 1 11.6	275 46.7
17	Betelgeuse	0.1–1.2	5 15	N 7 24.5	271 01.7
18	Canopus	–0.7	5 43	S52 41.9	263 56.2
19	Sirius	–1.5	6 04	S16 43.9	258 34.1
20	Adhara	1.5	6 18	S28 59.2	255 12.8
21	Castor	1.6	6 54	N31 51.4	246 08.6
22	Procyon	0.4	6 58	N 5 11.5	245 00.2
23	Pollux	1.1	7 04	N27 59.5	243 28.3
24	Avior	1.9	7 41	S59 32.8	234 18.3
25	Suhail	2.2	8 27	S43 28.9	222 53.0
26	Miaplacidus	1.7	8 31	S69 46.0	221 40.1
27	Alphard	2.0	8 46	S 8 42.8	217 56.7
28	Regulus	1.4	9 27	N11 54.2	207 44.2
29	Dubhe	1.8	10 22	N61 40.7	193 52.8
30	Denebola	2.1	11 07	N14 30.0	182 34.4
31	Gienah	2.6	11 34	S17 36.7	175 53.1
32	Acrux	1.3	11 45	S63 10.1	173 10.5
33	Gacrux	1.6	11 49	S57 11.0	172 02.0
34	Mimosa	1.3	12 06	S59 45.5	167 53.1
35	Alioth	1.8	12 12	N55 53.4	166 21.7
36	Spica	1.0	12 43	S11 13.6	158 32.1
37	Alkaid	1.9	13 05	N49 15.0	152 59.8
38	Hadar	0.6	13 22	S60 26.1	148 49.2
39	Menkent	2.1	13 25	S36 25.9	148 08.5
40	Arcturus	0.0	13 33	N19 07.1	145 56.5
41	Rigil Kent	–0.3	13 58	S60 53.2	139 53.0
42	Zuben'ubi	2.8	14 09	S16 05.6	137 06.3
43	Kochab	2.1	14 08	N74 06.3	137 21.1
44	Alphecca	2.2	14 52	N26 40.5	126 11.7
45	Antares	1.0	15 47	S26 27.5	112 27.1
46	Atria	1.9	16 07	S69 03.1	107 29.7
47	Sabik	2.4	16 28	S15 44.3	102 13.3
48	Shaula	1.6	16 51	S37 06.7	96 22.8
49	Rasalhague	2.1	16 52	N12 33.3	96 07.1
50	Eltanin	2.2	17 14	N51 29.6	90 46.6
51	Kaus Aust.	1.9	17 42	S34 22.6	83 44.6
52	Vega	0.0	17 54	N38 48.1	80 39.4
53	Nunki	2.0	18 13	S26 16.7	75 59.0
54	Altair	0.8	19 08	N 8 54.5	62 08.7
55	Peacock	1.9	19 43	S56 41.7	53 19.9
56	Deneb	1.3	19 58	N45 20.0	49 31.7
57	Enif	2.4	21 01	N 9 56.3	33 47.4
58	Al Na'ir	1.7	21 25	S46 53.9	27 44.0
59	Fomalhaut	1.2	22 14	S29 33.1	15 24.2
60	Markab	2.5	22 21	N15 16.7	13 38.5

SUN

Lat 52°N

Yr	Day of Mth	Week	Transit h m	Semi-Diam	Twilight h m	Sunrise h m	Sunset h m	Twilight h m
275	1	Mon	11 50	16.0	05 28	06 01	17 37	18 10
276	2	Tu	11 49	16.0	05 29	06 03	17 34	18 08
277	3	Wed	11 49	16.0	05 31	06 05	17 32	18 06
278	4	Th	11 49	16.0	05 33	06 06	17 30	18 03
279	5	Fri	11 48	16.0	05 34	06 08	17 28	18 01
280	6	Sat	11 48	16.0	05 36	06 10	17 25	17 59
281	7	Sun	11 48	16.0	05 38	06 11	17 23	17 57
282	8	Mon	11 47	16.0	05 39	06 13	17 21	17 54
283	9	Tu	11 47	16.0	05 41	06 15	17 18	17 52
284	10	Wed	11 47	16.0	05 43	06 17	17 16	17 50
285	11	Th	11 47	16.0	05 44	06 18	17 14	17 48
286	12	Fri	11 46	16.1	05 46	06 20	17 12	17 46
287	13	Sat	11 46	16.1	05 48	06 22	17 10	17 44
288	14	Sun	11 46	16.1	05 49	06 24	17 07	17 41
289	15	Mon	11 46	16.1	05 51	06 25	17 05	17 39
290	16	Tu	11 45	16.1	05 53	06 27	17 03	17 37
291	17	Wed	11 45	16.1	05 54	06 29	17 01	17 35
292	18	Th	11 45	16.1	05 56	06 30	16 59	17 33
293	19	Fri	11 45	16.1	05 58	06 32	16 57	17 31
294	20	Sat	11 45	16.1	06 00	06 34	16 55	17 29
295	21	Sun	11 45	16.1	06 01	06 36	16 53	17 27
296	22	Mon	11 44	16.1	06 03	06 38	16 50	17 25
297	23	Tu	11 44	16.1	06 05	06 39	16 48	17 23
298	24	Wed	11 44	16.1	06 06	06 41	16 46	17 21
299	25	Th	11 44	16.1	06 08	06 43	16 44	17 19
300	26	Fri	11 44	16.1	06 10	06 45	16 42	17 17
301	27	Sat	11 44	16.1	06 11	06 46	16 40	17 16
302	28	Sun	11 44	16.1	06 13	06 48	16 39	17 14
303	29	Mon	11 44	16.1	06 15	06 50	16 37	17 12
304	30	Tu	11 44	16.1	06 16	06 52	16 35	17 10
305	31	Wed	11 44	16.1	06 18	06 54	16 33	17 08

Lat Corr to Sunrise, Sunset etc.

Lat °	Twilight h m	Sunrise h m	Sunset h m	Twilight h m
N70	+0 19	+0 49	–0 51	–0 20
68	+0 16	+0 40	–0 41	–0 16
66	+0 13	+0 32	–0 33	–0 14
64	+0 10	+0 25	–0 27	–0 11
62	+0 08	+0 20	–0 21	–0 09
N60	+0 06	+0 14	–0 16	–0 07
58	+0 05	+0 10	–0 11	–0 05
56	+0 03	+0 06	–0 07	–0 04
54	+0 01	+0 03	–0 04	–0 02
50	–0 02	–0 03	+0 03	+0 01
N45	–0 05	–0 10	+0 09	+0 05
40	–0 08	–0 15	+0 14	+0 07
35	–0 11	–0 19	+0 19	+0 10
30	–0 14	–0 23	+0 23	+0 13
20	–0 18	–0 30	+0 30	+0 18
N10	–0 24	–0 46	+0 37	+0 23
0	–0 29	–0 42	+0 43	+0 29
S10	–0 36	–0 49	+0 49	+0 36
20	–0 43	–0 55	+0 57	+0 44
30	–0 53	–1 03	+1 05	+0 54
S35	–1 00	–1 08	+1 10	+1 00
40	–1 07	–1 13	+1 15	+1 08
45	–1 16	–1 19	+1 21	+1 16
S50	–1 27	–1 27	+1 28	+1 26

NOTES
The corrections to sunrise etc. are for middle of October.

MOON

Yr	Day of Mth	Week	Age days	Transit (Upper) h m	Diff m	Semi-diam ′	Hor Par ′	Lat 52°N Moonrise h m	Lat 52°N Moonset h m
275	1	Mon	15	00 22	45	15.0	54.9	17 46	07 33
276	2	Tu	16	01 07	46	14.9	54.5	18 13	08 40
277	3	Wed	17	01 53	47	14.8	54.3	18 43	09 44
278	4	Th	18	02 40	47	14.8	54.1	19 20	10 44
279	5	Fri	19	03 27	49	14.8	54.1	20 03	11 39
280	6	Sat	20	04 16	48	14.8	54.3	20 53	12 28
281	7	Sun	21	05 04	48	14.9	54.6	21 50	13 10
282	8	Mon	22	05 52	49	15.0	55.1	22 53	13 46
283	9	Tu	23	06 41	47	15.2	55.7	– –	14 16
284	10	Wed	24	07 28	48	15.4	56.5	00 01	14 43
285	11	Th	25	08 16	49	15.6	57.4	01 12	15 07
286	12	Fri	26	09 05	49	15.9	58.3	02 26	15 30
287	13	Sat	27	09 54	52	16.1	59.2	03 42	15 53
288	14	Sun	28	10 46	54	16.3	59.9	05 02	16 18
289	15	Mon	29	11 40	58	16.5	60.5	06 24	16 46
290	16	Tu	01	12 38	60	16.6	60.8	07 47	17 19
291	17	Wed	02	13 38	61	16.6	60.8	09 09	18 01
292	18	Th	03	14 39	61	16.5	60.5	10 24	18 52
293	19	Fri	04	15 40	60	16.4	60.0	11 29	19 53
294	20	Sat	05	16 40	56	16.2	59.4	12 21	21 02
295	21	Sun	06	17 36	53	16.0	58.7	13 03	22 16
296	22	Mon	07	18 29	49	15.8	58.0	13 36	23 30
297	23	Tu	08	19 18	47	15.6	57.3	14 03	– –
298	24	Wed	09	20 05	45	15.4	56.7	14 26	00 44
299	25	Th	10	20 50	44	15.3	56.1	14 47	01 56
300	26	Fri	11	21 34	45	15.1	55.6	15 07	03 06
301	27	Sat	12	22 19	44	15.0	55.1	15 28	04 14
302	28	Sun	13	23 03	46	14.9	54.8	15 50	05 22
303	29	Mon	14	23 49	46	14.8	54.5	16 15	06 29
304	30	Tu	15	24 35	–	14.8	54.2	16 44	07 33
305	31	Wed	16	00 35	48	14.7	54.1	17 19	08 35

Phases of the Moon

	d	h	m
☽ Last Quarter	8	07	33
● New Moon	15	12	03
☾ First Quarter	22	03	32
○ Full Moon	29	19	49

	d	h
Apogee	5	01
Perigee	17	01

NOVEMBER 2012

SUN AND ARIES

GMT	SUN GHA ° '	SUN Dec ° '	ARIES GHA ° '
Thursday, 1st November			
00	184 06.2	**S14 29.2**	40 41.2
02	214 06.2	**14 30.7**	70 46.1
04	244 06.3	**14 32.3**	100 51.0
06	274 06.3	**14 33.9**	130 55.9
08	304 06.3	**14 35.5**	161 00.9
10	334 06.3	**14 37.1**	191 05.8
12	4 06.4	**14 38.7**	221 10.7
14	34 06.4	**14 40.3**	251 15.7
16	64 06.4	**14 41.9**	281 20.6
18	94 06.4	**14 43.5**	311 25.5
20	124 06.5	**14 45.0**	341 30.4
22	154 06.5	**S14 46.6**	11 35.4
Friday, 2nd November			
00	184 06.5	**S14 48.2**	41 40.3
02	214 06.5	**14 49.8**	71 45.2
04	244 06.5	**14 51.4**	101 50.2
06	274 06.5	**14 52.9**	131 55.1
08	304 06.5	**14 54.5**	162 00.0
10	334 06.6	**14 56.1**	192 04.9
12	4 06.6	**14 57.6**	222 09.9
14	34 06.6	**14 59.2**	252 14.8
16	64 06.6	**15 00.8**	282 19.7
18	94 06.6	**15 02.3**	312 24.7
20	124 06.6	**15 03.9**	342 29.6
22	154 06.6	**S15 05.5**	12 34.5
Saturday, 3rd November			
00	184 06.6	**S15 07.0**	42 39.4
02	214 06.6	**15 08.6**	72 44.4
04	244 06.6	**15 10.1**	102 49.3
06	274 06.6	**15 11.7**	132 54.2
08	304 06.6	**15 13.2**	162 59.1
10	334 06.6	**15 14.8**	193 04.1
12	4 06.6	**15 16.3**	223 09.0
14	34 06.5	**15 17.9**	253 13.9
16	64 06.5	**15 19.4**	283 18.9
18	94 06.5	**15 21.0**	313 23.8
20	124 06.5	**15 22.5**	343 28.7
22	154 06.5	**S15 24.0**	13 33.6
Sunday, 4th November			
00	184 06.5	**S15 25.6**	43 38.6
02	214 06.5	**15 27.1**	73 43.5
04	244 06.4	**15 28.6**	103 48.4
06	274 06.4	**15 30.2**	133 53.4
08	304 06.4	**15 31.7**	163 58.3
10	334 06.4	**15 33.2**	194 03.2
12	4 06.3	**15 34.8**	224 08.1
14	34 06.3	**15 36.3**	254 13.1
16	64 06.3	**15 37.8**	284 18.0
18	94 06.3	**15 39.3**	314 22.9
20	124 06.2	**15 40.9**	344 27.9
22	154 06.2	**S15 42.4**	14 32.8
Monday, 5th November			
00	184 06.2	**S15 43.9**	44 37.7
02	214 06.1	**15 45.4**	74 42.6
04	244 06.1	**15 46.9**	104 47.6
06	274 06.0	**15 48.4**	134 52.5
08	304 06.0	**15 49.9**	164 57.4
10	334 06.0	**15 51.4**	195 02.4
12	4 05.9	**15 52.9**	225 07.3
14	34 05.9	**15 54.5**	255 12.2
16	64 05.8	**15 56.0**	285 17.1
18	94 05.8	**15 57.5**	315 22.1
20	124 05.7	**15 59.0**	345 27.0
22	154 05.7	**S16 00.4**	15 31.9

GMT	SUN GHA ° '	SUN Dec ° '	ARIES GHA ° '
Tuesday, 6th November			
00	184 05.6	**S16 01.9**	45 36.9
02	214 05.6	**16 03.4**	75 41.8
04	244 05.5	**16 04.9**	105 46.7
06	274 05.5	**16 06.4**	135 51.6
08	304 05.4	**16 07.9**	165 56.6
10	334 05.3	**16 09.4**	196 01.5
12	4 05.3	**16 10.9**	226 06.4
14	34 05.2	**16 12.4**	256 11.4
16	64 05.1	**16 13.8**	286 16.3
18	94 05.1	**16 15.3**	316 21.2
20	124 05.0	**16 16.8**	346 26.1
22	154 04.9	**S16 18.3**	16 31.1
Wednesday, 7th November			
00	184 04.9	**S16 19.7**	46 36.0
02	214 04.8	**16 21.2**	76 40.9
04	244 04.7	**16 22.7**	106 45.9
06	274 04.7	**16 24.1**	136 50.8
08	304 04.6	**16 25.6**	166 55.7
10	334 04.5	**16 27.1**	197 00.6
12	4 04.4	**16 28.5**	227 05.6
14	34 04.3	**16 30.0**	257 10.5
16	64 04.3	**16 31.4**	287 15.4
18	94 04.2	**16 32.9**	317 20.3
20	124 04.1	**16 34.3**	347 25.3
22	154 04.0	**S16 35.8**	17 30.2
Thursday, 8th November			
00	184 03.9	**S16 37.2**	47 35.1
02	214 03.8	**16 38.7**	77 40.1
04	244 03.7	**16 40.1**	107 45.0
06	274 03.6	**16 41.6**	137 49.9
08	304 03.5	**16 43.0**	167 54.8
10	334 03.4	**16 44.5**	197 59.8
12	4 03.3	**16 45.9**	228 04.7
14	34 03.2	**16 47.3**	258 09.6
16	64 03.1	**16 48.8**	288 14.6
18	94 03.0	**16 50.2**	318 19.5
20	124 02.9	**16 51.6**	348 24.4
22	154 02.8	**S16 53.1**	18 29.3
Friday, 9th November			
00	184 02.7	**S16 54.5**	48 34.3
02	214 02.6	**16 55.9**	78 39.2
04	244 02.5	**16 57.3**	108 44.1
06	274 02.4	**16 58.7**	138 49.1
08	304 02.3	**17 00.2**	168 54.0
10	334 02.2	**17 01.6**	198 58.9
12	4 02.1	**17 03.0**	229 03.8
14	34 01.9	**17 04.4**	259 08.8
16	64 01.8	**17 05.8**	289 13.7
18	94 01.7	**17 07.2**	319 18.6
20	124 01.6	**17 08.6**	349 23.6
22	154 01.5	**S17 10.0**	19 28.5
Saturday, 10th November			
00	184 01.3	**S17 11.4**	49 33.4
02	214 01.2	**17 12.8**	79 38.3
04	244 01.1	**17 14.2**	109 43.3
06	274 01.0	**17 15.6**	139 48.2
08	304 00.8	**17 17.0**	169 53.1
10	334 00.7	**17 18.4**	199 58.0
12	4 00.6	**17 19.8**	230 03.0
14	34 00.4	**17 21.2**	260 07.9
16	64 00.3	**17 22.6**	290 12.8
18	94 00.2	**17 23.9**	320 17.8
20	124 00.0	**17 25.3**	350 22.7
22	153 59.9	**S17 26.7**	20 27.6

GMT	SUN GHA ° '	SUN Dec ° '	ARIES GHA ° '
Sunday, 11th November			
00	183 59.7	**S17 28.1**	50 32.5
02	213 59.6	**17 29.5**	80 37.5
04	243 59.4	**17 30.8**	110 42.4
06	273 59.3	**17 32.2**	140 47.3
08	303 59.1	**17 33.6**	170 52.3
10	333 59.0	**17 34.9**	200 57.2
12	3 58.8	**17 36.3**	231 02.1
14	33 58.7	**17 37.7**	261 07.0
16	63 58.5	**17 39.0**	291 12.0
18	93 58.4	**17 40.4**	321 16.9
20	123 58.2	**17 41.7**	351 21.8
22	153 58.1	**S17 43.1**	21 26.8
Monday, 12th November			
00	183 57.9	**S17 44.4**	51 31.7
02	213 57.7	**17 45.8**	81 36.6
04	243 57.6	**17 47.1**	111 41.5
06	273 57.4	**17 48.5**	141 46.5
08	303 57.2	**17 49.8**	171 51.4
10	333 57.1	**17 51.2**	201 56.3
12	3 56.9	**17 52.5**	232 01.3
14	33 56.7	**17 53.8**	262 06.2
16	63 56.6	**17 55.2**	292 11.1
18	93 56.4	**17 56.5**	322 16.0
20	123 56.2	**17 57.8**	352 21.0
22	153 56.0	**S17 59.2**	22 25.9
Tuesday, 13th November			
00	183 55.9	**S18 00.5**	52 30.8
02	213 55.7	**18 01.8**	82 35.7
04	243 55.5	**18 03.1**	112 40.7
06	273 55.3	**18 04.4**	142 45.6
08	303 55.1	**18 05.8**	172 50.5
10	333 54.9	**18 07.1**	202 55.5
12	3 54.8	**18 08.4**	233 00.4
14	33 54.6	**18 09.7**	263 05.3
16	63 54.4	**18 11.0**	293 10.2
18	93 54.2	**18 12.3**	323 15.2
20	123 54.0	**18 13.6**	353 20.1
22	153 53.8	**S18 14.9**	23 25.0
Wednesday, 14th November			
00	183 53.6	**S18 16.2**	53 30.0
02	213 53.4	**18 17.5**	83 34.9
04	243 53.2	**18 18.8**	113 39.8
06	273 53.0	**18 20.1**	143 44.7
08	303 52.8	**18 21.4**	173 49.7
10	333 52.6	**18 22.7**	203 54.6
12	3 52.4	**18 24.0**	233 59.5
14	33 52.2	**18 25.2**	264 04.5
16	63 52.0	**18 26.5**	294 09.4
18	93 51.8	**18 27.8**	324 14.3
20	123 51.6	**18 29.1**	354 19.2
22	153 51.4	**S18 30.4**	24 24.2
Thursday, 15th November			
00	183 51.1	**S18 31.6**	54 29.1
02	213 50.9	**18 32.9**	84 34.0
04	243 50.7	**18 34.2**	114 39.0
06	273 50.5	**18 35.4**	144 43.9
08	303 50.3	**18 36.7**	174 48.8
10	333 50.0	**18 38.0**	204 53.7
12	3 49.8	**18 39.2**	234 58.7
14	33 49.6	**18 40.5**	265 03.6
16	63 49.4	**18 41.7**	295 08.5
18	93 49.1	**18 43.0**	325 13.5
20	123 48.9	**18 44.2**	355 18.4
22	153 48.7	**S18 45.5**	25 23.3

SUN AND ARIES

GMT	SUN GHA ° '	SUN Dec ° '	ARIES GHA ° '
Friday, 16th November			
00	183 48.5	**S18 46.7**	55 28.2
02	213 48.2	**18 48.0**	85 33.2
04	243 48.0	**18 49.2**	115 38.1
06	273 47.8	**18 50.4**	145 43.0
08	303 47.5	**18 51.7**	175 48.0
10	333 47.3	**18 52.9**	205 52.9
12	3 47.0	**18 54.1**	235 57.8
14	33 46.8	**18 55.4**	266 02.7
16	63 46.6	**18 56.6**	296 07.7
18	93 46.3	**18 57.8**	326 12.6
20	123 46.1	**18 59.0**	356 17.5
22	153 45.8	**S19 00.3**	26 22.5
Saturday, 17th November			
00	183 45.6	**S19 01.5**	56 27.4
02	213 45.3	**19 02.7**	86 32.3
04	243 45.1	**19 03.9**	116 37.2
06	273 44.8	**19 05.1**	146 42.2
08	303 44.6	**19 06.3**	176 47.1
10	333 44.3	**19 07.5**	206 52.0
12	3 44.1	**19 08.7**	236 57.0
14	33 43.8	**19 09.9**	267 01.9
16	63 43.5	**19 11.1**	297 06.8
18	93 43.3	**19 12.3**	327 11.7
20	123 43.0	**19 13.5**	357 16.7
22	153 42.8	**S19 14.7**	27 21.6
Sunday, 18th November			
00	183 42.5	**S19 15.9**	57 26.5
02	213 42.2	**19 17.1**	87 31.5
04	243 42.0	**19 18.2**	117 36.4
06	273 41.7	**19 19.4**	147 41.3
08	303 41.4	**19 20.6**	177 46.2
10	333 41.1	**19 21.8**	207 51.2
12	3 40.9	**19 23.0**	237 56.1
14	33 40.6	**19 24.1**	268 01.0
16	63 40.3	**19 25.3**	298 05.9
18	93 40.0	**19 26.5**	328 10.9
20	123 39.8	**19 27.6**	358 15.8
22	153 39.5	**S19 28.8**	28 20.7
Monday, 19th November			
00	183 39.2	**S19 29.9**	58 25.7
02	213 38.9	**19 31.1**	88 30.6
04	243 38.6	**19 32.3**	118 35.5
06	273 38.3	**19 33.4**	148 40.4
08	303 38.1	**19 34.6**	178 45.4
10	333 37.8	**19 35.7**	208 50.3
12	3 37.5	**19 36.8**	238 55.2
14	33 37.2	**19 38.0**	269 00.2
16	63 36.9	**19 39.1**	299 05.1
18	93 36.6	**19 40.3**	329 10.0
20	123 36.3	**19 41.4**	359 14.9
22	153 36.0	**S19 42.5**	29 19.9
Tuesday, 20th November			
00	183 35.7	**S19 43.7**	59 24.8
02	213 35.4	**19 44.8**	89 29.7
04	243 35.1	**19 45.9**	119 34.7
06	273 34.8	**19 47.0**	149 39.6
08	303 34.5	**19 48.2**	179 44.5
10	333 34.2	**19 49.3**	209 49.4
12	3 33.9	**19 50.4**	239 54.4
14	33 33.6	**19 51.5**	269 59.3
16	63 33.3	**19 52.6**	300 04.2
18	93 33.0	**19 53.7**	330 09.2
20	123 32.7	**19 54.8**	0 14.1
22	153 32.3	**S19 55.9**	30 19.0

GMT	SUN GHA ° '	SUN Dec ° '	ARIES GHA ° '
Wednesday, 21st November			
00	183 32.0	**S19 57.0**	60 23.9
02	213 31.7	**19 58.1**	90 28.9
04	243 31.4	**19 59.2**	120 33.8
06	273 31.1	**20 00.3**	150 38.7
08	303 30.8	**20 01.4**	180 43.7
10	333 30.4	**20 02.5**	210 48.6
12	3 30.1	**20 03.6**	240 53.5
14	33 29.8	**20 04.6**	270 58.4
16	63 29.5	**20 05.7**	301 03.4
18	93 29.1	**20 06.8**	331 08.3
20	123 28.8	**20 07.9**	1 13.2
22	153 28.5	**S20 08.9**	31 18.1
Thursday, 22nd November			
00	183 28.1	**S20 10.0**	61 23.1
02	213 27.8	**20 11.1**	91 28.0
04	243 27.5	**20 12.1**	121 32.9
06	273 27.1	**20 13.2**	151 37.9
08	303 26.8	**20 14.3**	181 42.8
10	333 26.5	**20 15.3**	211 47.7
12	3 26.1	**20 16.4**	241 52.6
14	33 25.8	**20 17.4**	271 57.6
16	63 25.4	**20 18.5**	302 02.5
18	93 25.1	**20 19.5**	332 07.4
20	123 24.8	**20 20.5**	2 12.4
22	153 24.4	**S20 21.6**	32 17.3
Friday, 23rd November			
00	183 24.1	**S20 22.6**	62 22.2
02	213 23.7	**20 23.7**	92 27.1
04	243 23.4	**20 24.7**	122 32.1
06	273 23.0	**20 25.7**	152 37.0
08	303 22.7	**20 26.7**	182 41.9
10	333 22.3	**20 27.8**	212 46.9
12	3 22.0	**20 28.8**	242 51.8
14	33 21.6	**20 29.8**	272 56.7
16	63 21.2	**20 30.8**	303 01.6
18	93 20.9	**20 31.8**	333 06.6
20	123 20.5	**20 32.9**	3 11.5
22	153 20.2	**S20 33.9**	33 16.4
Saturday, 24th November			
00	183 19.8	**S20 34.9**	63 21.4
02	213 19.4	**20 35.9**	93 26.3
04	243 19.1	**20 36.9**	123 31.2
06	273 18.7	**20 37.9**	153 36.1
08	303 18.3	**20 38.9**	183 41.1
10	333 18.0	**20 39.9**	213 46.0
12	3 17.6	**20 40.8**	243 50.9
14	33 17.2	**20 41.8**	273 55.9
16	63 16.9	**20 42.8**	304 00.8
18	93 16.5	**20 43.8**	334 05.7
20	123 16.1	**20 44.8**	4 10.6
22	153 15.7	**S20 45.8**	34 15.6
Sunday, 25th November			
00	183 15.3	**S20 46.7**	64 20.5
02	213 15.0	**20 47.7**	94 25.4
04	243 14.6	**20 48.7**	124 30.3
06	273 14.2	**20 49.6**	154 35.3
08	303 13.8	**20 50.6**	184 40.2
10	333 13.4	**20 51.6**	214 45.1
12	3 13.1	**20 52.5**	244 50.1
14	33 12.7	**20 53.5**	274 55.0
16	63 12.3	**20 54.4**	304 59.9
18	93 11.9	**20 55.4**	335 04.8
20	123 11.5	**20 56.3**	5 09.8
22	153 11.1	**S20 57.3**	35 14.7

GMT	SUN GHA ° '	SUN Dec ° '	ARIES GHA ° '
Monday, 26th November			
00	183 10.7	**S20 58.2**	65 19.6
02	213 10.3	**20 59.1**	95 24.6
04	243 09.9	**21 00.1**	125 29.5
06	273 09.5	**21 01.0**	155 34.4
08	303 09.1	**21 01.9**	185 39.3
10	333 08.7	**21 02.9**	215 44.3
12	3 08.3	**21 03.8**	245 49.2
14	33 07.9	**21 04.7**	275 54.1
16	63 07.5	**21 05.6**	305 59.1
18	93 07.1	**21 06.6**	336 04.0
20	123 06.7	**21 07.5**	6 08.9
22	153 06.3	**S21 08.4**	36 13.8
Tuesday, 27th November			
00	183 05.9	**S21 09.3**	66 18.8
02	213 05.5	**21 10.2**	96 23.7
04	243 05.1	**21 11.1**	126 28.6
06	273 04.7	**21 12.0**	156 33.6
08	303 04.2	**21 12.9**	186 38.5
10	333 03.8	**21 13.8**	216 43.4
12	3 03.4	**21 14.7**	246 48.3
14	33 03.0	**21 15.6**	276 53.3
16	63 02.6	**21 16.5**	306 58.2
18	93 02.2	**21 17.3**	337 03.1
20	123 01.7	**21 18.2**	7 08.1
22	153 01.3	**S21 19.1**	37 13.0
Wednesday, 28th November			
00	183 00.9	**S21 20.0**	67 17.9
02	213 00.5	**21 20.8**	97 22.8
04	243 00.0	**21 21.7**	127 27.8
06	272 59.6	**21 22.6**	157 32.7
08	302 59.2	**21 23.4**	187 37.6
10	332 58.8	**21 24.3**	217 42.5
12	2 58.3	**21 25.2**	247 47.5
14	32 57.9	**21 26.0**	277 52.4
16	62 57.5	**21 26.9**	307 57.3
18	92 57.0	**21 27.7**	338 02.3
20	122 56.6	**21 28.6**	8 07.2
22	152 56.2	**S21 29.4**	38 12.1
Thursday, 29th November			
00	182 55.7	**S21 30.3**	68 17.0
02	212 55.3	**21 31.1**	98 22.0
04	242 54.8	**21 31.9**	128 26.9
06	272 54.4	**21 32.8**	158 31.8
08	302 54.0	**21 33.6**	188 36.8
10	332 53.5	**21 34.4**	218 41.7
12	2 53.1	**21 35.2**	248 46.6
14	32 52.6	**21 36.1**	278 51.5
16	62 52.2	**21 36.9**	308 56.5
18	92 51.7	**21 37.7**	339 01.4
20	122 51.3	**21 38.5**	9 06.3
22	152 50.8	**S21 39.3**	39 11.3
Friday, 30th November			
00	182 50.4	**S21 40.1**	69 16.2
02	212 49.9	**21 40.9**	99 21.1
04	242 49.5	**21 41.7**	129 26.0
06	272 49.0	**21 42.5**	159 31.0
08	302 48.6	**21 43.3**	189 35.9
10	332 48.1	**21 44.1**	219 40.8
12	2 47.6	**21 44.9**	249 45.8
14	32 47.2	**21 45.7**	279 50.7
16	62 46.7	**21 46.5**	309 55.6
18	92 46.3	**21 47.3**	340 00.5
20	122 45.8	**21 48.0**	10 05.5
22	152 45.3	**S21 48.8**	40 10.4

Day	GMT hr	GHA ° '	Mean Var/hr 14°+	Dec ° '	Mean Var/hr '
1 Th	0	340 00.1	31.1	N20 05.1	2.8
	6	67 06.4	31.0	N20 22.0	2.2
	12	154 12.3	30.9	N20 35.4	1.7
	18	241 17.8	30.8	N20 45.4	1.1
2 Fri	0	328 23.0	30.8	N20 52.0	0.5
	6	55 28.1	30.8	N20 55.0	0.2
	12	142 33.1	30.8	N20 54.6	0.7
	18	229 38.1	30.8	N20 50.7	1.3
3 Sat	0	316 43.2	30.9	N20 43.3	1.9
	6	43 48.5	31.0	N20 32.4	2.4
	12	130 54.0	31.0	N20 18.0	3.0
	18	218 00.0	31.1	N20 00.3	3.5
4 Sun	0	305 06.3	31.1	N19 39.2	4.1
	6	32 13.1	31.2	N19 14.7	4.7
	12	119 20.4	31.4	N18 47.0	5.2
	18	206 28.2	31.4	N18 16.2	5.7
5 Mon	0	293 36.6	31.5	N17 42.1	6.3
	6	20 45.4	31.5	N17 05.1	6.7
	12	107 54.8	31.6	N16 25.0	7.2
	18	195 04.7	31.7	N15 42.1	7.6
6 Tu	0	282 15.0	31.7	N14 56.3	8.1
	6	9 25.6	31.8	N14 07.8	8.5
	12	96 36.6	31.8	N13 16.8	9.0
	18	183 47.8	31.9	N12 23.2	9.4
7 Wed	0	270 59.0	31.9	N11 27.2	9.7
	6	358 10.3	31.9	N10 28.9	10.1
	12	85 21.5	31.8	N 9 28.5	10.5
	18	172 32.4	31.7	N 8 26.0	10.8
8 Th	0	259 42.8	31.6	N 7 21.5	11.1
	6	346 52.8	31.5	N 6 15.3	11.3
	12	74 02.0	31.4	N 5 07.5	11.6
	18	161 10.3	31.2	N 3 58.2	11.8
9 Fri	0	248 17.5	31.0	N 2 47.7	11.9
	6	335 23.5	30.7	N 1 36.0	12.1
	12	62 28.1	30.5	N 0 23.3	12.2
	18	149 31.0	30.2	S 0 50.0	12.3
10 Sat	0	236 32.1	29.9	S 2 03.8	12.4
	6	323 31.2	29.4	S 3 17.9	12.4
	12	50 28.2	29.1	S 4 31.9	12.3
	18	137 22.7	28.6	S 5 45.7	12.2
11 Sun	0	224 14.8	28.2	S 6 58.9	12.0
	6	311 04.1	27.7	S 8 11.2	11.9
	12	37 50.6	27.2	S 9 22.3	11.6
	18	124 34.2	26.7	S10 31.9	11.2
12 Mon	0	211 14.7	26.2	S11 39.5	10.8
	6	297 52.2	25.7	S12 45.0	10.4
	12	24 26.5	25.1	S13 47.7	9.9
	18	110 57.7	24.7	S14 47.5	9.3
13 Tu	0	197 25.8	24.1	S15 43.9	8.7
	6	283 51.0	23.7	S16 36.6	8.1
	12	10 13.4	23.2	S17 25.2	7.3
	18	Eclipse of the Sun occurs today			
14 Wed	0	182 50.6	22.6	S18 48.8	5.7
	6	269 06.0	22.2	S19 23.3	4.8
	12	355 19.8	22.1	S19 52.5	3.9
	18	81 32.2	22.0	S20 16.3	2.9
15 Th	0	167 43.9	21.9	S20 34.5	2.0
	6	253 55.1	21.9	S20 46.9	1.1
	12	340 06.5	22.0	S20 53.7	0.1
	18	66 18.4	22.2	S20 54.7	0.8
16 Fri	0	152 31.3	22.4	S20 50.0	1.8
	6	238 45.7	22.8	S20 39.7	2.7
	12	325 01.9	23.1	S20 24.0	3.5
	18	51 20.4	23.6	S20 03.0	4.4
17 Sat	0	137 41.5	24.0	S19 37.0	5.2
	6	224 05.5	24.5	S19 06.2	6.0
	12	310 32.5	25.1	S18 30.9	6.6
	18	37 02.7	25.7	S17 51.4	7.3
18 Sun	0	123 36.4	26.2	S17 07.9	7.9
	6	210 13.5	26.8	S16 20.9	8.5
	12	296 54.0	27.4	S15 30.6	8.9
	18	23 38.1	27.9	S14 37.3	9.4
19 Mon	0	110 25.5	28.5	S13 41.4	9.7
	6	197 16.2	29.0	S12 43.1	10.1
	12	284 10.2	29.5	S11 42.7	10.4
	18	11 07.2	30.1	S10 40.5	10.6
20 Tu	0	98 07.0	30.5	S 9 36.7	10.9
	6	185 09.7	30.9	S 8 31.6	11.0
	12	272 14.8	31.3	S 7 25.5	11.1
	18	359 22.3	31.6	S 6 18.6	11.3
21 Wed	0	86 32.0	32.0	S 5 11.0	11.4
	6	173 43.6	32.3	S 4 03.0	11.3
	12	260 56.9	32.5	S 2 54.8	11.4
	18	348 11.8	32.7	S 1 46.6	11.3
22 Th	0	75 28.0	32.9	S 0 38.5	11.3
	6	162 45.3	33.1	N 0 29.3	11.2
	12	250 03.6	33.2	N 1 36.7	11.1
	18	337 22.6	33.2	N 2 43.4	11.0
23 Fri	0	64 42.2	33.3	N 3 49.4	10.9
	6	152 02.2	33.4	N 4 54.4	10.7
	12	239 22.5	33.4	N 5 58.5	10.5
	18	326 42.8	33.3	N 7 01.3	10.3
24 Sat	0	54 03.0	33.3	N 8 02.8	10.0
	6	141 22.9	33.3	N 9 02.9	9.8
	12	228 42.6	33.2	N10 01.4	9.4
	18	316 01.7	33.1	N10 58.1	9.2
25 Sun	0	43 20.2	33.0	N11 53.1	8.8
	6	130 38.0	32.8	N12 46.0	8.5
	12	217 55.1	32.7	N13 36.9	8.1
	18	305 11.3	32.5	N14 25.5	7.6
26 Mon	0	32 26.6	32.4	N15 11.8	7.3
	6	119 41.0	32.2	N15 55.6	6.8
	12	206 54.3	32.0	N16 36.9	6.4
	18	294 06.7	31.9	N17 15.5	5.9
27 Tu	0	21 18.1	31.7	N17 51.2	5.4
	6	108 28.5	31.6	N18 24.1	4.9
	12	195 38.0	31.4	N18 53.9	4.4
	18	282 46.5	31.3	N19 20.7	3.8
28 Wed	0	9 54.3	31.2	N19 44.2	3.3
	6	97 01.2	31.1	N20 04.6	2.8
	12	184 07.5	31.0	N20 21.5	2.2
	18	271 13.2	30.8	N20 35.2	1.6
29 Th	0	358 18.4	30.8	N20 45.3	1.1
	6	85 23.3	30.8	N20 52.1	0.5
	12	172 27.9	30.7	N20 55.3	0.1
	18	259 32.4	30.7	N20 55.0	0.6
30 Fri	0	346 36.9	30.7	N20 51.2	1.3
	6	73 41.5	30.8	N20 43.9	1.9
	12	160 46.3	30.9	N20 33.1	2.4
	18	247 51.4	30.9	N20 18.9	3.0

VENUS Mer Pass h m	GHA ° '	Mean Var/hr 14°+	Dec ° '	Mean Var/hr '	Day	JUPITER GHA ° '	Mean Var/hr 15°+	Dec ° '	Mean Var/hr '	Mer Pass h m
09 35	216 12.6	59.7	S 0 11.9	1.2	1 Th	326 43.1	2.7	N21 46.9	0.0	02 13
09 36	216 04.3	59.7	S 0 39.6	1.2	2 Fri	327 48.0	2.7	N21 46.3	0.0	02 08
09 36	215 55.9	59.7	S 1 07.2	1.2	3 Sat	328 53.0	2.7	N21 45.7	0.0	02 04
09 37	215 47.5	59.6	S 1 34.9	1.2	4 SUN	329 58.2	2.7	N21 45.1	0.0	02 00
09 38	215 39.0	59.6	S 2 02.7	1.2	5 Mon	331 03.6	2.7	N21 44.5	0.0	01 55
09 38	215 30.4	59.6	S 2 30.4	1.2	6 Tu	332 09.1	2.7	N21 43.8	0.0	01 51
09 39	215 21.7	59.6	S 2 58.2	1.2	7 Wed	333 14.8	2.7	N21 43.2	0.0	01 47
09 39	215 12.9	59.6	S 3 26.0	1.2	8 Th	334 20.7	2.8	N21 42.5	0.0	01 42
09 40	215 03.9	59.6	S 3 53.7	1.2	9 Fri	335 26.7	2.8	N21 41.8	0.0	01 38
09 41	214 54.9	59.6	S 4 21.4	1.2	10 Sat	336 32.8	2.8	N21 41.1	0.0	01 34
09 41	214 45.7	59.6	S 4 49.1	1.1	11 SUN	337 39.1	2.8	N21 40.3	0.0	01 29
09 42	214 36.4	59.6	S 5 16.7	1.1	12 Mon	338 45.5	2.8	N21 39.6	0.0	01 25
09 42	214 26.9	59.6	S 5 44.2	1.1	13 Tu	339 52.1	2.8	N21 38.8	0.0	01 20
09 43	214 17.3	59.6	S 6 11.7	1.1	14 Wed	340 58.7	2.8	N21 38.0	0.0	01 16
09 44	214 07.6	59.6	S 6 39.1	1.1	15 Th	342 05.5	2.8	N21 37.2	0.0	01 11
09 44	213 57.7	59.6	S 7 06.3	1.1	16 Fri	343 12.4	2.8	N21 36.4	0.0	01 07
09 45	213 47.6	59.6	S 7 33.5	1.1	17 Sat	344 19.5	2.8	N21 35.6	0.0	01 03
09 46	213 37.3	59.6	S 8 00.5	1.1	18 SUN	345 26.6	2.8	N21 34.8	0.0	00 58
09 46	213 26.8	59.6	S 8 27.4	1.1	19 Mon	346 33.8	2.8	N21 33.9	0.0	00 54
09 47	213 16.2	59.5	S 8 54.1	1.1	20 Tu	347 41.1	2.8	N21 33.0	0.0	00 49
09 48	213 05.3	59.5	S 9 20.6	1.1	21 Wed	348 48.5	2.8	N21 32.2	0.0	00 45
09 49	212 54.3	59.5	S 9 46.9	1.1	22 Th	349 55.9	2.8	N21 31.3	0.0	00 40
09 49	212 43.0	59.5	S10 13.1	1.1	23 Fri	351 03.4	2.8	N21 30.4	0.0	00 36
09 50	212 31.5	59.5	S10 39.0	1.1	24 Sat	352 11.0	2.8	N21 29.4	0.0	00 31
09 51	212 19.8	59.5	S11 04.8	1.1	25 SUN	353 18.7	2.8	N21 28.5	0.0	00 27
09 52	212 07.8	59.5	S11 30.2	1.1	26 Mon	354 26.4	2.8	N21 27.6	0.0	00 22
09 53	211 55.6	59.5	S11 55.5	1.0	27 Tu	355 34.1	2.8	N21 26.6	0.0	00 18
09 53	211 43.1	59.5	S12 20.4	1.0	28 Wed	356 41.9	2.8	N21 25.7	0.0	00 13
09 54	211 30.4	59.5	S12 45.1	1.0	29 Th	357 49.7	2.8	N21 24.7	0.0	00 09
09 55	211 17.4	59.5	S13 09.5	1.0	30 Fri	358 57.5	2.8	N21 23.7	0.0	00 04

VENUS, Av. Mag. –3.9
SHA November 5 171; 10 165; 15 160; 20 154; 25 148; 30 142

JUPITER, Av. Mag. –2.8
SHA November 5 286; 10 287; 15 288; 20 288; 25 289; 30 290

MARS Mer Pass h m	GHA ° '	Mean Var/hr 15°+	Dec ° '	Mean Var/hr '	Day	SATURN GHA ° '	Mean Var/hr 15°+	Dec ° '	Mean Var/hr '	Mer Pass h m
14 24	143 51.7	0.5	S23 55.9	0.2	1 Th	188 59.3	2.2	S10 27.5	0.1	11 22
14 23	144 02.6	0.4	S24 00.1	0.2	2 Fri	189 51.5	2.2	S10 30.0	0.1	11 19
14 23	144 13.3	0.4	S24 04.0	0.2	3 Sat	190 43.8	2.2	S10 32.4	0.1	11 15
14 22	144 24.0	0.4	S24 07.8	0.1	4 SUN	191 36.0	2.2	S10 34.8	0.1	11 12
14 21	144 34.5	0.4	S24 11.2	0.1	5 Mon	192 28.2	2.2	S10 37.2	0.1	11 09
14 21	144 44.9	0.4	S24 14.5	0.1	6 Tu	193 20.5	2.2	S10 39.6	0.1	11 05
14 20	144 55.2	0.4	S24 17.5	0.1	7 Wed	194 12.7	2.2	S10 42.0	0.1	11 02
14 19	145 05.4	0.4	S24 20.2	0.1	8 Th	195 05.0	2.2	S10 44.4	0.1	10 58
14 19	145 15.5	0.4	S24 22.6	0.1	9 Fri	195 57.3	2.2	S10 46.7	0.1	10 55
14 18	145 25.5	0.4	S24 24.9	0.1	10 Sat	196 49.5	2.2	S10 49.1	0.1	10 51
14 17	145 35.3	0.4	S24 26.8	0.1	11 SUN	197 41.8	2.2	S10 51.4	0.1	10 48
14 17	145 45.1	0.4	S24 28.5	0.1	12 Mon	198 34.2	2.2	S10 53.7	0.1	10 44
14 16	145 54.8	0.4	S24 29.9	0.0	13 Tu	199 26.5	2.2	S10 56.1	0.1	10 41
14 15	146 04.4	0.4	S24 31.1	0.0	14 Wed	200 18.8	2.2	S10 58.4	0.1	10 37
14 15	146 13.9	0.4	S24 32.0	0.0	15 Th	201 11.2	2.2	S11 00.7	0.1	10 34
14 14	146 23.4	0.4	S24 32.6	0.0	16 Fri	202 03.6	2.2	S11 02.9	0.1	10 30
14 13	146 32.8	0.4	S24 33.0	0.0	17 Sat	202 56.0	2.2	S11 05.2	0.1	10 27
14 13	146 42.1	0.4	S24 33.1	0.0	18 SUN	203 48.4	2.2	S11 07.4	0.1	10 23
14 12	146 51.3	0.4	S24 32.9	0.0	19 Mon	204 40.9	2.2	S11 09.7	0.1	10 20
14 12	147 00.5	0.4	S24 32.5	0.0	20 Tu	205 33.4	2.2	S11 11.9	0.1	10 16
14 11	147 09.7	0.4	S24 31.8	0.0	21 Wed	206 25.9	2.2	S11 14.1	0.1	10 13
14 10	147 18.8	0.4	S24 30.8	0.1	22 Th	207 18.4	2.2	S11 16.3	0.1	10 09
14 10	147 27.8	0.4	S24 29.6	0.1	23 Fri	208 10.9	2.2	S11 18.4	0.1	10 06
14 09	147 36.9	0.4	S24 28.1	0.1	24 Sat	209 03.5	2.2	S11 20.6	0.1	10 02
14 09	147 45.8	0.4	S24 26.3	0.1	25 SUN	209 56.1	2.2	S11 22.7	0.1	09 59
14 08	147 54.8	0.4	S24 24.2	0.1	26 Mon	210 48.8	2.2	S11 24.8	0.1	09 55
14 07	148 03.7	0.4	S24 21.9	0.1	27 Tu	211 41.5	2.2	S11 26.9	0.1	09 52
14 07	148 12.6	0.4	S24 19.3	0.1	28 Wed	212 34.2	2.2	S11 29.0	0.1	09 48
14 06	148 21.5	0.4	S24 16.4	0.1	29 Th	213 26.9	2.2	S11 31.1	0.1	09 45
14 06	148 30.4	0.4	S24 13.2	0.1	30 Fri	214 19.7	2.2	S11 33.1	0.1	09 41

MARS, Av. Mag. +1.2
SHA November 5 100; 10 96; 15 92; 20 88; 25 83; 30 79

SATURN, Av. Mag. +0.6
SHA November 5 148; 10 147; 15 147; 20 146; 25 146; 30 145

STARS

No.	Name	Mag	Transit h m	Dec ° ′	SHA ° ′
	Oh GMT November 1				
Ψ	ARIES	–	21 14	–	–
1	Alpheratz	2.1	21 23	N29 10.0	357 43.7
2	Ankaa	2.4	21 41	S42 14.2	353 15.8
3	Schedar	2.2	21 55	N56 36.8	349 40.5
4	Diphda	2.0	21 58	S17 54.9	348 56.1
5	Achernar	0.5	22 52	S57 10.4	335 26.6
6	POLARIS	2.0	0 07	N89 19.2	317 31.4
7	Hamal	2.0	23 21	N23 31.5	328 00.9
8	Acamar	3.2	0 16	S40 15.2	315 18.2
9	Menkar	2.5	0 20	N 4 08.5	314 15.2
10	Mirfak	1.8	0 42	N49 54.4	308 40.4
11	Aldebaran	0.9	1 54	N16 32.1	290 49.5
12	Rigel	0.1	2 32	S 8 11.2	281 12.1
13	Capella	0.1	2 35	N46 00.5	280 34.6
14	Bellatrix	1.6	2 43	N 6 21.6	278 32.2
15	Elnath	1.7	2 44	N28 37.0	278 12.8
16	Alnilam	1.7	2 54	S 1 11.7	275 46.5
17	Betelgeuse	0.1–1.2	3 13	N 7 24.5	271 01.5
18	Canopus	–0.7	3 41	S52 42.1	263 55.9
19	Sirius	–1.5	4 02	S16 44.0	258 33.8
20	Adhara	1.5	4 16	S28 59.3	255 12.6
21	Castor	1.6	4 52	N31 51.3	246 08.3
22	Procyon	0.4	4 56	N 5 11.4	245 00.0
23	Pollux	1.1	5 03	N27 59.5	243 28.1
24	Avior	1.9	5 39	S59 32.9	234 17.9
25	Suhail	2.2	6 25	S43 28.9	222 52.7
26	Miaplacidus	1.7	6 30	S69 46.0	221 39.6
27	Alphard	2.0	6 44	S 8 42.9	217 56.5
28	Regulus	1.4	7 25	N11 54.1	207 44.0
29	Dubhe	1.8	8 20	N61 40.5	193 52.5
30	Denebola	2.1	9 05	N14 29.9	182 34.3
31	Gienah	2.6	9 32	S17 36.7	175 52.9
32	Acrux	1.3	9 43	S63 10.0	173 10.2
33	Gacrux	1.6	9 48	S57 10.9	172 01.8
34	Mimosa	1.3	10 04	S59 45.4	167 52.9
35	Alioth	1.8	10 10	N55 53.2	166 21.5
36	Spica	1.0	10 41	S11 13.6	158 32.0
37	Alkaid	1.9	11 03	N49 14.9	152 59.7
38	Hadar	0.6	11 20	S60 25.9	148 49.1
39	Menkent	2.1	11 23	S36 25.8	148 08.4
40	Arcturus	0.0	11 32	N19 07.0	145 56.4
41	Rigil Kent	–0.3	11 56	S60 53.1	139 52.9
42	Zuben'ubi	2.8	12 07	S16 05.6	137 06.2
43	Kochab	2.1	12 06	N74 06.1	137 21.2
44	Alphecca	2.2	12 50	N26 40.4	126 11.7
45	Antares	1.0	13 45	S26 27.5	112 27.1
46	Atria	1.9	14 05	S69 03.0	107 29.8
47	Sabik	2.4	14 26	S15 44.3	102 13.3
48	Shaula	1.6	14 49	S37 06.7	96 22.8
49	Rasalhague	2.1	14 50	N12 33.3	96 07.1
50	Eltanin	2.2	15 12	N51 29.5	90 46.8
51	Kaus Aust.	1.9	15 40	S34 22.6	83 44.7
52	Vega	0.0	15 52	N38 48.1	80 39.5
53	Nunki	2.0	16 11	S26 16.7	75 59.1
54	Altair	0.8	17 06	N 8 54.4	62 08.8
55	Peacock	1.9	17 41	S56 41.7	53 20.1
56	Deneb	1.3	17 56	N45 20.0	49 31.9
57	Enif	2.4	18 59	N 9 56.3	33 47.5
58	Al Na'ir	1.7	19 23	S46 54.0	27 44.2
59	Fomalhaut	1.2	20 12	S29 33.2	15 24.3
60	Markab	2.5	20 19	N15 16.7	13 38.6

SUN

Yr	Day of Mth	Day of Week	Transit h m	Semi-Diam	Lat 52°N Twilight h m	Lat 52°N Sunrise h m	Lat 52°N Sunset h m	Lat 52°N Twilight h m
306	1	Th	11 44	16.1	06 20	06 55	16 31	17 07
307	2	Fri	11 44	16.1	06 21	06 57	16 29	17 05
308	3	Sat	11 44	16.2	06 23	06 59	16 27	17 03
309	4	Sun	11 44	16.2	06 25	07 01	16 26	17 02
310	5	Mon	11 44	16.2	06 26	07 03	16 24	17 00
311	6	Tu	11 44	16.2	06 28	07 04	16 22	16 58
312	7	Wed	11 44	16.2	06 30	07 06	16 20	16 57
313	8	Th	11 44	16.2	06 31	07 08	16 19	16 55
314	9	Fri	11 44	16.2	06 33	07 10	16 17	16 54
315	10	Sat	11 44	16.2	06 35	07 12	16 16	16 52
316	11	Sun	11 44	16.2	06 36	07 13	16 14	16 51
317	12	Mon	11 44	16.2	06 38	07 15	16 13	16 50
318	13	Tu	11 44	16.2	06 40	07 17	16 11	16 48
319	14	Wed	11 45	16.2	06 41	07 19	16 10	16 47
320	15	Th	11 45	16.2	06 43	07 20	16 08	16 46
321	16	Fri	11 45	16.2	06 45	07 22	16 07	16 45
322	17	Sat	11 45	16.2	06 46	07 24	16 06	16 43
323	18	Sun	11 45	16.2	06 48	07 26	16 04	16 42
324	19	Mon	11 45	16.2	06 49	07 27	16 03	16 41
325	20	Tu	11 46	16.2	06 51	07 29	16 02	16 40
326	21	Wed	11 46	16.2	06 52	07 31	16 01	16 39
327	22	Th	11 46	16.2	06 54	07 32	16 00	16 38
328	23	Fri	11 47	16.2	06 55	07 34	15 59	16 37
329	24	Sat	11 47	16.2	06 57	07 36	15 58	16 36
330	25	Sun	11 47	16.2	06 58	07 37	15 57	16 36
331	26	Mon	11 47	16.2	07 00	07 39	15 56	16 35
332	27	Tu	11 48	16.2	07 01	07 40	15 55	16 34
333	28	Wed	11 48	16.2	07 02	07 42	15 54	16 33
334	29	Th	11 48	16.2	07 04	07 43	15 53	16 33
335	30	Fri	11 49	16.2	07 05	07 45	15 53	16 32

Lat Corr to Sunrise, Sunset etc.

Lat °	Twilight h m	Sunrise h m	Sunset h m	Twilight h m
N70	+1 27	+2 27	–2 27	–1 27
68	+1 09	+1 52	–1 52	–1 10
66	+0 56	+1 27	–1 26	–0 57
64	+0 44	+1 07	–1 07	–0 45
62	+0 34	+0 51	–0 51	–0 35
N60	+0 26	+0 38	–0 38	–0 26
58	+0 19	+0 26	–0 26	–0 18
56	+0 12	+0 17	–0 16	–0 12
54	+0 06	+0 08	–0 08	–0 06
50	–0 06	–0 07	+0 07	+0 06
N45	–0 17	–0 22	+0 22	+0 16
40	–0 26	–0 34	+0 34	+0 26
35	–0 34	–0 45	+0 45	+0 34
30	–0 42	–0 54	+0.54	+0 42
20	–0 56	–1 10	+1 10	+0 57
N10	–1 09	–1 24	+1 25	+1 10
0	–1 21	–1 37	+1 38	+1 22
S10	–1 36	–1 50	+1 51	+1 35
20	–1 52	–2 05	+2 06	+1 51
30	–2 10	–2 21	+2 23	+2 10
S35	–2 22	–2 31	+2 33	+2 22
40	–2 36	–2 42	+2 44	+2 36
45	–2 52	–2 55	+2 57	+2 53
S50	–3 14	–3 12	+3 13	+3 13

NOTES

The corrections to sunrise etc. are for middle of November.

MOON

Yr	Day of Mth	Day of Week	Age days	Transit (Upper) h m	Diff m	Semi-diam ′	Hor Par ′	Lat 52°N Moonrise h m	Lat 52°N Moonset h m
306	1	Th	17	01 23	48	14.7	54.0	17 59	09 32
307	2	Fri	18	02 11	48	14.7	54.0	18 47	10 23
308	3	Sat	19	02 59	48	14.8	54.2	19 41	11 07
309	4	Sun	20	03 47	47	14.9	54.5	20 42	11 45
310	5	Mon	21	04 34	47	15.0	55.0	21 46	12 17
311	6	Tu	22	05 21	47	15.1	55.6	22 54	12 44
312	7	Wed	23	06 08	46	15.3	56.3	– –	13 08
313	8	Th	24	06 54	48	15.6	57.2	00 04	13 31
314	9	Fri	25	07 42	49	15.8	58.2	01 17	13 53
315	10	Sat	26	08 31	52	16.1	59.1	02 33	14 16
316	11	Sun	27	09 23	55	16.4	60.0	03 52	14 42
317	12	Mon	28	10 18	59	16.6	60.7	05 15	15 12
318	13	Tu	29	11 17	63	16.7	61.2	06 38	15 49
319	14	Wed	01	12 20	63	16.7	61.4	07 58	16 37
320	15	Th	02	13 23	63	16.7	61.2	09 11	17 35
321	16	Fri	03	14 26	60	16.5	60.7	10 11	18 44
322	17	Sat	04	15 26	56	16.3	60.0	10 59	19 59
323	18	Sun	05	16 22	52	16.1	59.1	11 36	21 16
324	19	Mon	06	17 14	49	15.9	58.2	12 06	22 32
325	20	Tu	07	18 03	46	15.6	57.3	12 31	23 46
326	21	Wed	08	18 49	44	15.4	56.5	12 53	– –
327	22	Th	09	19 33	44	15.2	55.8	13 13	00 57
328	23	Fri	10	20 17	44	15.1	55.3	13 34	02 06
329	24	Sat	11	21 01	45	14.9	54.8	13 55	03 13
330	25	Sun	12	21 46	46	14.8	54.4	14 19	04 20
331	26	Mon	13	22 32	47	14.8	54.2	14 47	05 25
332	27	Tu	14	23 19	48	14.7	54.0	15 19	06 27
333	28	Wed	15	24 07	–	14.7	54.0	15 57	07 26
334	29	Th	16	00 07	48	14.7	54.0	16 43	08 19
335	30	Fri	17	00 55	49	14.7	54.1	17 35	09 06

Phases of the Moon

	Phase	d	h	m
☽	Last Quarter	7	00	36
●	New Moon	13	22	08
☾	First Quarter	20	14	31
○	Full Moon	28	14	46

	d	h
Apogee	1	15
Perigee	14	10
Apogee	28	20

Saturday, 1st December

GMT	SUN GHA ° '	SUN Dec ° '	ARIES GHA ° '
00	182 44.9	S21 49.6	70 15.3
02	212 44.4	21 50.4	100 20.3
04	242 43.9	21 51.1	130 25.2
06	272 43.5	21 51.9	160 30.1
08	302 43.0	21 52.6	190 35.0
10	332 42.5	21 53.4	220 40.0
12	2 42.0	21 54.2	250 44.9
14	32 41.6	21 54.9	280 49.8
16	62 41.1	21 55.7	310 54.8
18	92 40.6	21 56.4	340 59.7
20	122 40.1	21 57.2	11 04.6
22	152 39.7	S21 57.9	41 09.5

Sunday, 2nd December

GMT	SUN GHA ° '	SUN Dec ° '	ARIES GHA ° '
00	182 39.2	S21 58.6	71 14.5
02	212 38.7	21 59.4	101 19.4
04	242 38.2	22 00.1	131 24.3
06	272 37.7	22 00.8	161 29.3
08	302 37.3	22 01.5	191 34.2
10	332 36.8	22 02.3	221 39.1
12	2 36.3	22 03.0	251 44.0
14	32 35.8	22 03.7	281 49.0
16	62 35.3	22 04.4	311 53.9
18	92 34.8	22 05.1	341 58.8
20	122 34.3	22 05.8	12 03.8
22	152 33.8	S22 06.5	42 08.7

Monday, 3rd December

GMT	SUN GHA ° '	SUN Dec ° '	ARIES GHA ° '
00	182 33.4	S22 07.2	72 13.6
02	212 32.9	22 07.9	102 18.5
04	242 32.4	22 08.6	132 23.5
06	272 31.9	22 09.3	162 28.4
08	302 31.4	22 10.0	192 33.3
10	332 30.9	22 10.7	222 38.3
12	2 30.4	22 11.4	252 43.2
14	32 29.9	22 12.1	282 48.1
16	62 29.4	22 12.8	312 53.0
18	92 28.9	22 13.4	342 58.0
20	122 28.4	22 14.1	13 02.9
22	152 27.9	S22 14.8	43 07.8

Tuesday, 4th December

GMT	SUN GHA ° '	SUN Dec ° '	ARIES GHA ° '
00	182 27.4	S22 15.4	73 12.7
02	212 26.9	22 16.1	103 17.7
04	242 26.4	22 16.8	133 22.6
06	272 25.9	22 17.4	163 27.5
08	302 25.3	22 18.1	193 32.5
10	332 24.8	22 18.7	223 37.4
12	2 24.3	22 19.4	253 42.3
14	32 23.8	22 20.0	283 47.2
16	62 23.3	22 20.7	313 52.2
18	92 22.8	22 21.3	343 57.1
20	122 22.3	22 21.9	14 02.0
22	152 21.8	S22 22.6	44 07.0

Wednesday, 5th December

GMT	SUN GHA ° '	SUN Dec ° '	ARIES GHA ° '
00	182 21.2	S22 23.2	74 11.9
02	212 20.7	22 23.8	104 16.8
04	242 20.2	22 24.5	134 21.7
06	272 19.7	22 25.1	164 26.7
08	302 19.2	22 25.7	194 31.6
10	332 18.6	22 26.3	224 36.5
12	2 18.1	22 26.9	254 41.5
14	32 17.6	22 27.5	284 46.4
16	62 17.1	22 28.1	314 51.3
18	92 16.6	22 28.7	344 56.2
20	122 16.0	22 29.3	15 01.2
22	152 15.5	S22 29.9	45 06.1

Thursday, 6th December

GMT	SUN GHA ° '	SUN Dec ° '	ARIES GHA ° '
00	182 15.0	S22 30.5	75 11.0
02	212 14.4	22 31.1	105 16.0
04	242 13.9	22 31.7	135 20.9
06	272 13.4	22 32.3	165 25.8
08	302 12.8	22 32.9	195 30.7
10	332 12.3	22 33.5	225 35.7
12	2 11.8	22 34.0	255 40.6
14	32 11.2	22 34.6	285 45.5
16	62 10.7	22 35.2	315 50.5
18	92 10.2	22 35.7	345 55.4
20	122 09.6	22 36.3	16 00.3
22	152 09.1	S22 36.9	46 05.2

Friday, 7th December

GMT	SUN GHA ° '	SUN Dec ° '	ARIES GHA ° '
00	182 08.6	S22 37.4	76 10.2
02	212 08.0	22 38.0	106 15.1
04	242 07.5	22 38.5	136 20.0
06	272 06.9	22 39.1	166 24.9
08	302 06.4	22 39.6	196 29.9
10	332 05.9	22 40.2	226 34.8
12	2 05.3	22 40.7	256 39.7
14	32 04.8	22 41.2	286 44.7
16	62 04.2	22 41.8	316 49.6
18	92 03.7	22 42.3	346 54.5
20	122 03.1	22 42.8	16 59.4
22	152 02.6	S22 43.4	47 04.4

Saturday, 8th December

GMT	SUN GHA ° '	SUN Dec ° '	ARIES GHA ° '
00	182 02.0	S22 43.9	77 09.3
02	212 01.5	22 44.4	107 14.2
04	242 00.9	22 44.9	137 19.2
06	272 00.4	22 45.4	167 24.1
08	301 59.8	22 45.9	197 29.0
10	331 59.3	22 46.4	227 33.9
12	1 58.7	22 46.9	257 38.9
14	31 58.2	22 47.4	287 43.8
16	61 57.6	22 47.9	317 48.7
18	91 57.0	22 48.4	347 53.7
20	121 56.5	22 48.9	17 58.6
22	151 55.9	S22 49.4	48 03.5

Sunday, 9th December

GMT	SUN GHA ° '	SUN Dec ° '	ARIES GHA ° '
00	181 55.4	S22 49.9	78 08.4
02	211 54.8	22 50.3	108 13.4
04	241 54.2	22 50.8	138 18.3
06	271 53.7	22 51.3	168 23.2
08	301 53.1	22 51.8	198 28.2
10	331 52.6	22 52.2	228 33.1
12	1 52.0	22 52.7	258 38.0
14	31 51.4	22 53.2	288 42.9
16	61 50.9	22 53.6	318 47.9
18	91 50.3	22 54.1	348 52.8
20	121 49.7	22 54.5	18 57.7
22	151 49.2	S22 55.0	49 02.6

Monday, 10th December

GMT	SUN GHA ° '	SUN Dec ° '	ARIES GHA ° '
00	181 48.6	S22 55.4	79 07.6
02	211 48.0	22 55.9	109 12.5
04	241 47.4	22 56.3	139 17.4
06	271 46.9	22 56.7	169 22.4
08	301 46.3	22 57.2	199 27.3
10	331 45.7	22 57.6	229 32.2
12	1 45.2	22 58.0	259 37.1
14	31 44.6	22 58.4	289 42.1
16	61 44.0	22 58.9	319 47.0
18	91 43.4	22 59.3	349 51.9
20	121 42.9	22 59.7	19 56.9
22	151 42.3	S23 00.1	50 01.8

Tuesday, 11th December

GMT	SUN GHA ° '	SUN Dec ° '	ARIES GHA ° '
00	181 41.7	S23 00.5	80 06.7
02	211 41.1	23 00.9	110 11.6
04	241 40.5	23 01.3	140 16.6
06	271 40.0	23 01.7	170 21.5
08	301 39.4	23 02.1	200 26.4
10	331 38.8	23 02.5	230 31.4
12	1 38.2	23 02.9	260 36.3
14	31 37.6	23 03.3	290 41.2
16	61 37.1	23 03.7	320 46.1
18	91 36.5	23 04.0	350 51.1
20	121 35.9	23 04.4	20 56.0
22	151 35.3	S23 04.8	51 00.9

Wednesday, 12th December

GMT	SUN GHA ° '	SUN Dec ° '	ARIES GHA ° '
00	181 34.7	S23 05.2	81 05.9
02	211 34.1	23 05.5	111 10.8
04	241 33.6	23 05.9	141 15.7
06	271 33.0	23 06.2	171 20.6
08	301 32.4	23 06.6	201 25.6
10	331 31.8	23 07.0	231 30.5
12	1 31.2	23 07.3	261 35.4
14	31 30.6	23 07.7	291 40.4
16	61 30.0	23 08.0	321 45.3
18	91 29.4	23 08.3	351 50.2
20	121 28.8	23 08.7	21 55.1
22	151 28.2	S23 09.0	52 00.1

Thursday, 13th December

GMT	SUN GHA ° '	SUN Dec ° '	ARIES GHA ° '
00	181 27.7	S23 09.3	82 05.0
02	211 27.1	23 09.7	112 09.9
04	241 26.5	23 10.0	142 14.9
06	271 25.9	23 10.3	172 19.8
08	301 25.3	23 10.6	202 24.7
10	331 24.7	23 10.9	232 29.6
12	1 24.1	23 11.3	262 34.6
14	31 23.5	23 11.6	292 39.5
16	61 22.9	23 11.9	322 44.4
18	91 22.3	23 12.2	352 49.4
20	121 21.7	23 12.5	22 54.3
22	151 21.1	S23 12.8	52 59.2

Friday, 14th December

GMT	SUN GHA ° '	SUN Dec ° '	ARIES GHA ° '
00	181 20.5	S23 13.1	83 04.1
02	211 19.9	23 13.4	113 09.1
04	241 19.3	23 13.6	143 14.0
06	271 18.7	23 13.9	173 18.9
08	301 18.1	23 14.2	203 23.9
10	331 17.5	23 14.5	233 28.8
12	1 16.9	23 14.8	263 33.7
14	31 16.3	23 15.0	293 38.6
16	61 15.7	23 15.3	323 43.6
18	91 15.1	23 15.6	353 48.5
20	121 14.5	23 15.8	23 53.4
22	151 13.9	S23 16.1	53 58.4

Saturday, 15th December

GMT	SUN GHA ° '	SUN Dec ° '	ARIES GHA ° '
00	181 13.3	S23 16.3	84 03.3
02	211 12.7	23 16.6	114 08.2
04	241 12.1	23 16.8	144 13.1
06	271 11.5	23 17.1	174 18.1
08	301 10.9	23 17.3	204 23.0
10	331 10.3	23 17.5	234 27.9
12	1 09.7	23 17.8	264 32.9
14	31 09.0	23 18.0	294 37.8
16	61 08.4	23 18.2	324 42.7
18	91 07.8	23 18.5	354 47.6
20	121 07.2	23 18.7	24 52.6
22	151 06.6	S23 18.9	54 57.5

Sunday, 16th December

GMT	SUN GHA ° '	SUN Dec ° '	ARIES GHA ° '
00	181 06.0	S23 19.1	85 02.4
02	211 05.4	23 19.3	115 07.4
04	241 04.8	23 19.5	145 12.3
06	271 04.2	23 19.8	175 17.2
08	301 03.6	23 20.0	205 22.1
10	331 03.0	23 20.2	235 27.1
12	1 02.3	23 20.3	265 32.0
14	31 01.7	23 20.5	295 36.9
16	61 01.1	23 20.7	325 41.8
18	91 00.5	23 20.9	355 46.8
20	120 59.9	23 21.1	25 51.7
22	150 59.3	S23 21.3	55 56.6

Monday, 17th December

GMT	SUN GHA ° '	SUN Dec ° '	ARIES GHA ° '
00	180 58.7	S23 21.5	86 01.6
02	210 58.1	23 21.6	116 06.5
04	240 57.4	23 21.8	146 11.4
06	270 56.8	23 22.0	176 16.3
08	300 56.2	23 22.1	206 21.3
10	330 55.6	23 22.3	236 26.2
12	0 55.0	23 22.4	266 31.1
14	30 54.4	23 22.6	296 36.1
16	60 53.8	23 22.7	326 41.0
18	90 53.1	23 22.9	356 45.9
20	120 52.5	23 23.0	26 50.8
22	150 51.9	S23 23.2	56 55.8

Tuesday, 18th December

GMT	SUN GHA ° '	SUN Dec ° '	ARIES GHA ° '
00	180 51.3	S23 23.3	87 00.7
02	210 50.7	23 23.5	117 05.6
04	240 50.1	23 23.6	147 10.6
06	270 49.5	23 23.7	177 15.5
08	300 48.8	23 23.8	207 20.4
10	330 48.2	23 24.0	237 25.3
12	0 47.6	23 24.1	267 30.3
14	30 47.0	23 24.2	297 35.2
16	60 46.4	23 24.3	327 40.1
18	90 45.8	23 24.4	357 45.1
20	120 45.1	23 24.5	27 50.0
22	150 44.5	S23 24.6	57 54.9

Wednesday, 19th December

GMT	SUN GHA ° '	SUN Dec ° '	ARIES GHA ° '
00	180 43.9	S23 24.7	87 59.8
02	210 43.3	23 24.8	118 04.8
04	240 42.7	23 24.9	148 09.7
06	270 42.0	23 25.0	178 14.6
08	300 41.4	23 25.1	208 19.6
10	330 40.8	23 25.2	238 24.5
12	0 40.2	23 25.2	268 29.4
14	30 39.6	23 25.3	298 34.3
16	60 38.9	23 25.4	328 39.3
18	90 38.3	23 25.5	358 44.2
20	120 37.7	23 25.5	28 49.1
22	150 37.1	S23 25.6	58 54.0

Thursday, 20th December

GMT	SUN GHA ° '	SUN Dec ° '	ARIES GHA ° '
00	180 36.5	S23 25.6	88 59.0
02	210 35.8	23 25.7	119 03.9
04	240 35.2	23 25.7	149 08.8
06	270 34.6	23 25.8	179 13.8
08	300 34.0	23 25.8	209 18.7
10	330 33.4	23 25.9	239 23.6
12	0 32.7	23 25.9	269 28.5
14	30 32.1	23 26.0	299 33.5
16	60 31.5	23 26.0	329 38.4
18	90 30.9	23 26.0	359 43.3
20	120 30.3	23 26.1	29 48.3
22	150 29.6	S23 26.1	59 53.2

Friday, 21st December

GMT	SUN GHA ° '	SUN Dec ° '	ARIES GHA ° '
00	180 29.0	S23 26.1	89 58.1
02	210 28.4	23 26.1	120 03.0
04	240 27.8	23 26.1	150 08.0
06	270 27.2	23 26.1	180 12.9
08	300 26.5	23 26.1	210 17.8
10	330 25.9	23 26.1	240 22.8
12	0 25.3	23 26.1	270 27.7
14	30 24.7	23 26.1	300 32.6
16	60 24.0	23 26.1	330 37.5
18	90 23.4	23 26.1	0 42.5
20	120 22.8	23 26.1	30 47.4
22	150 22.2	S23 26.1	60 52.3

Saturday, 22nd December

GMT	SUN GHA ° '	SUN Dec ° '	ARIES GHA ° '
00	180 21.6	S23 26.1	90 57.3
02	210 20.9	23 26.1	121 02.2
04	240 20.3	23 26.0	151 07.1
06	270 19.7	23 26.0	181 12.0
08	300 19.1	23 26.0	211 17.0
10	330 18.5	23 25.9	241 21.9
12	0 17.8	23 25.9	271 26.8
14	30 17.2	23 25.9	301 31.8
16	60 16.6	23 25.8	331 36.7
18	90 16.0	23 25.8	1 41.6
20	120 15.4	23 25.7	31 46.5
22	150 14.7	S23 25.7	61 51.5

Sunday, 23rd December

GMT	SUN GHA ° '	SUN Dec ° '	ARIES GHA ° '
00	180 14.1	S23 25.6	91 56.4
02	210 13.5	23 25.5	122 01.3
04	240 12.9	23 25.5	152 06.2
06	270 12.2	23 25.4	182 11.2
08	300 11.6	23 25.3	212 16.1
10	330 11.0	23 25.3	242 21.0
12	0 10.4	23 25.2	272 26.0
14	30 09.8	23 25.1	302 30.9
16	60 09.1	23 25.0	332 35.8
18	90 08.5	23 24.9	2 40.7
20	120 07.9	23 24.8	32 45.7
22	150 07.3	S23 24.7	62 50.6

Monday, 24th December

GMT	SUN GHA ° '	SUN Dec ° '	ARIES GHA ° '
00	180 06.7	S23 24.6	92 55.5
02	210 06.0	23 24.5	123 00.5
04	240 05.4	23 24.4	153 05.4
06	270 04.8	23 24.3	183 10.3
08	300 04.2	23 24.2	213 15.2
10	330 03.6	23 24.1	243 20.2
12	0 02.9	23 24.0	273 25.1
14	30 02.3	23 23.9	303 30.0
16	60 01.7	23 23.7	333 35.0
18	90 01.1	23 23.6	3 39.9
20	120 00.5	23 23.5	33 44.8
22	149 59.9	S23 23.4	63 49.7

Tuesday, 25th December

GMT	SUN GHA ° '	SUN Dec ° '	ARIES GHA ° '
00	179 59.2	S23 23.2	93 54.7
02	209 58.6	23 23.1	123 59.6
04	239 58.0	23 22.9	154 04.5
06	269 57.4	23 22.8	184 09.5
08	299 56.8	23 22.6	214 14.4
10	329 56.1	23 22.5	244 19.3
12	359 55.5	23 22.3	274 24.2
14	29 54.9	23 22.2	304 29.2
16	59 54.3	23 22.0	334 34.1
18	89 53.7	23 21.8	4 39.0
20	119 53.1	23 21.7	34 44.0
22	149 52.4	S23 21.5	64 48.9

Wednesday, 26th December

GMT	SUN GHA ° '	SUN Dec ° '	ARIES GHA ° '
00	179 51.8	S23 21.3	94 53.8
02	209 51.2	23 21.1	124 58.7
04	239 50.6	23 21.0	155 03.7
06	269 50.0	23 20.8	185 08.6
08	299 49.4	23 20.6	215 13.5
10	329 48.8	23 20.4	245 18.5
12	359 48.1	23 20.2	275 23.4
14	29 47.5	23 20.0	305 28.3
16	59 46.9	23 19.8	335 33.2
18	89 46.3	23 19.6	5 38.2
20	119 45.7	23 19.4	35 43.1
22	149 45.1	S23 19.2	65 48.0

Thursday, 27th December

GMT	SUN GHA ° '	SUN Dec ° '	ARIES GHA ° '
00	179 44.5	S23 19.0	95 53.0
02	209 43.8	23 18.7	125 57.9
04	239 43.2	23 18.5	156 02.8
06	269 42.6	23 18.3	186 07.7
08	299 42.0	23 18.1	216 12.7
10	329 41.4	23 17.8	246 17.6
12	359 40.8	23 17.6	276 22.5
14	29 40.2	23 17.4	306 27.5
16	59 39.6	23 17.1	336 32.4
18	89 39.0	23 16.9	6 37.3
20	119 38.3	23 16.6	36 42.2
22	149 37.7	S23 16.4	66 47.2

Friday, 28th December

GMT	SUN GHA ° '	SUN Dec ° '	ARIES GHA ° '
00	179 37.1	S23 16.1	96 52.1
02	209 36.5	23 15.9	126 57.0
04	239 35.9	23 15.6	157 01.9
06	269 35.3	23 15.3	187 06.9
08	299 34.7	23 15.1	217 11.8
10	329 34.1	23 14.8	247 16.7
12	359 33.5	23 14.5	277 21.7
14	29 32.9	23 14.3	307 26.6
16	59 32.3	23 14.0	337 31.5
18	89 31.7	23 13.7	7 36.4
20	119 31.1	23 13.4	37 41.4
22	149 30.5	S23 13.1	67 46.3

Saturday, 29th December

GMT	SUN GHA ° '	SUN Dec ° '	ARIES GHA ° '
00	179 29.8	S23 12.8	97 51.2
02	209 29.2	23 12.5	127 56.2
04	239 28.6	23 12.2	158 01.1
06	269 28.0	23 11.9	188 06.0
08	299 27.4	23 11.6	218 10.9
10	329 26.8	23 11.3	248 15.9
12	359 26.2	23 11.0	278 20.8
14	29 25.6	23 10.7	308 25.7
16	59 25.0	23 10.4	338 30.7
18	89 24.4	23 10.0	8 35.6
20	119 23.8	23 09.7	38 40.5
22	149 23.2	S23 09.4	68 45.4

Sunday, 30th December

GMT	SUN GHA ° '	SUN Dec ° '	ARIES GHA ° '
00	179 22.6	S23 09.1	98 50.4
02	209 22.0	23 08.7	128 55.3
04	239 21.4	23 08.4	159 00.2
06	269 20.8	23 08.0	189 05.2
08	299 20.2	23 07.7	219 10.1
10	329 19.6	23 07.4	249 15.0
12	359 19.0	23 07.0	279 19.9
14	29 18.4	23 06.7	309 24.9
16	59 17.8	23 06.3	339 29.8
18	89 17.2	23 05.9	9 34.7
20	119 16.6	23 05.6	39 39.7
22	149 16.0	S23 05.2	69 44.6

Monday, 31st December

GMT	SUN GHA ° '	SUN Dec ° '	ARIES GHA ° '
00	179 15.4	S23 04.8	99 49.5
02	209 14.9	23 04.5	129 54.4
04	239 14.3	23 04.1	159 59.4
06	269 13.7	S23 03.7	190 04.3
08	299 13.1	S23 03.3	220 09.2
10	329 12.5	23 02.9	250 14.2
12	359 11.9	23 02.6	280 19.1
14	29 11.3	S23 02.2	310 24.0
16	59 10.7	S23 01.8	340 28.9
18	89 10.1	23 01.4	10 33.9
20	119 09.5	23 01.0	40 38.8
22	149 08.9	S23 00.6	70 43.7

DECEMBER 2012

MOON

Day	GMT hr	GHA ° ′	Mean Var/hr 14°+	Dec ° ′	Mean Var/hr ′
1 Sat	0	334 57.0	31.0	N20 01.2	3.6
	6	62 03.1	31.1	N19 40.2	4.1
	12	149 09.7	31.3	N19 15.9	4.7
	18	236 17.1	31.3	N18 48.4	5.2
2 Sun	0	323 25.1	31.5	N18 17.7	5.6
	6	50 33.9	31.6	N17 43.9	6.2
	12	137 43.4	31.7	N17 07.1	6.6
	18	224 53.7	31.8	N16 27.4	7.1
3 Mon	0	312 04.8	32.0	N15 44.9	7.6
	6	39 16.6	32.1	N14 59.8	8.0
	12	126 29.1	32.2	N14 12.0	8.4
	18	213 42.2	32.3	N13 21.8	8.8
4 Tu	0	300 55.8	32.4	N12 29.2	9.2
	6	28 09.9	32.4	N11 34.4	9.5
	12	115 24.4	32.5	N10 37.5	9.8
	18	202 39.1	32.4	N 9 38.5	10.1
5 Wed	0	289 53.9	32.5	N 8 37.7	10.4
	6	17 08.7	32.5	N 7 35.2	10.7
	12	104 23.3	32.4	N 6 31.0	11.0
	18	191 37.6	32.3	N 5 25.4	11.2
6 Th	0	278 51.3	32.2	N 4 18.4	11.3
	6	6 04.4	32.0	N 3 10.3	11.6
	12	93 16.6	31.8	N 2 01.2	11.7
	18	180 27.8	31.6	N 0 51.2	11.8
7 Fri	0	267 37.7	31.4	S 0 19.4	11.9
	6	354 46.1	31.1	S 1 30.5	11.9
	12	81 52.8	30.8	S 2 42.0	11.9
	18	168 57.7	30.4	S 3 53.5	11.9
8 Sat	0	256 00.5	30.1	S 5 04.8	11.8
	6	343 01.0	29.6	S 6 15.8	11.7
	12	69 59.0	29.2	S 7 26.2	11.6
	18	156 54.3	28.7	S 8 35.7	11.3
9 Sun	0	243 46.8	28.2	S 9 44.0	11.1
	6	330 36.3	27.7	S10 50.8	10.8
	12	57 22.6	27.2	S11 55.9	10.5
	18	144 05.6	26.6	S12 58.8	10.1
10 Mon	0	230 45.3	26.0	S13 59.3	9.6
	6	317 21.6	25.4	S14 57.1	9.1
	12	43 54.5	24.9	S15 51.7	8.5
	18	130 24.0	24.4	S16 42.8	7.8
11 Tu	0	216 50.3	23.8	S17 30.1	7.2
	6	303 13.3	23.3	S18 13.3	6.3
	12	29 33.5	22.9	S18 51.9	5.6
	18	115 50.9	22.5	S19 25.7	4.7
12 Wed	0	202 06.0	22.1	S19 54.5	3.9
	6	288 19.0	21.8	S20 18.0	3.0
	12	14 30.4	21.7	S20 35.9	1.9
	18	100 40.6	21.6	S20 48.2	1.0
13 Th	0	186 50.2	21.5	S20 54.7	0.1
	6	272 59.6	21.7	S20 55.3	1.0
	12	359 09.2	21.7	S20 50.2	1.9
	18	85 19.7	22.0	S20 39.3	2.9
14 Fri	0	171 31.5	22.3	S20 22.7	3.8
	6	257 45.0	22.7	S20 00.6	4.6
	12	344 00.7	23.1	S19 33.2	5.5
	18	70 18.8	23.6	S19 00.7	6.3
15 Sat	0	156 39.8	24.1	S18 23.5	7.0
	6	243 03.8	24.6	S17 41.8	7.7
	12	329 31.1	25.1	S16 56.0	8.4
	18	56 01.9	25.8	S16 06.4	8.9
16 Sun	0	142 36.1	26.3	S15 13.3	9.5
	6	229 13.8	26.9	S14 17.1	9.8
	12	315 55.1	27.5	S13 18.1	10.3
	18	42 39.8	28.0	S12 16.7	10.6
17 Mon	0	129 28.0	28.6	S11 13.2	10.9
	6	216 19.4	29.1	S10 08.0	11.2
	12	303 13.9	29.7	S 9 01.2	11.3
	18	30 11.5	30.1	S 7 53.3	11.5
18 Tu	0	117 11.8	30.5	S 6 44.5	11.6
	6	204 14.7	31.0	S 5 35.0	11.7
	12	291 20.1	31.3	S 4 25.1	11.7
	18	18 27.7	31.7	S 3 15.0	11.7
19 Wed	0	105 37.3	31.9	S 2 04.9	11.7
	6	192 48.6	32.2	S 0 55.1	11.6
	12	280 01.6	32.4	N 0 14.3	11.4
	18	7 15.9	32.6	N 1 23.2	11.3
20 Th	0	94 31.5	32.8	N 2 31.3	11.2
	6	181 48.0	32.9	N 3 38.5	11.1
	12	269 05.3	33.0	N 4 44.7	10.9
	18	356 23.2	33.0	N 5 49.7	10.6
21 Fri	0	83 41.5	33.1	N 6 53.4	10.3
	6	171 00.1	33.1	N 7 55.6	10.1
	12	258 18.7	33.1	N 8 56.2	9.9
	18	345 37.3	33.0	N 9 55.2	9.5
22 Sat	0	72 55.7	33.0	N10 52.4	9.1
	6	160 13.8	32.9	N11 47.6	8.8
	12	247 31.4	32.8	N12 40.8	8.4
	18	334 48.5	32.8	N13 31.9	8.1
23 Sun	0	62 04.8	32.6	N14 20.7	7.8
	6	149 20.5	32.4	N15 07.2	7.3
	12	236 35.3	32.3	N15 51.1	6.8
	18	323 49.2	32.2	N16 32.6	6.4
24 Mon	0	51 02.2	32.0	N17 11.3	6.0
	6	138 14.3	31.9	N17 47.2	5.5
	12	225 25.4	31.6	N18 20.3	5.0
	18	312 35.5	31.5	N18 50.4	4.5
25 Tu	0	39 44.8	31.4	N19 17.5	4.0
	6	126 53.1	31.3	N19 41.4	3.4
	12	214 00.6	31.2	N20 02.1	2.9
	18	301 07.3	31.0	N20 19.5	2.2
26 Wed	0	28 13.3	30.9	N20 33.6	1.8
	6	115 18.7	30.8	N20 44.2	1.1
	12	202 23.6	30.7	N20 51.4	0.6
	18	289 28.1	30.7	N20 55.2	0.0
27 Th	0	16 32.3	30.7	N20 55.4	0.6
	6	103 36.3	30.6	N20 52.1	1.1
	12	190 40.2	30.7	N20 45.3	1.8
	18	277 44.2	30.7	N20 35.0	2.3
28 Fri	0	4 48.4	30.8	N20 21.2	3.0
	6	91 52.9	30.8	N20 03.9	3.5
	12	178 57.8	30.9	N19 43.2	4.0
	18	266 03.2	31.0	N19 19.2	4.6
29 Sat	0	353 09.1	31.1	N18 51.9	5.2
	6	80 15.7	31.2	N18 21.3	5.7
	12	167 23.0	31.3	N17 47.6	6.1
	18	254 31.1	31.5	N17 10.9	6.6
30 Sun	0	341 40.0	31.6	N16 31.3	7.2
	6	68 49.6	31.7	N15 48.8	7.6
	12	156 00.1	31.9	N15 03.6	8.0
	18	243 11.4	32.1	N14 15.8	8.4
31 Mon	0	330 23.5	32.1	N13 25.5	8.8
	6	57 36.2	32.2	N12 32.9	9.2
	12	144 49.7	32.3	N11 38.1	9.5
	18	232 03.7	32.4	N10 41.3	9.8

PLANETS

VENUS Mer Pass h m	VENUS GHA ° ′	VENUS Mean Var/hr 14°+	VENUS Dec ° ′	VENUS Mean Var/hr ′	Day	JUPITER GHA ° ′	JUPITER Mean Var/hr 15°+	JUPITER Dec ° ′	JUPITER Mean Var/hr ′	JUPITER Mer Pass h m
09 56	211 04.2	59.4	S13 33.6	1.0	1 Sat	0 05.3	2.8	N21 22.8	0.0	00 00
09 57	210 50.6	59.4	S13 57.4	1.0	2 SUN	1 13.2	2.8	N21 21.8	0.0	23 51
09 58	210 36.8	59.4	S14 20.9	1.0	3 Mon	2 21.1	2.8	N21 20.8	0.0	23 46
09 59	210 22.7	59.4	S14 44.0	0.9	4 Tu	3 28.9	2.8	N21 19.8	0.0	23 42
10 00	210 08.4	59.4	S15 06.8	0.9	5 Wed	4 36.8	2.8	N21 18.8	0.0	23 37
10 01	209 53.7	59.4	S15 29.1	0.9	6 Th	5 44.6	2.8	N21 17.8	0.0	23 33
10 02	209 38.7	59.4	S15 51.2	0.9	7 Fri	6 52.4	2.8	N21 16.8	0.0	23 28
10 03	209 23.5	59.3	S16 12.8	0.9	8 Sat	8 00.2	2.8	N21 15.8	0.0	23 24
10 04	209 07.9	59.3	S16 34.0	0.9	9 SUN	9 08.0	2.8	N21 14.8	0.0	23 19
10 05	208 52.1	59.3	S16 54.8	0.8	10 Mon	10 15.7	2.8	N21 13.8	0.0	23 15
10 06	208 36.0	59.3	S17 15.1	0.8	11 Tu	11 23.3	2.8	N21 12.8	0.0	23 10
10 07	208 19.5	59.3	S17 35.0	0.8	12 Wed	12 30.9	2.8	N21 11.8	0.0	23 06
10 08	208 02.8	59.3	S17 54.4	0.8	13 Th	13 38.5	2.8	N21 10.8	0.0	23 01
10 09	207 45.7	59.3	S18 13.4	0.8	14 Fri	14 45.9	2.8	N21 09.9	0.0	22 57
10 11	207 28.4	59.3	S18 31.9	0.8	15 Sat	15 53.3	2.8	N21 08.9	0.0	22 52
10 12	207 10.8	59.3	S18 49.9	0.7	16 SUN	17 00.7	2.8	N21 07.9	0.0	22 48
10 13	206 52.9	59.2	S19 07.3	0.7	17 Mon	18 07.9	2.8	N21 07.0	0.0	22 43
10 14	206 34.7	59.2	S19 24.3	0.7	18 Tu	19 15.0	2.8	N21 06.0	0.0	22 39
10 15	206 16.3	59.2	S19 40.7	0.7	19 Wed	20 22.0	2.8	N21 05.1	0.0	22 34
10 17	205 57.5	59.2	S19 56.6	0.6	20 Th	21 29.0	2.8	N21 04.2	0.0	22 30
10 18	205 38.5	59.2	S20 11.9	0.6	21 Fri	22 35.8	2.8	N21 03.2	0.0	22 25
10 19	205 19.2	59.2	S20 26.6	0.6	22 Sat	23 42.4	2.8	N21 02.3	0.0	22 21
10 21	204 59.7	59.2	S20 40.8	0.6	23 SUN	24 49.0	2.8	N21 01.5	0.0	22 17
10 22	204 39.9	59.2	S20 54.4	0.5	24 Mon	25 55.4	2.8	N21 00.6	0.0	22 12
10 23	204 19.9	59.2	S21 07.4	0.5	25 Tu	27 01.7	2.8	N20 59.7	0.0	22 08
10 25	203 59.6	59.1	S21 19.7	0.5	26 Wed	28 07.9	2.8	N20 58.9	0.0	22 03
10 26	203 39.1	59.1	S21 31.5	0.5	27 Th	29 13.9	2.7	N20 58.1	0.0	21 59
10 27	203 18.3	59.1	S21 42.6	0.4	28 Fri	30 19.8	2.7	N20 57.3	0.0	21 55
10 29	202 57.4	59.1	S21 53.1	0.4	29 Sat	31 25.5	2.7	N20 56.5	0.0	21 50
10 30	202 36.2	59.1	S22 03.0	0.4	30 SUN	32 31.1	2.7	N20 55.8	0.0	21 46
10 32	202 14.9	59.1	S22 12.1	0.4	31 Mon	33 36.5	2.7	N20 55.0	0.0	21 42

VENUS, Av. Mag. –3.9
SHA December 5 136; 10 130; 15 123; 20 117; 25 110; 30 10

JUPITER, Av. Mag. –2.8
SHA December 5 290; 10 291; 15 292; 20 292; 25 293; 30 294

MARS Mer Pass h m	MARS GHA ° ′	MARS Mean Var/hr 15°+	MARS Dec ° ′	MARS Mean Var/hr ′	Day	SATURN GHA ° ′	SATURN Mean Var/hr 15°+	SATURN Dec ° ′	SATURN Mean Var/hr ′	SATURN Mer Pass h m
14 05	148 39.2	0.4	S24 09.8	0.2	1 Sat	215 12.5	2.2	S11 35.2	0.1	09 38
14 04	148 48.1	0.4	S24 06.1	0.2	2 SUN	216 05.4	2.2	S11 37.2	0.1	09 34
14 04	148 56.9	0.4	S24 02.1	0.2	3 Mon	216 58.3	2.2	S11 39.1	0.1	09 31
14 03	149 05.8	0.4	S23 57.8	0.2	4 Tu	217 51.2	2.2	S11 41.1	0.1	09 27
14 03	149 14.7	0.4	S23 53.3	0.2	5 Wed	218 44.2	2.2	S11 43.1	0.1	09 24
14 02	149 23.5	0.4	S23 48.5	0.2	6 Th	219 37.2	2.2	S11 45.0	0.1	09 20
14 01	149 32.4	0.4	S23 43.5	0.2	7 Fri	220 30.3	2.2	S11 46.9	0.1	09 17
14 01	149 41.3	0.4	S23 38.1	0.2	8 Sat	221 23.4	2.2	S11 48.8	0.1	09 13
14 00	149 50.2	0.4	S23 32.5	0.2	9 SUN	222 16.5	2.2	S11 50.6	0.1	09 10
14 00	149 59.2	0.4	S23 26.6	0.3	10 Mon	223 09.7	2.2	S11 52.5	0.1	09 06
13 59	150 08.1	0.4	S23 20.5	0.3	11 Tu	224 03.0	2.2	S11 54.3	0.1	09 02
13 59	150 17.1	0.4	S23 14.1	0.3	12 Wed	224 56.3	2.2	S11 56.1	0.1	08 59
13 58	150 26.2	0.4	S23 07.4	0.3	13 Th	225 49.6	2.2	S11 57.9	0.1	08 55
13 57	150 35.3	0.4	S23 00.4	0.3	14 Fri	226 43.0	2.2	S11 59.6	0.1	08 52
13 57	150 44.4	0.4	S22 53.2	0.3	15 Sat	227 36.5	2.2	S12 01.4	0.1	08 48
13 56	150 53.6	0.4	S22 45.8	0.3	16 SUN	228 30.0	2.2	S12 03.1	0.1	08 45
13 55	151 02.8	0.4	S22 38.1	0.3	17 Mon	229 23.5	2.2	S12 04.7	0.1	08 41
13 55	151 12.1	0.4	S22 30.1	0.3	18 Tu	230 17.2	2.2	S12 06.4	0.1	08 38
13 54	151 21.4	0.4	S22 21.8	0.3	19 Wed	231 10.8	2.2	S12 08.0	0.1	08 34
13 54	151 30.8	0.4	S22 13.4	0.4	20 Th	232 04.6	2.2	S12 09.6	0.1	08 30
13 53	151 40.3	0.4	S22 04.6	0.4	21 Fri	232 58.4	2.2	S12 11.2	0.1	08 27
13 52	151 49.9	0.4	S21 55.6	0.4	22 Sat	233 52.2	2.2	S12 12.8	0.1	08 23
13 52	151 59.5	0.4	S21 46.4	0.4	23 SUN	234 46.1	2.3	S12 14.3	0.1	08 20
13 51	152 09.2	0.4	S21 36.9	0.4	24 Mon	235 40.1	2.3	S12 15.8	0.1	08 16
13 50	152 19.0	0.4	S21 27.2	0.4	25 Tu	236 34.2	2.3	S12 17.3	0.1	08 12
13 50	152 28.8	0.4	S21 17.2	0.4	26 Wed	237 28.3	2.3	S12 18.8	0.1	08 09
13 49	152 38.8	0.4	S21 07.0	0.4	27 Th	238 22.5	2.3	S12 20.2	0.1	08 05
13 48	152 48.8	0.4	S20 56.6	0.4	28 Fri	239 16.7	2.3	S12 21.6	0.1	08 02
13 48	152 58.9	0.4	S20 45.9	0.5	29 Sat	240 11.0	2.3	S12 23.0	0.1	07 58
13 47	153 09.1	0.4	S20 35.0	0.5	30 SUN	241 05.4	2.3	S12 24.3	0.1	07 54
13 46	153 19.4	0.4	S20 23.9	0.5	31 Mon	241 59.9	2.3	S12 25.6	0.1	07 51

MARS, Av. Mag. +1.2
SHA December 5 75; 10 71; 15 67; 20 63; 25 58; 30 54

SATURN, Av. Mag. +0.7
SHA December 5 145; 10 144; 15 144; 20 143; 25 143; 30 142

STARS

No.	Name	Mag	Transit h m	Dec ° ′	SHA ° ′
	Oh GMT December 1				
Ψ	ARIES	–	19 16	–	–
1	Alpheratz	2.1	19 25	N29 10.0	357 43.8
2	Ankaa	2.4	19 43	S42 14.3	353 16.0
3	Schedar	2.2	19 57	N56 36.9	349 40.7
4	Diphda	2.0	20 00	S17 55.0	348 56.1
5	Achernar	0.5	20 54	S57 10.5	335 26.7
6	POLARIS	2.0	22 05	N89 19.4	317 35.3
7	Hamal	2.0	21 23	N23 31.5	328 00.9
8	Acamar	3.2	22 14	S40 15.3	315 18.3
9	Menkar	2.5	22 18	N 4 08.4	314 15.1
10	Mirfak	1.8	22 41	N49 54.5	308 40.4
11	Aldebaran	0.9	23 52	N16 32.0	290 49.4
12	Rigel	0.1	0 34	S 8 11.3	281 12.0
13	Capella	0.1	0 37	N46 00.5	280 34.4
14	Bellatrix	1.6	0 45	N 6 21.6	278 32.0
15	Elnath	1.7	0 46	N28 37.0	278 12.7
16	Alnilam	1.7	0 56	S 1 11.7	275 46.4
17	Betelgeuse	0.1–1.2	1 15	N 7 24.4	271 01.3
18	Canopus	–0.7	1 43	S52 42.2	263 55.7
19	Sirius	–1.5	2 04	S16 44.2	258 33.7
20	Adhara	1.5	2 18	S28 59.5	255 12.4
21	Castor	1.6	2 54	N31 51.3	246 08.1
22	Procyon	0.4	2 59	N 5 11.3	244 59.8
23	Pollux	1.1	3 05	N27 59.4	243 27.8
24	Avior	1.9	3 41	S59 33.0	234 17.6
25	Suhail	2.2	4 27	S43 29.1	222 52.4
26	Miaplacidus	1.7	4 32	S69 46.1	221 39.1
27	Alphard	2.0	4 46	S 8 43.0	217 56.2
28	Regulus	1.4	5 27	N11 54.0	207 43.8
29	Dubhe	1.8	6 22	N61 40.4	193 52.1
30	Denebola	2.1	7 08	N14 29.8	182 34.0
31	Gienah	2.6	7 34	S17 36.8	175 52.7
32	Acrux	1.3	7 45	S63 10.0	173 09.8
33	Gacrux	1.6	7 50	S57 10.9	172 01.4
34	Mimosa	1.3	8 06	S59 45.3	167 52.5
35	Alioth	1.8	8 12	N55 53.1	166 21.2
36	Spica	1.0	8 43	S11 13.7	158 31.8
37	Alkaid	1.9	9 06	N49 14.7	152 59.5
38	Hadar	0.6	9 22	S60 25.9	148 48.7
39	Menkent	2.1	9 25	S36 25.8	148 08.2
40	Arcturus	0.0	9 34	N19 06.8	145 56.3
41	Rigil Kent	–0.3	9 58	S60 53.0	139 52.6
42	Zuben'ubi	2.8	10 09	S16 05.6	137 06.1
43	Kochab	2.1	10 08	N74 05.9	137 21.0
44	Alphecca	2.2	10 52	N26 40.3	126 11.6
45	Antares	1.0	11 47	S26 27.5	112 27.0
46	Atria	1.9	12 07	S69 02.8	107 29.7
47	Sabik	2.4	12 28	S15 44.3	102 13.3
48	Shaula	1.6	12 51	S37 06.6	96 22.8
49	Rasalhague	2.1	12 52	N12 33.2	96 07.1
50	Eltanin	2.2	13 14	N51 29.4	90 46.8
51	Kaus Aust.	1.9	13 42	S34 22.6	83 44.7
52	Vega	0.0	13 54	N38 48.0	80 39.6
53	Nunki	2.0	14 13	S26 16.7	75 59.1
54	Altair	0.8	15 08	N 8 54.4	62 08.9
55	Peacock	1.9	15 43	S56 41.6	53 20.3
56	Deneb	1.3	15 58	N45 20.0	49 32.0
57	Enif	2.4	17 01	N 9 56.3	33 47.6
58	Al Na'ir	1.7	17 25	S46 54.0	27 44.4
59	Fomalhaut	1.2	18 14	S29 33.2	15 24.4
60	Markab	2.5	18 21	N15 16.7	13 38.7

SUN

Yr	Day of Mth	Day of Week	Transit h m	Semi-Diam	Lat 52°N Twilight h m	Lat 52°N Sunrise h m	Lat 52°N Sunset h m	Lat 52°N Twilight h m
336	1	Sat	11 49	16.2	07 06	07 46	15 52	16 32
337	2	Sun	11 50	16.3	07 08	07 47	15 51	16 31
338	3	Mon	11 50	16.3	07 09	07 49	15 51	16 31
339	4	Tu	11 50	16.3	07 10	07 50	15 50	16 30
340	5	Wed	11 51	16.3	07 11	07 51	15 50	16 30
341	6	Th	11 51	16.3	07 12	07 53	15 49	16 30
342	7	Fri	11 52	16.3	07 14	07 54	15 49	16 29
343	8	Sat	11 52	16.3	07 15	07 55	15 49	16 29
344	9	Sun	11 53	16.3	07 16	07 56	15 49	16 29
345	10	Mon	11 53	16.3	07 17	07 57	15 48	16 29
346	11	Tu	11 53	16.3	07 18	07 58	15 48	16 29
347	12	Wed	11 54	16.3	07 19	07 59	15 48	16 29
348	13	Th	11 54	16.3	07 19	08 00	15 48	16 29
349	14	Fri	11 55	16.3	07 20	08 01	15 48	16 29
350	15	Sat	11 55	16.3	07 21	08 02	15 49	16 29
351	16	Sun	11 56	16.3	07 22	08 03	15 49	16 30
352	17	Mon	11 56	16.3	07 23	08 04	15 49	16 30
353	18	Tu	11 57	16.3	07 23	08 04	15 49	16 30
354	19	Wed	11 57	16.3	07 24	08 05	15 50	16 31
355	20	Th	11 58	16.3	07 24	08 05	15 50	16 31
356	21	Fri	11 58	16.3	07 25	08 06	15 51	16 32
357	22	Sat	11 59	16.3	07 25	08 06	15 51	16 32
358	23	Sun	11 59	16.3	07 26	08 07	15 52	16 33
359	24	Mon	12 00	16.3	07 26	08 07	15 52	16 33
360	25	Tu	12 00	16.3	07 27	08 08	15 53	16 34
361	26	Wed	12 01	16.3	07 27	08 08	15 54	16 35
362	27	Th	12 01	16.3	07 27	08 08	15 55	16 35
363	28	Fri	12 02	16.3	07 27	08 08	15 56	16 36
364	29	Sat	12 02	16.3	07 27	08 08	15 56	16 37
365	30	Sun	12 03	16.3	07 28	08 08	15 57	16 38
366	31	Mon	12 03	16.3	07 28	08 08	15 58	16 39

Lat Corr to Sunrise, Sunset etc.

Lat °	Twilight h m	Sunrise h m	Sunset h m	Twilight h m
N70	+2 26	SBH	SBH	–2 26
68	+1 52	SBH	SBH	–1 52
66	+1 27	+2 25	–2 26	–1 27
64	+1 08	+1 44	–1 45	–1 08
62	+0 52	+1 16	–1 17	–0 52
N60	+0 38	+0 55	–0 56	–0 38
58	+0 27	+0 38	–0 39	–0 27
56	+0 17	+0 23	–0 24	–0 17
54	+0 08	+0 11	–0 11	–0 08
50	–0 08	–0 10	+0 10	+0 08
N45	–0 23	–0 30	+0 30	+0 23
40	–0 37	–0 47	+0 47	+0 37
35	–0 48	–1 01	+1 01	+0 48
30	–0 59	–1 13	+1 13	+0 59
20	–1 17	–1 34	+1 34	+1 18
N10	–1 34	–1 53	+1 53	+1 35
0	–1 51	–2 10	+2 10	+1 52
S10	–2 09	–2 27	+2 27	+2 10
20	–2 30	–2 46	+2 46	+2 31
30	–2 54	–3 08	+3 08	+2 51
S35	–3 09	–3 21	+3 21	+3 16
40	–3 27	–3 36	+3 36	+3 29
45	–3 49	–3 53	+3 54	+3 51
S50	–4 18	–4 16	+4 16	+4 20

NOTES
The corrections to sunrise etc. are for middle of December. SBH means Sun below Horizon.

MOON

Yr	Day of Mth	Day of Week	Age days	Transit (Upper) h m	Diff m	Semi-diam ′	Hor Par ′	Lat 52°N Moonrise h m	Lat 52°N Moonset h m
336	1	Sat	18	01 44	47	14.8	54.3	18 34	09 46
337	2	Sun	19	02 31	47	14.9	54.6	19 36	10 19
338	3	Mon	20	03 18	46	15.0	55.0	20 42	10 48
339	4	Tu	21	04 04	45	15.1	55.5	21 50	11 12
340	5	Wed	22	04 49	46	15.3	56.2	23 00	11 35
341	6	Th	23	05 35	47	15.5	57.0	– –	11 56
342	7	Fri	24	06 22	48	15.8	57.8	00 13	12 18
343	8	Sat	25	07 10	52	16.0	58.7	01 27	12 41
344	9	Sun	26	08 02	55	16.3	59.7	02 45	13 08
345	10	Mon	27	08 57	60	16.5	60.4	04 06	13 40
346	11	Tu	28	09 57	62	16.6	61.1	05 26	14 21
347	12	Wed	29	10 59	65	16.7	61.4	06 43	15 13
348	13	Th	00	12 04	63	16.7	61.4	07 51	16 17
349	14	Fri	01	13 07	60	16.6	61.0	08 47	17 31
350	15	Sat	02	14 07	56	16.5	60.4	09 31	18 51
351	16	Sun	03	15 03	52	16.2	59.5	10 05	20 10
352	17	Mon	04	15 55	49	16.0	58.6	10 33	21 28
353	18	Tu	05	16 44	46	15.7	57.6	10 57	22 42
354	19	Wed	06	17 30	45	15.4	56.7	11 19	23 54
355	20	Th	07	18 15	44	15.2	55.9	11 40	– –
356	21	Fri	08	18 59	45	15.0	55.2	12 01	01 03
357	22	Sat	09	19 44	45	14.9	54.7	12 24	02 10
358	23	Sun	10	20 29	47	14.8	54.3	12 50	03 16
359	24	Mon	11	21 16	47	14.7	54.1	13 20	04 19
360	25	Tu	12	22 03	49	14.7	54.0	13 56	05 19
361	26	Wed	13	22 52	48	14.7	54.0	14 39	06 15
362	27	Th	14	23 40	48	14.7	54.1	15 29	07 04
363	28	Fri	15	24 28	–	14.8	54.3	16 26	07 46
364	29	Sat	16	00 28	48	14.9	54.6	17 28	08 22
365	30	Sun	17	01 16	46	15.0	54.9	18 33	08 52
366	31	Mon	18	02 02	46	15.1	55.3	19 41	09 18

Phases of the Moon

	Phase	d	h	m
☽	Last Quarter	6	15	31
●	New Moon	13	08	42
☾	First Quarter	20	05	19
○	Full Moon	28	10	21

	d	h
Perigee	12	23
Apogee	25	21

POLARIS (POLE STAR) TABLE – 2012

FOR DETERMINING THE LATITUDE FROM A SEXTANT ALTITUDE

LHA Aries	Q	LHA Aries	Q	LHA Aries	Q	LHA Aries	Q	LHA Aries	Q	LHA Aries	Q	LHA Aries	Q	LHA Aries	Q
° ′		° ′		° ′		° ′		° ′		° ′		° ′		° ′	
358 20		**85 31**		**122 26**		**155 25**		**206 49**		**283 45**		**317 11**		**352 35**	
	– 30		– 29		– 6		+ 17		+ 40		+ 19		– 4		– 27
0 23		**87 30**		**123 51**		**156 57**		**213 52**		**285 20**		**318 35**		**354 26**	
	– 31		– 28		– 5		+ 18		+ 41		+ 18		– 5		– 28
2 31		**89 25**		**125 16**		**158 31**		**229 59**		**286 54**		**320 00**		**356 21**	
	– 32		– 27		– 4		+ 19		+ 40		+ 17		– 6		– 29
4 45		**91 16**		**126 40**		**160 06**		**237 02**		**288 26**		**321 25**		**358 20**	
	– 33		– 26		– 3		+ 20		+ 39		+ 16		– 7		– 30
7 07		**93 05**		**128 04**		**161 43**		**241 45**		**289 58**		**322 50**		**0 23**	
	– 34		– 25		– 2		+ 21		+ 38		+ 15		– 8		– 31
9 38		**94 51**		**129 28**		**163 21**		**245 35**		**291 28**		**324 16**		**2 31**	
	– 35		– 24		– 1		+ 22		+ 37		+ 14		– 9		– 32
12 19		**96 34**		**130 53**		**165 01**		**248 54**		**292 58**		**325 42**		**4 45**	
	– 36		– 23		0		+ 23		+ 36		+ 13		– 10		– 33
15 16		**98 15**		**132 18**		**166 43**		**251 52**		**294 27**		**327 08**		**7 07**	
	– 37		– 22		+ 1		+ 24		+ 35		+ 12		– 11		– 34
18 32		**99 55**		**133 42**		**168 28**		**254 35**		**295 55**		**328 35**		**9 38**	
	– 38		– 21		+ 2		+ 25		+ 34		+ 11		– 12		– 35
22 19		**101 33**		**135 06**		**170 14**		**257 07**		**297 22**		**330 03**		**12 19**	
	– 39		– 20		+ 3		+ 26		+ 33		+ 10		– 13		– 36
26 59		**103 09**		**136 30**		**172 03**		**259 30**		**298 49**		**331 32**		**15 16**	
	– 40		– 19		+ 4		+ 27		+ 32		+ 9		– 14		– 37
33 58		**104 43**		**137 55**		**173 56**		**261 46**		**300 15**		**333 01**		**18 32**	
	– 41		– 18		+ 5		+ 28		+ 31		+ 8		– 15		– 38
49 53		**106 17**		**139 19**		**175 52**		**263 55**		**301 41**		**334 31**		**22 19**	
	– 40		– 17		+ 6		+ 29		+ 30		+ 7		– 16		– 39
56 52		**107 49**		**140 44**		**177 51**		**266 00**		**303 07**		**336 02**		**26 59**	
	– 39		– 16		+ 7		+ 30		+ 29		+ 6		– 17		– 40
61 32		**109 20**		**142 10**		**179 56**		**267 59**		**304 32**		**337 34**		**33 58**	
	– 38		– 15		+ 8		+ 31		+ 28		+ 5		– 18		– 41
65 19		**110 50**		**143 36**		**182 05**		**269 55**		**305 56**		**339 08**		**49 53**	
	– 37		– 14		+ 9		+ 32		+ 27		+ 4		– 19		– 40
68 35		**112 19**		**145 02**		**184 21**		**271 48**		**307 21**		**340 42**		**56 52**	
	– 36		– 13		+ 10		+ 33		+ 26		+ 3		– 20		– 39
71 32		**113 48**		**146 29**		**186 44**		**273 37**		**308 45**		**342 18**		**61 32**	
	– 35		– 12		+ 11		+ 34		+ 25		+ 2		– 21		– 38
74 13		**115 16**		**147 56**		**189 16**		**275 23**		**310 09**		**343 56**		**65 19**	
	– 34		– 11		+ 12		+ 35		+ 24		+ 1		– 22		– 37
76 44		**116 43**		**149 24**		**191 59**		**277 08**		**311 33**		**345 36**		**68 35**	
	– 33		– 10		+ 13		+ 36		+ 23		0		– 23		– 36
79 06		**118 09**		**150 53**		**194 57**		**278 50**		**312 58**		**347 17**		**71 32**	
	– 32		– 9		+ 14		+ 37		+ 22		– 1		– 24		– 35
81 20		**119 35**		**152 23**		**198 16**		**280 30**		**314 23**		**349 00**		**74 13**	
	– 31		– 8		+ 15		+ 38		+ 21		– 2		– 25		– 34
83 28		**121 01**		**153 53**		**202 06**		**282 08**		**315 47**		**350 46**		**76 44**	
	– 30		– 7		+ 16		+ 39		+ 20		– 3		– 26		– 33
85 31		**122 26**		**155 25**		**206 49**		**283 45**		**317 11**		**352 35**		**79 06**	

In critical cases, ascend

Q, which does not include refraction, is to be applied to the corrected sextant altitude of Polaris.
Polaris: Mag. 2.1, SHA 318 04, Dec N89 19.1

The Pole Star table, above, changes annually and is the last of the ephemeral tables.

ECLIPSES 2012

There are four eclipses, two solar and two lunar. All times are UT (GMT).

20th May
Annular eclipse of the Sun, starting at 2056 and ending at 0249 on 21st May. Visible in Asia, Pacific Ocean, North America.

4th June
Partial eclipse of the Moon, starting at 0847 and ending at 1320. Visible in Asia, Antarctica, North America, South America and Pacific Ocean.

13th November
Total eclipse of the Sun, starting at 1938 and ending at 0046 on 14th November. Visible in Australia, New Zealand, south Pacific Ocean and southern part of South America. Totality passes through extreme north of Australia and thence in Pacific Ocean.

28th November
Penumbral eclipse of the Moon, starting at 1213 and ending at 1653. Visible in eastern Africa, Europe, Asia, Australia, North America, Greenland, Indian Ocean, Pacific Ocean.

USABILITY IN 2013
The almanac may be used for Sun and Stars in 2013. It cannot be used for Moon or Planets.

For the Sun, take out the GHA and declination for the same date but, for January and February, for a time 18 hours 12 minutes later and, for March to December, for a time 5 hours 48 minutes earlier than the time of observation. In both cases add 87 degrees to the GHA so obtained.

For the Stars calculate the GHA and declination for the same date and the same time but, for January and February, add 44.0 minutes of arc and, for March to December, subtract 15.1 minutes of arc from the GHA so found.

In all cases the error is unlikely to exceed 4.0 minutes of arc.

SUN ALTITUDE TOTAL CORRECTION TABLE

FOR CORRECTING THE OBSERVED ALTITUDE OF THE SUN'S LOWER LIMB

ALWAYS ADDITIVE

Height of the eye above the sea. Top line metres, lower line feet

Obs Alt	0.9 3	1.8 6	2.4 8	3.0 10	3.7 12	4.3 14	4.9 16	5.5 18	6.0 20	7.6 25	9.0 30	12 40	15 50	18 60	21 70	24 80
°	'	'	'	'	'	'	'	'	'	'	'	'	'	'	'	'
9	8.6	8.0	7.6	7.2	6.9	6.6	6.4	6.2	5.9	5.4	4.9	4.1	3.4	2.7	2.1	1.5
10	9.1	8.5	8.1	7.9	7.5	7.2	7.0	6.7	6.6	6.0	5.5	4.7	3.9	3.3	2.7	2.1
11	9.6	9.0	8.6	8.3	8.0	7.7	7.4	7.2	7.0	6.4	6.0	5.2	4.4	3.7	3.1	2.5
12	10.0	9.4	9.0	8.7	8.4	8.1	7.8	7.6	7.4	6.8	6.4	5.6	4.8	4.1	3.5	2.9
13	10.3	9.7	9.3	9.0	8.7	8.4	8.2	7.9	7.7	7.2	6.7	5.9	5.2	4.5	3.9	3.3
14	10.6	10.0	9.6	9.3	9.0	8.7	8.5	8.2	8.0	7.5	7.0	6.2	5.5	4.8	4.2	3.6
15	10.9	10.2	9.9	9.5	9.2	9.0	8.7	8.5	8.2	7.7	7.2	6.4	5.7	5.0	4.4	3.8
16	11.1	10.5	10.1	9.7	9.5	9.2	8.9	8.7	8.5	7.9	7.5	6.7	5.9	5.2	4.6	4.1
17	11.3	10.7	10.3	10.0	9.7	9.4	9.1	8.9	8.7	8.2	7.7	6.9	6.1	5.5	4.9	4.3
18	11.5	10.8	10.5	10.1	9.9	9.6	9.3	9.1	8.9	8.3	7.9	7.0	6.3	5.6	5.0	4.5
19	11.6	11.0	10.6	10.3	10.0	9.7	9.5	9.2	9.0	8.5	8.0	7.2	6.5	5.8	5.2	4.6
20	11.8	11.2	10.8	10.4	10.2	9.9	9.6	9.4	9.2	8.6	8.2	7.4	6.6	5.9	5.3	4.8
21	11.9	11.3	10.9	10.6	10.3	10.0	9.8	9.5	9.3	8.8	8.3	7.5	6.8	6.1	5.5	4.9
22	12.0	11.4	11.0	10.7	10.4	10.1	9.9	9.7	9.4	8.9	8.4	7.6	6.9	6.2	5.6	5.0
23	12.1	11.5	11.1	10.8	10.5	10.2	10.0	9.8	9.5	9.0	8.5	7.7	7.0	6.3	5.7	5.1
24	12.2	11.6	11.2	10.9	10.6	10.3	10.1	9.9	9.6	9.1	8.6	7.8	7.1	6.4	5.8	5.2
25	12.3	11.7	11.3	11.0	10.7	10.4	10.2	10.0	9.7	9.2	8.7	7.9	7.2	6.5	5.9	5.3
26	12.4	11.8	11.4	11.1	10.8	10.5	10.3	10.1	9.8	9.3	8.8	8.0	7.3	6.6	6.0	5.4
27	12.5	11.9	11.5	11.2	10.9	10.6	10.4	10.1	9.9	9.4	8.9	8.1	7.4	6.7	6.1	5.5
28	12.6	12.0	11.6	11.3	11.0	10.7	10.4	10.2	10.0	9.5	9.0	8.2	7.4	6.8	6.2	5.6
30	12.7	12.1	11.7	11.4	11.1	10.8	10.6	10.4	10.1	9.6	9.1	8.3	7.6	6.9	6.3	5.7
32	12.9	12.2	11.9	11.5	11.2	11.0	10.7	10.5	10.2	9.7	9.3	8.4	7.7	7.0	6.4	5.8
34	13.0	12.3	12.0	11.6	11.3	11.1	10.8	10.6	10.3	9.8	9.4	8.5	7.8	7.1	6.5	5.9
36	13.1	12.4	12.1	11.7	11.4	11.2	10.9	10.7	10.4	9.9	9.5	8.6	7.9	7.2	6.6	6.0
38	13.2	12.5	12.1	11.8	11.5	11.2	11.0	10.8	10.5	10.0	9.5	8.7	8.0	7.3	6.7	6.1
40	13.3	12.6	12.2	11.9	11.6	11.3	11.1	10.8	10.6	10.1	9.6	8.8	8.1	7.4	6.8	6.2
42	13.4	12.7	12.3	12.0	11.7	11.4	11.2	10.9	10.7	10.2	9.7	8.9	8.2	7.5	6.9	6.3
44	13.4	12.7	12.4	12.0	11.7	11.5	11.2	11.0	10.7	10.2	9.8	8.9	8.2	7.5	6.9	6.3
46	13.5	12.8	12.4	12.1	11.8	11.5	11.3	11.0	10.8	10.3	9.8	9.0	8.3	7.6	7.0	6.4
48	13.6	12.9	12.5	12.2	11.9	11.6	11.3	11.1	10.9	10.4	9.9	9.1	8.3	7.7	7.1	6.4
50	13.6	12.9	12.5	12.2	11.9	11.6	11.4	11.1	10.9	10.4	9.9	9.1	8.4	7.7	7.1	6.5
52	13.6	13.0	12.6	12.3	12.0	11.7	11.4	11.2	11.0	10.5	10.0	9.2	8.4	7.8	7.2	6.5
54	13.7	13.0	12.6	12.3	12.0	11.7	11.5	11.3	11.0	10.5	10.0	9.2	8.5	7.8	7.2	6.6
56	13.7	13.1	12.7	12.4	12.1	11.8	11.5	11.3	11.1	10.6	10.1	9.3	8.5	7.9	7.3	6.7
58	13.8	13.1	12.7	12.4	12.1	11.8	11.6	11.3	11.1	10.6	10.1	9.3	8.6	7.9	7.3	6.8
60	13.8	13.1	12.8	12.4	12.1	11.9	11.6	11.4	11.1	10.6	10.2	9.3	8.6	7.9	7.3	6.8
62	13.9	13.2	12.8	12.5	12.2	11.9	11.7	11.4	11.2	10.7	10.2	9.4	8.7	8.0	7.4	6.8
64	13.9	13.2	12.8	12.5	12.2	11.9	11.7	11.5	11.2	10.7	10.2	9.4	8.7	8.0	7.4	6.9
66	14.0	13.2	12.9	12.5	12.3	12.0	11.7	11.5	11.3	10.7	10.3	9.5	8.7	8.1	7.5	7.0
70	14.1	13.3	12.9	12.6	12.3	12.0	11.8	11.6	11.3	10.8	10.3	9.5	8.8	8.1	7.5	7.0
80	14.2	13.5	13.1	12.8	12.5	12.2	11.9	11.7	11.5	11.0	10.5	9.7	8.9	8.3	7.7	7.1
90	14.3	13.6	13.2	12.9	12.6	12.3	12.1	11.9	11.6	11.1	10.6	9.8	9.1	8.4	7.8	7.2

MONTHLY CORRECTION

Jan	Feb	Mar	Apr	May	Jun	Jul	Aug	Sep	Oct	Nov	Dec
+0.3'	+0.2'	+0.1'	0.01'	–0.11'	–0.2'	–0.2'	–0.2'	–0.11'	+0.1'	+0.2'	+0.3'

STAR OR PLANET ALTITUDE TOTAL CORRECTION TABLE

ALWAYS SUBRTRACTIVE

Height of the eye above the sea. Top line metres, lower line feet

Obs Alt	1.5 5	3.0 10	4.6 15	6.0 20	7.6 25	9.0 30	10.7 35	12 40	13.7 45	15 50	16.8 55	18 60	21.3 70
°	'	'	'	'	'	'	'	'	'	'	'	'	'
9	8.0	8.9	9.6	10.3	10.7	11.2	11.6	12.0	12.4	12.8	13.1	13.5	14.1
10	7.4	8.4	9.1	9.7	10.2	10.6	11.1	11.5	11.8	12.2	12.5	12.9	13.5
11	7.0	7.9	8.6	9.2	9.7	10.2	10.6	11.0	11.4	11.8	12.0	12.4	13.0
12	6.6	7.5	8.2	8.8	9.3	9.8	10.2	10.6	11.0	11.4	11.6	12.0	12.6
13	6.2	7.2	7.9	8.4	9.0	9.4	9.9	10.3	10.6	11.0	11.3	11.6	12.3
14	5.9	6.9	7.6	8.1	8.6	9.2	9.6	10.0	10.3	10.7	11.0	11.3	12.0
15	5.7	6.6	7.3	7.9	8.4	8.9	9.3	9.7	10.1	10.4	10.8	11.1	11.7
16	5.5	6.4	7.1	7.7	8.2	8.7	9.1	9.5	9.9	10.2	10.5	10.9	11.5
17	5.3	6.2	6.9	7.5	8.0	8.5	8.9	9.3	9.7	10.0	10.3	10.7	11.3
18	5.1	6.0	6.7	7.3	7.8	8.3	8.7	9.1	9.5	9.8	10.2	10.5	11.1
19	4.9	5.8	6.5	7.1	7.6	8.1	8.5	8.9	9.3	9.7	10.0	10.3	11.0
20	4.8	5.7	6.4	7.0	7.5	8.0	8.4	8.8	9.2	9.6	9.9	10.2	10.8
25	4.2	5.1	5.8	6.4	6.9	7.4	7.8	8.2	8.6	9.0	9.3	9.6	10.2
30	3.8	4.7	5.4	6.0	6.5	7.0	7.4	7.8	8.2	8.6	8.9	9.2	9.8
35	3.5	4.4	5.1	5.7	6.3	6.7	7.2	7.6	7.9	8.3	8.6	8.9	9.5
40	3.3	4.2	4.9	5.5	6.0	6.5	6.9	7.3	7.7	8.1	8.4	8.7	9.3
50	3.0	3.9	4.6	5.2	5.7	6.2	6.6	7.0	7.4	7.7	8.1	8.4	9.0
60	2.7	3.6	4.4	4.9	5.5	5.9	6.4	6.8	7.1	7.5	7.8	8.1	8.8
70	2.5	3.4	4.1	4.7	5.3	5.7	6.2	6.6	6.9	7.3	7.6	7.9	8.6
80	2.3	3.3	4.0	4.6	5.1	5.5	6.0	6.4	6.7	7.1	7.4	7.8	8.4
90	2.2	3.1	3.8	4.4	4.9	5.4	5.8	6.2	6.6	6.9	7.3	7.6	8.2

The above table contains the combined effects of Dip of the Horizon and Refraction and is therefore a total correction table for a Star or Planet. It is always subtractive.

45

MOON ALTITUDE TOTAL CORRECTION TABLE

Upper Limb ADD/SUBRACT									Lower Limb ADD								
	Horizontal Parallax									Horizontal Parallax							
Obs Alt	54'	55'	56'	57'	58'	59'	60'	61'	Obs Alt	54'	55'	56'	57'	58'	59'	60'	61'
10	23.4	24.0	24.6	25.5	26.0	26.7	27.5	28.3	**10**	52.7	54.0	55.3	56.5	57.7	59.0	60.2	61.5
12	23.8	24.6	25.2	26.0	26.5	27.2	28.0	28.7	**12**	53.2	54.5	55.7	57.0	58.4	59.5	60.7	62.0
14	24.0	24.8	25.4	26.1	26.7	27.5	28.3	29.0	**14**	53.5	54.7	56.0	57.3	58.5	59.8	61.0	62.3
16	24.0	24.8	25.5	26.1	26.7	27.5	28.3	28.8	**16**	53.5	54.6	56.0	57.3	58.5	59.8	61.0	62.2
18	23.8	24.6	25.2	26.0	26.5	27.3	28.0	28.6	**18**	53.4	54.6	55.7	57.0	58.4	59.5	60.6	62.0
20	23.6	24.2	25.0	25.5	26.2	27.0	27.5	28.2	**20**	53.0	54.4	55.5	56.8	58.0	59.0	60.4	61.5
22	23.2	23.8	24.6	25.0	25.7	26.5	27.0	27.8	**22**	52.5	53.7	55.0	56.3	57.5	58.8	60.0	61.0
24	22.7	23.2	24.0	24.5	25.3	25.8	26.5	27.0	**24**	52.0	53.3	54.5	55.5	56.7	58.0	59.4	60.5
26	22.0	22.6	23.4	24.0	24.5	25.0	25.7	26.5	**26**	51.5	52.5	53.7	55.0	56.3	57.5	58.0	59.8
28	21.4	22.0	22.6	23.3	23.8	24.5	25.0	25.5	**28**	50.7	52.0	53.0	54.4	55.5	56.5	57.8	59.0
30	20.6	21.2	21.8	22.3	23.0	23.5	24.3	24.7	**30**	50.0	51.0	52.3	53.5	54.5	55.7	57.0	58.0
32	19.8	20.2	21.0	21.3	22.0	22.5	23.2	23.7	**32**	49.3	50.4	51.3	52.5	53.7	54.8	56.0	57.0
34	19.0	19.4	20.0	20.5	21.0	21.5	22.2	22.7	**34**	48.3	49.5	50.5	51.5	52.7	53.7	55.0	56.0
36	18.0	18.4	19.0	19.5	20.0	20.5	21.0	21.7	**36**	47.3	48.5	49.5	50.5	51.7	52.7	54.0	55.0
38	16.8	17.4	17.8	18.5	19.0	19.5	20.0	20.4	**38**	46.4	47.4	48.5	49.5	50.5	51.5	52.7	53.8
40	15.8	16.2	16.8	17.3	17.7	18.2	18.8	19.2	**40**	45.3	46.3	47.3	48.3	49.5	50.5	51.5	52.5
42	14.7	15.2	15.6	16.0	16.5	17.0	17.5	18.0	**42**	44.0	45.0	46.0	47.0	48.0	49.0	50.0	51.0
44	13.5	13.8	14.2	14.6	15.0	15.5	16.0	16.5	**44**	42.7	43.7	44.7	45.7	46.7	47.7	48.7	49.7
46	12.0	12.6	13.0	13.4	13.8	14.2	14.5	15.0	**46**	41.5	42.5	43.5	44.5	45.5	46.5	47.5	48.5
48	10.5	11.2	11.6	12.0	12.4	12.8	13.2	13.5	**48**	40.2	41.2	42.2	43.0	44.0	45.0	46.0	47.0
50	9.3	10.0	10.2	10.6	11.0	11.3	11.7	12.0	**50**	39.0	40.0	41.0	41.8	42.6	43.6	44.5	45.5
52	8.0	8.4	8.6	9.2	9.5	9.7	10.0	10.5	**52**	37.5	38.5	39.3	40.2	41.0	42.0	42.8	43.7
54	6.7	6.8	7.2	7.5	7.8	8.2	8.5	8.7	**54**	36.0	37.0	38.0	38.8	39.5	40.5	41.3	42.0
56	5.2	5.5	5.6	6.0	6.3	6.5	7.0	7.0	**56**	34.5	35.5	36.2	37.0	38.0	38.7	39.5	40.5
58	3.7	3.7	4.2	4.5	4.5	5.0	5.0	5.5	**58**	33.0	34.0	34.7	35.5	36.3	37.0	38.0	38.8
60	2.0	2.2	2.5	2.7	3.0	3.2	3.5	3.5	**60**	31.5	32.4	33.0	34.0	34.5	35.5	36.0	37.0
62	+0.5	+0.7	+0.8	+1.0	+1.2	+1.5	+1.5	+1.7	**62**	30.0	30.5	31.5	32.0	33.0	33.5	34.5	35.0
64	–1.2	–1.0	–1.0	–0.8	–0.6	–0.5	–0.3	–0.1	**64**	28.3	29.0	29.6	30.5	31.0	31.8	32.5	33.3
66	3.0	2.8	2.6	2.5	2.4	2.3	2.0	2.0	**66**	26.5	27.3	28.0	28.5	29.3	30.0	30.7	31.5
68	4.5	4.5	4.4	4.3	4.2	4.0	4.0	4.0	**68**	25.0	25.5	26.3	26.8	27.5	28.0	28.8	29.5
70	6.3	6.2	6.2	6.1	6.0	6.0	5.8	5.8	**70**	23.3	23.8	24.5	25.0	25.5	26.2	27.0	27.5
72	8.0	8.0	8.0	8.0	8.0	8.0	7.8	7.8	**72**	21.5	22.0	22.5	23.3	23.8	24.5	25.0	25.5
74	9.7	9.7	9.7	9.7	9.7	9.7	9.7	9.7	**74**	19.7	20.3	20.7	21.2	22.0	22.5	23.0	23.5
76	11.5	11.5	11.5	11.5	11.6	11.7	11.7	11.7	**76**	18.0	18.5	19.0	19.5	20.0	20.5	21.0	21.5
78	13.5	13.5	13.5	13.6	13.6	13.7	13.7	13.7	**78**	16.0	16.5	17.0	17.5	18.0	18.5	19.0	19.5
80	15.4	15.4	15.4	15.5	15.6	15.7	15.7	16.0	**80**	14.2	14.7	15.3	15.5	16.0	16.5	17.0	17.5
82	17.0	17.0	17.2	17.3	17.5	17.7	17.8	18.0	**82**	12.5	13.0	13.3	13.5	14.0	14.5	15.0	15.5
84	18.8	19.0	19.2	19.3	19.5	19.7	19.9	20.0	**84**	10.5	11.0	11.5	11.7	12.0	12.5	13.0	13.4
86	20.8	21.0	21.0	21.2	21.5	21.7	22.0	22.0	**86**	8.8	9.0	9.5	9.8	10.0	10.5	11.0	11.3
88	22.6	22.8	23.0	23.2	23.4	23.7	24.0	24.2	**88**	7.0	7.2	7.5	8.0	8.3	8.5	8.7	9.0
90									**90**								

HEIGHT OF EYE CORRECTION (ADD)

Height of eye (m)	0	1.5	3	4.6	6	7.6	9	10.7	12	14	15	17	18	20	21	23	24	26	30
(ft)	0	5	10	15	20	25	30	35	40	45	50	55	60	65	70	75	80	85	100
Correction +	9.8'	7.6'	6.7'	6.0'	5.5'	5.0'	4.5'	4.0'	3.5'	3.2'	3.0'	2.5'	2.3'	2.0'	1.7'	1.3'	1.0'	0.8'	0.0'

MEAN REFRACTION

ALWAYS SUBTRACTIVE

App. Alt.	Refr.	App. Alt.	Refr.	App. Alt.	Refr.	App. Alt.	Refr.
° '	'	° '	'	° '	'	° '	'
0 00	**34.9**	**5 00**	9.8	**10 00**	5.3	**16 30**	3.2
10	32.8	10	9.5	10	5.2	**17 00**	3.1
20	30.9	20	9.3	20	5.1	30	3.0
30	29.1	30	9.0	30	5.0	**18 00**	2.9
40	27.4	40	8.8	40	5.0	30	2.9
50	25.8	50	8.6	50	4.9	**19 00**	2.8
1 00	**24.4**	**6 00**	8.4	**11 00**	4.8	**20 00**	2.6
10	23.1	10	8.2	10	4.7	**21 00**	2.5
20	21.9	20	8.0	20	4.7	**22 00**	2.4
30	20.9	30	7.8	30	4.6	**23 00**	2.3
40	19.9	40	7.7	40	4.5	**24 00**	2.2
50	19.0	50	7.5	50	4.5	**26 00**	2.0
2 00	**18.1**	**7 00**	7.3	**12 00**	4.4	**28 00**	1.8
10	17.4	10	7.2	10	4.4	**30 00**	1.7
20	16.7	20	7.0	20	4.3	**32 00**	1.5
30	16.0	30	6.9	30	4.2	**34 00**	1.4
40	15.4	40	6.8	40	4.2	**36 00**	1.3
50	14.8	50	6.6	50	4.1	**38 00**	1.2
3 00	**14.2**	**8 00**	6.5	**13 00**	4.1	**40 00**	1.1
10	13.7	10	6.4	10	4.0	**43 00**	1.0
20	13.3	20	6.3	20	4.0	**46 00**	0.9
30	12.8	30	6.1	30	3.9	**50 00**	0.8
40	12.4	40	6.0	40	3.9	**55 00**	0.7
50	12.0	50	5.9	50	3.8	**60 00**	0.6
4 00	**11.7**	**9 00**	5.8	**14 00**	3.8	**65 00**	0.5
10	11.3	10	5.7	20	3.7	**70 00**	0.4
20	11.0	20	5.6	40	3.6	**75 00**	0.3
30	10.7	30	5.5	**15 00**	3.5	**80 00**	0.2
40	10.4	40	5.4	30	3.4	**85 00**	0.1
50	10.1	50	5.4	**16 00**	3.3	**90 00**	0.0

SUN GHA CORRECTION TABLE

Min. or Sec.	Add for Minutes	Add for 1 Hour + Minutes	Add for Secs.	Min. or Sec.	Add for Minutes	Add for 1 Hour + Minutes	Add for Secs.
	° '	° '	'		° '	° '	'
0	0 0.0	15 0.0	0.0	**30**	7 30.0	22 30.0	7.5
1	0 15.0	15 15.0	0.3	**31**	7 45.0	22 45.0	7.8
2	0 30.0	15 30.0	0.5	**32**	8 0.0	23 0.0	8.0
3	0 45.0	15 45.0	0.8	**33**	8 15.0	23 15.0	8.3
4	1 0.0	16 0.0	1.0	**34**	8 30.0	23 30.0	8.5
5	1 15.0	16 15.0	1.3	**35**	8 45.0	23 45.0	8.8
6	1 30.0	16 30.0	1.5	**36**	9 0.0	24 0.0	9.0
7	1 45.0	16 45.0	1.8	**37**	9 15.0	24 15.0	9.3
8	2 0.0	17 0.0	2.0	**38**	9 30.0	24 30.0	9.5
9	2 15.0	17 15.0	2.3	**39**	9 45.0	24 45.0	9.8
10	2 30.0	17 30.0	2.5	**40**	10 0.0	25 0.0	10.0
11	2 45.0	17 45.0	2.8	**41**	10 15.0	25 15.0	10.3
12	3 0.0	18 0.0	3.0	**42**	10 30.0	25 30.0	10.5
13	3 15.0	18 15.0	3.3	**43**	10 45.0	25 45.0	10.8
14	3 30.0	18 30.0	3.5	**44**	11 0.0	26 0.0	11.0
15	3 45.0	18 45.0	3.8	**45**	11 15.0	26 15.0	11.3
16	4 0.0	19 0.0	4.0	**46**	11 30.0	26 30.0	11.5
17	4 15.0	19 15.0	4.3	**47**	11 45.0	26 45.0	11.8
18	4 30.0	19 30.0	4.5	**48**	12 0.0	27 0.0	12.0
19	4 45.0	19 45.0	4.8	**49**	12 15.0	27 15.0	12.3
20	5 0.0	20 0.0	5.0	**50**	12 30.0	27 30.0	12.5
21	5 15.0	20 15.0	5.3	**51**	12 45.0	27 45.0	12.8
22	5 30.0	20 30.0	5.5	**52**	13 0.0	28 0.0	13.0
23	5 45.0	20 45.0	5.8	**53**	13 15.0	28 15.0	13.3
24	6 0.0	21 0.0	6.0	**54**	13 30.0	28 30.0	13.5
25	6 15.0	21 15.0	6.3	**55**	13 45.0	28 45.0	13.8
26	6 30.0	21 30.0	6.5	**56**	14 0.0	29 0.0	14.0
27	6 45.0	21 45.0	6.8	**57**	14 15.0	29 15.0	14.3
28	7 0.0	22 0.0	7.0	**58**	14 30.0	29 30.0	14.5
29	7 15.0	22 15.0	7.3	**59**	14 45.0	29 45.0	14.8
				60	15 0.0	30 0.0	15.0

ARIES GHA CORRECTION TABLE

Correction for 1 HOUR + MINS: 1 Hour + Mins		Correction		Correction for MINS: Mins		Correction		Correction for SECONDS: Seconds		Correction
0	+	15	2.5	**0**	+	0	0.0	**0**	+	0.0
1	+	15	17.5	**1**	+	0	15.0	**1**	+	0.3
2	+	15	32.6	**2**	+	0	30.1	**2**	+	0.5
3	+	15	47.6	**3**	+	0	45.1	**3**	+	0.8
4	+	16	2.7	**4**	+	1	0.2	**4**	+	1.0
5	+	16	17.7	**5**	+	1	15.2	**5**	+	1.3
6	+	16	32.7	**6**	+	1	30.2	**6**	+	1.5
7	+	16	47.8	**7**	+	1	45.3	**7**	+	1.8
8	+	17	2.8	**8**	+	2	0.3	**8**	+	2.0
9	+	17	17.9	**9**	+	2	15.4	**9**	+	2.3
10	+	17	32.9	**10**	+	2	30.4	**10**	+	2.5
11	+	17	48.0	**11**	+	2	45.5	**11**	+	2.8
12	+	18	3.0	**12**	+	3	0.5	**12**	+	3.0
13	+	18	18.0	**13**	+	3	15.5	**13**	+	3.3
14	+	18	33.1	**14**	+	3	30.6	**14**	+	3.5
15	+	18	48.1	**15**	+	3	45.6	**15**	+	3.8
16	+	19	3.2	**16**	+	4	0.7	**16**	+	4.0
17	+	19	18.2	**17**	+	4	15.7	**17**	+	4.3
18	+	19	33.2	**18**	+	4	30.7	**18**	+	4.5
19	+	19	48.3	**19**	+	4	45.8	**19**	+	4.8
20	+	20	3.3	**20**	+	5	0.8	**20**	+	5.0
21	+	20	18.4	**21**	+	5	15.9	**21**	+	5.3
22	+	20	33.4	**22**	+	5	30.9	**22**	+	5.5
23	+	20	48.4	**23**	+	5	45.9	**23**	+	5.8
24	+	21	3.5	**24**	+	6	1.0	**24**	+	6.0
25	+	21	18.5	**25**	+	6	16.0	**25**	+	6.3
26	+	21	33.6	**26**	+	6	31.1	**26**	+	6.5
27	+	21	48.6	**27**	+	6	46.1	**27**	+	6.8
28	+	22	3.6	**28**	+	7	1.1	**28**	+	7.0
29	+	22	18.7	**29**	+	7	16.2	**29**	+	7.3
30	+	22	33.7	**30**	+	7	31.2	**30**	+	7.5
31	+	22	48.8	**31**	+	7	46.3	**31**	+	7.8
32	+	23	3.8	**32**	+	8	1.3	**32**	+	8.0
33	+	23	18.9	**33**	+	8	16.4	**33**	+	8.3
34	+	23	33.9	**34**	+	8	31.4	**34**	+	8.5
35	+	23	48.9	**35**	+	8	46.4	**35**	+	8.8
36	+	24	4.0	**36**	+	9	1.5	**36**	+	9.0
37	+	24	19.0	**37**	+	9	16.5	**37**	+	9.3
38	+	24	34.1	**38**	+	9	31.6	**38**	+	9.5
39	+	24	49.1	**39**	+	9	46.6	**39**	+	9.8
40	+	25	4.1	**40**	+	10	1.6	**40**	+	10.0
41	+	25	19.2	**41**	+	10	16.7	**41**	+	10.3
42	+	25	34.2	**42**	+	10	31.7	**42**	+	10.5
43	+	25	49.3	**43**	+	10	46.8	**43**	+	10.8
44	+	26	4.3	**44**	+	11	1.8	**44**	+	11.0
45	+	26	19.3	**45**	+	11	16.8	**45**	+	11.3
46	+	26	34.4	**46**	+	11	31.9	**46**	+	11.5
47	+	26	49.4	**47**	+	11	46.9	**47**	+	11.8
48	+	27	4.5	**48**	+	12	2.0	**48**	+	12.0
49	+	27	19.5	**49**	+	12	17.0	**49**	+	12.3
50	+	27	34.6	**50**	+	12	32.1	**50**	+	12.5
51	+	27	49.6	**51**	+	12	47.1	**51**	+	12.8
52	+	28	4.6	**52**	+	13	2.1	**52**	+	13.0
53	+	28	19.7	**53**	+	13	17.2	**53**	+	13.3
54	+	28	34.7	**54**	+	13	32.2	**54**	+	13.5
55	+	28	49.8	**55**	+	13	47.3	**55**	+	13.8
56	+	29	4.8	**56**	+	14	2.3	**56**	+	14.0
57	+	29	19.8	**57**	+	14	17.3	**57**	+	14.3
58	+	29	34.9	**58**	+	14	32.4	**58**	+	14.5
59	+	29	49.9	**59**	+	14	47.4	**59**	+	14.8
60	+	30	4.9	**60**	+	15	2.5	**60**	+	15.0

PLANETS GHA CORRECTION TABLE (HOURS)

ALWAYS ADD

HOURS

Var/hr	14°58.8'	14°59.0'	14°59.1'	14°59.3'	14°59.4'	14°59.6'	14°59.7'	14°59.9'	15°00.0'
	° ′	° ′	° ′	° ′	° ′	° ′	° ′	° ′	° ′
0	0 00.0	0 00.0	0 00.0	0 00.0	0 00.0	0 00.0	0 00.0	0 00.0	0 00.0
1	14 58.8	14 59.0	14 59.1	14 59.3	14 59.4	14 59.6	14 59.7	14 59.9	15 00.0
2	29 57.6	29 58.0	29 58.2	29 58.6	29 58.8	29 59.2	29 59.4	29 59.8	30 00.0
3	44 56.4	44 57.0	44 57.3	44 57.9	44 58.2	44 58.8	44 59.1	44 59.7	45 00.0
4	59 55.2	59 56.0	59 56.4	59 57.2	59 57.6	59 58.4	59 58.8	59 59.6	60 00.0
5	74 54.0	74 55.0	74 55.5	74 56.5	74 57.0	74 58.0	74 58.5	74 59.5	75 00.0
6	89 52.8	89 54.0	89 54.6	89 55.8	89 56.4	89 57.6	89 58.2	89 59.4	90 00.0
7	104 51.6	104 53.0	104 53.7	104 55.1	104 55.8	104 57.2	104 57.9	104 59.3	105 00.0
8	119 50.4	119 52.0	119 52.8	119 54.4	119 55.2	119 56.8	119 57.6	119 59.2	120 00.0
9	134 49.2	134 51.0	134 51.9	134 53.7	134 54.6	134 56.4	134 57.3	134 59.1	135 00.0
10	149 48.0	149 50.0	149 51.0	149 53.0	149 54.0	149 56.0	149 57.0	149 59.0	150 00.0
11	164 46.8	164 49.0	164 50.1	164 52.3	164 53.4	164 55.6	164 56.7	164 58.9	165 00.0
12	179 45.6	179 48.0	179 49.2	179 51.6	179 52.8	179 55.2	179 56.4	179 58.8	180 00.0
13	194 44.4	194 47.0	194 48.3	194 50.9	194 52.2	194 54.8	194 56.1	194 58.7	195 00.0
14	209 43.2	209 46.0	209 47.4	209 50.2	209 51.6	209 54.4	209 55.8	209 58.6	210 00.0
15	224 42.0	224 45.0	224 46.5	224 49.5	224 51.0	224 54.0	224 55.5	224 58.5	225 00.0
16	239 40.8	239 44.0	239 45.6	239 48.8	239 50.4	239 53.6	239 55.2	239 58.4	240 00.0
17	254 39.6	254 43.0	254 44.7	254 48.1	254 49.8	254 53.2	254 54.9	254 58.3	255 00.0
18	269 38.4	269 42.0	269 43.8	269 47.4	269 49.2	269 52.8	269 54.6	269 58.2	270 00.0
19	284 37.2	284 41.0	284 42.9	284 46.7	284 48.6	284 52.4	284 54.3	284 58.1	285 00.0
20	299 36.0	299 40.0	299 42.0	299 46.0	299 48.0	299 52.0	299 54.0	299 58.0	300 00.0
21	314 34.8	314 39.0	314 41.1	314 45.3	314 47.4	314 51.6	314 53.7	314 57.9	315 00.0
22	329 33.6	329 38.0	329 40.2	329 44.6	329 46.8	329 51.2	329 53.4	329 57.8	330 00.0
23	344 32.4	344 37.0	344 39.3	344 43.7	344 46.2	344 50.8	344 53.1	344 57.7	345 00.0
24	359 31.2	359 36.0	359 38.4	359 43.2	359 45.6	359 50.4	359 52.8	359 57.6	0 00.0

HOURS

Var/hr	15°00.2'	15°00.3'	15°00.5'	15°00.6'	15°00.8'	15°00.9'	15°01.1'	15°01.2'	15°01.4'
	° ′	° ′	° ′	° ′	° ′	° ′	° ′	° ′	° ′
0	0 00.0	0 00.0	0 00.0	0 00.0	0 00.0	0 00.0	0 00.0	0 00.0	0 00.0
1	15 00.2	15 00.3	15 00.5	15 00.6	15 00.8	15 00.9	15 01.1	15 01.2	15 01.4
2	30 00.4	30 00.6	30 01.0	30 01.2	30 01.6	30 01.8	30 02.2	30 02.4	30 02.8
3	45 00.6	45 00.9	45 01.5	45 01.8	45 02.4	45 02.7	45 03.3	45 03.6	45 04.2
4	60 00.8	60 01.2	60 02.0	60 02.4	60 03.2	60 03.6	60 04.4	60 04.8	60 05.6
5	75 01.0	75 01.5	75 02.5	75 03.0	75 04.0	75 04.5	75 05.5	75 06.0	75 07.0
6	90 01.2	90 01.8	90 03.0	90 03.6	90 04.8	90 05.4	90 06.6	90 07.2	90 08.4
7	105 01.4	105 02.1	105 03.5	105 04.2	105 05.6	105 06.3	105 07.7	105 08.4	105 09.8
8	120 01.6	120 02.4	120 04.0	120 04.8	120 06.4	120 07.2	120 08.8	120 09.6	120 11.2
9	135 01.8	135 02.7	135 04.5	135 05.4	135 07.2	135 08.1	135 09.9	135 10.8	135 12.6
10	150 02.0	150 03.0	150 05.0	150 06.0	150 08.0	150 09.0	150 11.0	150 12.0	150 14.0
11	165 02.2	165 03.3	165 05.5	165 06.6	165 08.8	165 09.9	165 12.1	165 13.2	165 15.4
12	180 02.4	180 03.6	180 06.0	180 07.2	180 09.6	180 10.8	180 13.2	180 14.4	180 16.8
13	195 02.6	195 03.9	190 06.5	195 07.8	195 10.4	195 11.7	195 14.3	195 15.6	195 18.2
14	210 02.8	210 04.2	210 07.0	210 08.4	210 11.2	210 12.6	210 15.4	210 16.8	210 19.6
15	225 03.0	225 04.5	225 07.5	225 09.0	225 12.0	225 13.5	225 16.5	225 18.5	225 21.0
16	240 03.2	240 04.8	240 08.0	240 09.6	240 12.8	240 14.4	240 17.6	240 19.2	240 22.4
17	255 03.4	255 05.1	255 08.5	255 10.2	255 13.6	255 15.3	255 18.7	255 20.4	255 23.8
18	270 03.6	270 05.4	270 09.0	27010.8	270 14.4	270 16.2	270 19.8	270 21.6	270 25.2
19	285 03.8	285 05.7	285 09.5	285 11.4	285 15.2	285 17.1	285 20.9	285 22.8	285 26.6
20	300 04.0	300 06.0	300 10.0	300 12.0	300 16.0	300 18.0	300 22.0	300 24.0	300 28.0
21	315 04.2	315 06.3	315 10.5	315 12.6	315 16.8	315 18.9	315 23.1	315 25.2	315 29.4
22	330 04.4	330 06.6	330 11.0	330 13.2	330 17.6	330 19.8	330 24.2	330 26.4	330 30.8
23	345 04.6	345 06.9	345 11.5	345 13.8	345 18.4	345 20.7	345 25.3	345 27.6	345 32.2
24	0 04.8	0 07.2	0 12.0	0.14.4	0 19.2	0 21.6	0 26.4	0 28.8	0 33.6

PLANETS GHA CORRECTION TABLE (HOURS)

ALWAYS ADD

HOURS

Var/hr	15°01.5'	15°01.7'	15°01.8'	15°02.0'	15°02.1'	15°02.3'	15°02.4'	15°02.6'	15°02.7'
	° ′	° ′	° ′	° ′	° ′	° ′	° ′	° ′	° ′
0	0 00.0	0 00.0	0 00.0	0 00.0	0 00.0	0 00.0	0 00.0	0 00.0	0 00.0
1	15 01.5	15 01.7	15 01.8	15 02.0	15 02.1	15 02.3	15 02.4	15 02.6	15 02.7
2	30 03.0	30 03.4	30 03.6	30 04.0	30 04.2	30 04.6	30 04.8	30 05.2	30 05.4
3	45 04.5	45 05.1	45 05.4	45 06.0	45 06.3	45 06.9	45 07.2	45 07.8	45 08.1
4	60 06.0	60 06.8	60 07.2	60 08.0	60 08.4	60 09.2	60 09.6	60 10.4	60 10.8
5	75 07.5	75 08.5	75 09.0	75 10.0	75 10.5	75 11.5	75 12.0	75 13.0	75 13.5
6	90 09.0	90 10.2	90 10.8	90 12.0	90 12.6	90 13.8	90 14.4	90 15.6	90 16.2
7	105 10.5	105 11.9	105 12.6	105 14.0	105 14.7	105 16.1	105 16.8	105 18.2	105 18.9
8	120 12.0	120 13.6	120 14.4	120 16.0	120 16.8	120 18.4	120 19.2	120 20.8	120 21.6
9	135 13.5	135 15.3	135 16.2	135 18.0	135 18.9	135 20.7	135 21.6	135 23.4	135 24.3
10	150 15.0	150 17.0	150 18.0	150 20.0	150 21.0	150 23.0	150 24.0	150 26.0	150 27.0
11	165 16.5	165 18.7	165 19.8	165 22.0	165 23.1	165 25.8	165 26.4	165 28.6	165 29.7
12	180 18.0	180 20.4	180 21.6	180 24.0	180 25.2	180 27.6	180 28.8	180 31.2	180 32.4
13	195 19.5	195 22.1	195 23.4	195 26.0	195 27.3	195 29.9	195 31.2	195 33.8	195 35.1
14	210 21.0	210 23.8	210 25.2	210 28.0	210 29.4	210 32.2	210 33.6	210 36.4	210 37.8
15	225 22.5	225 25.5	225 27.0	225 30.0	225 31.5	225 34.5	225 36.0	225 39.0	225 40.5
16	240 24.0	240 27.2	240 28.8	240 32.0	240 33.6	240 36.8	240 38.4	240 41.6	240 43.2
17	255 25.5	255 28.9	255 30.6	255 34.0	255 35.7	255 39.1	255 40.8	255 44.8	255 45.9
18	270 27.0	270 30.6	270 32.4	270 36.0	270 37.8	270 41.4	270 43.2	270 46.8	270 48.6
19	285 28.5	285 32.3	285 34.2	285 38.0	285 39.9	285 43.7	285 45.6	285 49.4	285 51.3
20	300 30.0	300 34.0	300 36.0	300 40.0	300 42.0	300 46.0	300 48.0	300 52.0	300 54.0
21	315 31.5	315 35.7	315 37.8	315 42.0	315 44.1	315 48.3	315 50.4	315 54.6	315 56.7
22	330 33.0	330 37.4	330 39.6	330 44.0	330 46.2	330 50.6	330 52.8	330 57.2	330 59.4
23	345 34.5	345 39.1	345 41.4	345 46.0	345 48.3	345 52.9	345 55.2	345 59.8	346 02.1
24	0 36.0	0 40.8	0 43.2	0 48.0	0 50.4	0 55.2	0 57.6	1 02.4	1 04.8

HOURS

Var/hr	15°02.9'	15°03.0'	15°03.2'	15°03.3'	15°03.5'	15°03.6'	15°03.8'	15°03.9'	15°04.1'
	° ′	° ′	° ′	° ′	° ′	° ′	° ′	° ′	° ′
0	0 00.0	0 00.0	0 00.0	0 00.0	0 00.0	0 00.0	0 00.0	0 00.0	0 00.0
1	15 02.9	15 03.0	15 03.2	15 03.3	15 03.5	15 03.6	15 03.8	15 03.9	15 04.1
2	30 05.8	30 06.0	30 06.4	30 06.6	30 07.0	15 07.2	30 07.6	30 07.8	30 08.2
3	45 08.7	45 09.0	45 09.6	45 09.9	45 10.5	45 10.8	45 11.4	45 11.7	45 12.3
4	60 11.6	60 12.0	60 12.8	60 13.2	60 14.0	60 14.4	60 15.2	60 15.6	60 16.4
5	75 14.5	75 15.0	75 16.0	75 16.5	75 17.5	75 18.0	75 19.0	75 19.5	75 20.5
6	90 17.4	90 18.0	90 19.2	90 19.8	90 21.0	90 21.6	90 22.8	90 23.4	90 24.6
7	105 20.3	105 21.0	105 22.4	105 23.1	105 24.5	105 25.2	105 26.6	105 27.3	105 28.7
8	120 23.2	120 24.0	120 25.6	120 26.4	120 28.0	120 28.8	120 30.4	120 31.2	120 32.8
9	135 26.1	135 27.0	135 28.8	135 29.7	135 31.5	135 32.4	135 34.2	135 35.1	135 36.9
10	150 29.0	150 30.0	150 32.0	150 33.0	150 35.0	150 36.0	150 38.0	150 39.0	150 41.0
11	165 31.9	165 33.0	165 35.2	165 36.3	165 38.5	165 39.6	165 41.8	165 42.9	165 45.1
12	180 34.8	180 36.0	180 38.4	180 39.6	180 42.0	180 43.2	180 45.6	180 46.8	180 49.2
13	195 37.7	195 39.0	195 41.6	195 42.9	195 45.5	195 46.8	195 49.4	195 50.7	195 53.3
14	210 40.6	210 42.0	210 44.8	210 46.2	210 49.0	210 50.4	210 53.2	210 54.6	210 57.4
15	225 43.5	225 45.0	225 48.0	225 49.5	225 52.5	225 54.0	225 57.0	225 58.5	226 01.5
16	240 46.4	240 48.0	240 51.2	240 52.8	240 56.0	240 57.6	241 00.8	241 02.4	241 05.6
17	255 49.3	255 51.0	255 54.4	255 56.1	255 59.5	256 01.2	256 04.6	256 06.3	256 09.7
18	270 52.2	270 54.0	270 57.6	270 59.4	271 03.0	271 04.8	271 08.4	271 10.2	271 13.8
19	285 55.1	285 57.0	286 00.8	286 02.7	286 06.5	286 08.4	286 12.2	286 14.1	286 17.9
20	300 58.0	301 00.0	301 04.0	301 06.0	301 10.0	301 12.0	301 16.0	301 18.0	301 22.0
21	316 00.9	316 03.0	316 07.2	316 09.3	316 13.5	316 15.6	316 19.8	316 21.9	316 26.1
22	331 03.8	331 06.0	331 10.4	331 12.6	331 17.0	331 19.2	331 23.6	331 25.8	331 30.2
23	346 06.7	346 09.0	346 13.6	346 15.9	346 20.5	346 22.8	346 27.4	346 29.7	346 34.3
24	001 09.6	001 12.0	001 16.8	001 19.2	001 24.0	001 26.4	001 31.2	001 33.6	001 38.4

PLANETS GHA CORRECTION TABLE (MINUTES)

var/hr	14°58.8'	14°59.4'	15°00.0'	15°00.6'	15°01.2'	15°01.8'	15°02.4'	15°03.0'	15°03.6'	Sec	'
	° '	° '	° '	° '	° '	° '	° '	° '	° '		
0	0 00.0	0 00.0	0 00.0	0 00.0	0 00.0	0 00.0	0 00.0	0 00.0	0 00.0	0	0.0
1	0 15.0	0 15.0	0 15.0	0 15.0	0 15.0	0 15.0	0 15.0	0 15.0	0 15.1	1	0.3
2	0 30.0	0 30.0	0 30.0	0 30.0	0 30.0	0 30.1	0 30.1	0 30.1	0 30.1	2	0.5
3	0 44.9	0 45.0	0 45.0	0 45.0	0 45.1	0 45.1	0 45.1	0 45.1	0 45.2	3	0.8
4	0 59.9	1 00.0	1 00.0	1 00.0	1 00.1	1 00.1	1 00.2	1 00.2	1 00.2	4	1.0
5	1 14.9	1 14.9	1 15.0	1 15.0	1 15.1	1 15.2	1 15.2	1 15.2	1 15.3	5	1.3
6	1 29.9	1 29.9	1 30.0	1 30.1	1 30.1	1 30.2	1 30.2	1 30.3	1 30.4	6	1.5
7	1 44.9	1 44.9	1 45.0	1 45.1	1 45.1	1 45.2	1 45.3	1 45.3	1 45.4	7	1.8
8	1 59.8	1 59.9	2 00.0	2 00.1	2 00.2	2 00.2	2 00.3	2 00.4	2 00.5	8	2.0
9	2 14.8	2 14.9	2 15.0	2 15.1	2 15.2	2 15.3	2 15.4	2 15.4	2 15.5	9	2.3
10	2 29.8	2 29.9	2 30.0	2 30.1	2 30.2	2 30.3	2 30.4	2 30.5	2 30.6	10	2.5
11	2 44.8	2 44.9	2 45.0	2 45.1	2 45.2	2 45.3	2 45.4	2 45.5	2 45.7	11	2.8
12	2 59.8	2 59.9	3 00.0	3 00.1	3 00.2	3 00.4	3 00.5	3 00.6	3 00.7	12	3.0
13	3 14.7	3 14.9	3 15.0	3 15.1	3 15.3	3 15.4	3 15.5	3 15.6	3 15.8	13	3.3
14	3 29.7	3 29.9	3 30.0	3 30.1	3 30.3	3 30.4	3 30.6	3 30.7	3 30.8	14	3.5
15	3 44.7	3 44.8	3 45.0	3 45.2	3 45.3	3 45.4	3 45.6	3 45.7	3 45.9	15	3.8
16	3 59.7	3 59.8	4 00.0	4 00.2	4 00.3	4 00.5	4 00.6	4 00.8	4 01.0	16	4.0
17	4 14.7	4 14.8	4 15.0	4 15.2	4 15.3	4 15.5	4 15.7	4 15.8	4 16.0	17	4.3
18	4 29.6	4 29.8	4 30.0	4 30.2	4 30.4	4 30.5	4 30.7	4 30.9	4 31.1	18	4.5
19	4 44.6	4 44.8	4 45.0	4 45.2	4 45.4	4 45.6	4 45.8	4 45.9	4 46.1	19	4.8
20	4 59.6	4 59.8	5 00.0	5 00.2	5 00.4	5 00.6	5 00.8	5 01.0	5 01.2	20	5.0
21	5 14.6	5 14.8	5 15.0	5 15.2	5 15.4	5 15.6	5 15.8	5 16.0	5 16.3	21	5.3
22	5 29.6	5 29.8	5 30.0	5 30.2	5 30.4	5 30.7	5 30.9	5 31.1	5 31.3	22	5.5
23	5 44.5	5 44.8	5 45.0	5 45.2	5 45.5	5 45.7	5 45.9	5 46.1	5 46.4	23	5.8
24	5 59.5	5 59.8	6 00.0	6 00.2	6 00.5	6 00.7	6 01.0	6 01.2	6 01.4	24	6.0
25	6 14.5	6 14.8	6 15.0	6 15.2	6 15.5	6 15.7	6 16.0	6 16.2	6 16.5	25	6.3
26	6 29.5	6 29.7	6 30.0	6 30.3	6 30.5	6 30.8	6 31.0	6 31.3	6 31.6	26	6.5
27	6 04.5	6 44.7	6 45.0	6 45.3	6 45.5	6 45.8	6 46.1	6 46.3	6 46.6	27	6.8
28	6 59.4	6 59.7	7 00.0	7 00.3	7 00.6	7 00.8	7 01.1	7 01.4	7 01.7	28	7.0
29	7 14.4	7 14.7	7 15.0	7 15.3	7 15.6	7 15.9	7 16.2	7 16.4	7 16.7	29	7.3
30	7 29.4	7 29.7	7 30.0	7 30.3	7 30.6	7 30.9	7 31.2	7 31.5	7 31.8	30	7.5
31	7 44.4	7 44.7	7 45.0	7 45.3	7 45.6	7 45.9	7 46.2	7 46.5	7 46.9	31	7.8
32	7 59.4	7 59.7	8 00.0	8 00.3	8 00.6	8 01.0	8 01.3	8 01.6	8 01.9	32	8.0
33	8 14.3	8 14.7	8 15.0	8 15.3	8 15.7	8 16.0	8 16.3	8 16.6	8 17.0	33	8.3
34	8 29.3	8 29.7	8 30.0	8 30.3	8 30.7	8 31.0	8 31.4	8 31.7	8 32.0	34	8.5
35	8 44.3	8 44.6	8 45.0	8 45.5	8 45.7	8 46.0	8 46.4	8 46.7	8 47.1	35	8.8
36	8 59.3	8 59.6	9 00.0	9 00.4	9 00.7	9 01.1	9 01.4	9 01.8	9 02.2	36	9.0
37	9 14.3	9 14.6	9 15.0	9 15.4	9 15.7	9 16.1	9 16.5	9 16.8	9 17.2	37	9.3
38	9 29.2	9 29.6	9 30.0	9 30.4	9 30.8	9 31.1	9 31.5	9 31.9	9 32.3	38	9.5
39	9 44.2	9 44.6	9 45.0	9 45.4	9 45.8	9 46.2	9 46.6	9 46.9	9 47.3	39	9.8
40	9 59.2	9 59.6	10 00.0	10 00.4	10 00.8	10 01.2	10 01.6	10 02.0	10 02.4	40	10.0
41	10 14.2	10 14.6	10 15.0	10 15.4	10 15.8	10 16.2	10 16.6	10 17.0	10 17.5	41	10.3
42	10 29.2	10 29.6	10 30.0	10 30.4	10 30.8	10 31.3	10 31.7	10 32.1	10 32.5	42	10.5
43	10 44.1	10 44.6	10 45.0	10 45.4	10 45.9	10 46.3	10 46.7	10 47.1	10 47.6	43	10.8
44	10 59.1	10 59.6	11 00.0	11 00.4	11 00.9	11 01.3	11 01.8	11 02.2	11 02.6	44	11.0
45	11 14.1	11 14.6	11 15.0	11 15.4	11 15.9	11 16.3	11 16.8	11 17.2	11 17.7	45	11.3
46	11 29.1	11 29.5	11 30.0	11 30.5	11 30.9	11 31.4	11 31.8	11 32.3	11 32.8	46	11.5
47	11 44.1	11 44.5	11 45.0	11 45.5	11 45.9	11 46.4	11 46.9	11 47.3	11 47.8	47	11.8
48	11 59.0	11 59.5	12 00.0	12 00.5	12 01.0	12 01.4	12 01.9	12 02.4	12 02.9	48	12.0
49	12 14.0	12 14.5	12 15.0	12 15.5	12 16.0	12 16.5	12 17.0	12 17.4	12 17.9	49	12.3
50	12 29.0	12 29.5	12 30.0	12 30.5	12 31.0	12 31.5	12 32.0	12 32.5	12 33.0	50	12.5
51	12 44.0	12 44.5	12 45.0	12 45.5	12 46.0	12 46.5	12 47.0	12 47.5	12 48.1	51	12.8
52	12 59.0	12 59.5	13 00.0	13 00.5	13 01.0	13 01.6	13 02.1	13 02.6	13 03.1	52	13.1
53	13 13.9	13 14.5	13 15.0	13 15.5	13 16.1	13 16.6	13 17.1	13 17.6	13 18.2	53	13.3
54	13 28.9	13 29.5	13 30.0	13 30.5	13 31.1	13 31.6	13 32.2	13 32.7	13 33.2	54	13.5
55	13 43.9	13 44.4	13 45.0	13 45.6	13 46.1	13 46.6	13 47.2	13 47.7	13 48.3	55	13.8
56	13 58.9	13 59.4	14 00.0	14 00.6	14 01.1	14 01.7	14 02.2	14 02.8	14 03.4	56	14.0
57	14 13.9	14 14.4	14 15.0	14 15.6	14 16.1	14 16.7	14 17.3	14 17.8	14 18.4	57	14.3
58	14 28.8	14 29.4	14 30.0	14 30.6	14 31.2	14 31.7	14 32.3	14 32.9	14 33.5	58	14.5
59	14 43.8	14 44.4	14 45.0	14 45.6	14 46.2	14 46.8	14 47.4	14 47.9	14 48.5	59	14.8
60	14 58.8	14 59.4	15 00.0	15 00.6	15 01.2	15 01.8	15 02.4	15 03.0	15 03.6	60	15.0

PLANETS NOTES 2012

VENUS is a brilliant object in the evening sky from the beginning of the year until the end of May, when it becomes too close to the Sun for observation. In mid-June it reappears in the morning sky, where it remains until the end of the year. Venus is in conjunction with Jupiter on 15th March and with Saturn on 27th November. In each case Venus is the brighter object.

JUPITER can be seen in the evening sky at the beginning of the year but in late April it becomes too close to the Sun for observation, reappearing in the morning sky in late May. Its westward elongation gradually increases until it is in opposition on 3rd December.

MARS rises shortly before midnight at the beginning of the year, and is at opposition on 3rd March, when it is visible throughout the night. From mid-June until the end of the year, as its eastward elongation gradually decreases, it is visible only in the evening sky. Mars is in conjunction with Saturn on 17th August, when Saturn is the brighter object.

SATURN rises shortly after midnight at the beginning of the year and is in opposition on 15th April when it can be seen throughout the night. From mid-July until early October it is visible only in the evening sky, and then becomes too close to the Sun for observation. From mid-November it reappears in the morning sky until the end of the year. Saturn is in conjunction with Mars on 17th August, when Saturn is the brighter object, and with Mercury on 6th October, when Mercury is the brighter object.

VISIBILITY OF PLANETS IN MORNING AND EVENING TWILIGHT

	Morning	*Evening*
VENUS	13th June to 31st December	1st January to 30th May
JUPITER	28th May to 3rd December	1st January to 29th April 3rd December to 31st December
MARS	1st January to 3rd March	3rd March to 31st December
SATURN	1st January to 15th April 12th November to 31st December	15th April to 8th October

PLANETS DECLINATION CORRECTION TABLE

Var/hr h m	0.0'	0.1'	0.2'	0.3'	0.4'	0.5'	0.6'	0.7'	0.8'	0.9'	1.0'	1.1'	1.2'	1.3'	1.4'	1.5'
0 00	0.0	0.0	0.0	0.0	0.0	0.0	0.0	0.0	0.0	0.0	0.0	0.0	0.0	0.0	0.0	0.0
12	0.0	0.0	0.0	0.1	0.1	0.1	0.1	0.1	0.2	0.2	02	0.2	0.2	0.3	0.3	0.3
24	0.0	0.0	0.1	0.1	0.2	0.2	0.2	0.3	0.3	0.4	0.4	0.4	0.5	0.6	0.6	0.6
36	0.0	0.1	0.1	0.2	0.2	0.3	0.4	0.4	0.5	0.5	0.6	0.7	0.7	0.8	0.8	0.9
48	0.0	0.1	0.2	0.2	0.3	0.4	0.5	0.6	0.6	0.7	0.8	0.9	1.0	1.0	1.1	1.2
1 00	0.0	0.1	0.2	0.3	0.4	0.5	0.6	0.7	0.8	0.9	1.0	1.1	1.2	1.3	1.4	1.5
12	0.0	0.1	0.2	0.4	0.5	0.6	0.7	0.8	1.0	1.1	1.2	1.3	1.4	1.6	1.7	1.8
24	0.0	0.1	0.3	0.4	0.6	0.7	0.8	1.0	1.1	1.3	1.4	1.5	1.7	1.8	2.0	2.1
36	0.0	0.2	0.3	0.5	0.6	0.8	1.0	1.1	1.3	1.4	1.6	1.8	1.9	2.1	2.2	2.4
48	0.0	0.2	0.4	0.5	0.7	0.9	1.1	1.3	1.4	1.6	1.8	2.0	2.2	2.3	2.5	2.7
2 00	0.0	0.2	0.4	0.6	0.8	1.0	1.2	1.4	1.6	1.8	2.0	2.2	2.4	2.6	2.8	3.0
12	0.0	0.2	0.4	0.7	0.9	1.1	1.3	1.5	1.8	2.0	2.2	2.4	2.6	2.9	3.1	3.3
24	0.0	02	0.5	0.7	1.0	1.2	1.4	1.7	1.9	2.2	2.4	2.6	2.9	3.1	3.4	3.6
36	0.0	0.3	0.5	0.8	1.0	1.3	1.6	1.8	2.1	2.3	2.6	2.9	3.1	3.4	3.6	3.9
48	0.0	0.3	0.6	0.8	1.1	1.4	1.7	2.0	2.2	2.5	2.8	3.1	3.4	3.6	3.9	4.2
3 00	0.0	0.3	0.6	0.9	1.2	1.5	1.8	2.1	2.4	2.7	3.0	3.3	3.6	3.9	4.2	4.5
12	0.0	0.3	0.6	1.0	1.3	1.6	1.9	2.2	2.6	2.9	3.2	3.5	3.8	4.2	4.5	4.8
24	0.0	0.3	0.7	1.0	1.4	1.7	2.0	2.4	2.7	3.1	3.4	3.7	4.1	4.4	4.8	5.1
36	0.0	0.4	0.7	1.1	1.4	1.8	2.2	2.5	2.9	3.2	3.6	4.0	4.3	4.7	5.0	5.4
48	0.0	0.4	0.8	1.1	1.5	1.9	2.3	2.7	3.0	3.4	3.8	4.2	4.6	4.9	5.3	5.7
4 00	0.0	0.4	0.8	1.2	1.6	2.0	2.4	2.8	3.2	3.6	4.0	4.4	4.8	5.2	5.6	6.0
12	0.0	0.4	0.8	1.3	1.7	2.1	2.5	2.9	3.4	3.8	4.2	4.6	5.0	5.5	5.9	6.3
24	0.0	0.4	0.9	1.3	1.8	2.2	2.6	3.1	3.5	4.0	4.4	4.8	5.3	5.7	6.2	6.6
36	0.0	0.5	0.9	1.4	1.8	2.3	2.8	3.2	3.7	4.1	4.6	5.1	5.5	6.0	6.4	6.9
48	0.0	0.5	1.0	1.4	1.9	2.4	2.9	3.4	3.8	4.3	4.8	5.3	5.8	6.2	6.7	7.2
5 00	0.0	0.5	1.0	1.5	2.0	2.5	3.0	3.5	4.0	4.5	5.0	5.5	6.0	6.5	7.0	7.5
12	0.0	0.5	1.0	1.6	2.1	2.6	3.1	3.6	4.2	4.7	5.2	5.7	6.2	6.8	7.3	7.8
24	0.0	0.5	1.1	1.6	2.2	2.7	3.2	3.8	4.3	4.9	5.4	5.9	6.5	7.0	7.6	8.1
36	0.0	0.6	1.1	1.7	2.2	2.8	3.4	3.9	4.5	5.0	5.6	6.2	6.7	7.3	7.8	8.4
48	0.0	0.6	1.2	1.7	2.3	2.9	3.5	4.1	4.6	5.2	5.8	6.4	7.0	7.5	8.1	8.7
6 00	0.0	0.6	1.2	1.8	2.4	3.0	3.6	4.2	4.8	5.4	6.0	6.6	7.2	7.8	8.4	9.0
12	0.0	0.6	1.2	1.9	2.5	3.1	3.7	4.3	5.0	5.6	6.2	6.8	7.4	8.1	8.7	9.3
24	0.0	0.6	1.3	1.9	2.6	3.2	3.8	4.5	5.1	5.8	6.4	7.0	7.7	8.3	9.0	9.6
36	0.0	0.7	1.3	2.0	2.6	3.3	4.0	4.6	5.3	5.9	6.6	7.3	7.9	8.6	9.2	9.9
48	0.0	0.7	1.4	2.0	2.7	3.4	4.1	4.8	5.4	6.1	6.8	7.5	8.2	B.8	9.5	10.2
7 00	0.0	0.7	1.4	2.1	2.8	3.5	4.2	4.9	5.6	6.3	7.0	7.7	8.4	9.1	9.8	10.5
12	0.0	0.7	1.4	2.2	2.9	3.6	4.3	5.0	5.8	6.5	7.2	7.9	8.6	9.4	10.1	10.8
24	0.0	0.7	1.5	2.2	3.0	3.7	4.4	5.2	5.9	6.7	7.4	8.1	8.9	9.6	10.4	11.1
36	0.0	0.8	1.5	2.3	3.0	3.8	4.6	5.3	6.1	6.8	7.6	8.4	9.1	9.9	10.6	11.4
48	0.0	0.8	1.6	2.3	3.1	3.9	4.7	5.5	6.2	7.0	7.8	8.6	9.4	10.1	10.9	11.7
8 00	0.0	0.8	1.6	2.4	3.2	4.0	4.8	5.6	6.4	7.2	8.0	8.8	9.6	10.4	11.2	12.0
12	0.0	0.8	1.6	2.5	3.3	4.1	4.9	5.7	6.6	7.4	8.2	9.0	9.8	10.7	11.5	12.3
24	0.0	0.8	1.7	2.5	3.4	4.2	5.0	5.9	6.7	7.6	8.4	9.2	10.1	10.9	11.8	12.6
36	0.0	0.9	1.7	2.6	3.4	4.3	5.2	6.0	6.9	7.7	8.6	9.5	10.3	11.2	12.0	12.9
48	0.0	0.9	1.8	2.6	3.5	4.4	5.3	6.2	7.0	7.9	8.8	9.7	10.6	11.4	12.3	13.2
9 00	0.0	0.9	1.8	2.7	3.6	4.5	5.4	6.3	7.2	8.1	9.0	9.9	10.8	11.7	12.6	13.5
12	0.0	0.9	1.8	2.8	3.7	4.6	5.5	6.4	7.4	8.3	9.2	10.1	11.0	12.0	12.9	13.8
24	0.0	0.9	1.9	2.8	3.8	4.7	5.6	6.6	7.5	8.5	9.4	10.3	11.3	12.2	13.2	14.1
36	0.0	1.0	1.9	2.9	3.8	4.8	5.8	6.7	7.7	8.6	9.6	10.6	11.5	12.5	13.4	14.4
48	0.0	1.0	2.0	2.9	3.9	4.9	5.9	6.9	7.8	8.8	9.8	10.8	11.8	12.7	13.7	14.7
10 00	0.0	1.0	2.0	3.0	4.0	5.0	6.0	7.0	8.0	9.0	10.0	11.0	12.0	13.0	14.0	15.0
12	0.0	1.0	2.0	3.1	4.1	5.1	6.1	7.1	8.2	9.2	10.2	11.2	12.2	13.2	14.3	15.3
24	0.0	1.0	2.1	3.1	4.2	5.2	6.2	7.3	8.3	9.4	10.4	11.4	12.5	13.5	14.6	15.6
36	0.0	1.1	2.2	3.2	4.2	5.3	6.4	7.4	8.5	9.5	10.6	11.7	12.7	13.8	14.8	15.9
48	0.0	1.1	2.2	3.2	4.3	5.4	6.5	7.6	8.6	9.7	10.8	11.9	13.0	14.0	15.1	16.2
11 00	0.0	1.1	2.2	3.3	4.4	5.5	6.6	7.7	8.8	9.9	11.0	12.1	13.2	14.3	15.4	16.5
12	0.0	1.1	2.2	3.4	4.5	5.6	6.7	7.8	9.0	10.1	11.2	12.3	13.4	14.6	15.7	16.8
24	0.0	1.1	2.3	3.4	4.6	5.7	6.8	8.0	9.1	10.3	11.4	12.5	13.7	14.8	16.0	17.1
36	0.0	1.2	2.3	3.5	4.6	5.8	7.0	8.1	9.3	10.4	11.6	12.8	13.9	15.1	16.2	17.4
48	0.0	1.2	2.4	3.5	4.7	5.9	7.1	8.3	9.4	10.6	11.8	13.0	14.2	15.3	16.5	17.7
12 00	0.0	1.2	2.4	3.6	4.8	6.0	7.2	8.4	9.6	10.8	12.0	13.2	14.4	15.6	16.8	18.0

PLANETS DECLINATION CORRECTION TABLE

Var/hr h m	0.0'	0.1'	0.2'	0.3'	0.4'	0.5'	0.6'	0.7'	0.8'	0.9'	1.0'	1.1'	1.2'	1.3'	1.4'	1.5'
12 00	0.0	1.2	2.4	3.6	4.8	6.0	7.2	8.4	9.6	10.8	12.0	13.2	14.4	15.6	16.8	18.0
12	0.0	1.2	2.4	3.7	4.9	6.1	7.3	8.5	9.8	11.0	12.2	13.4	14.6	15.9	17.1	18 3
24	0.0	1.2	2.5	3.7	5.0	6.2	7.4	8.7	9.9	11.2	12.4	13.6	14.9	16.1	17.4	18 6
36	0.0	1.3	2.5	3.8	5.0	6.3	7.6	8.8	10.1	11.3	12.6	13.9	15.1	16.4	17.6	18.9
48	0.0	1.3	2.6	3.8	5.1	6.4	7.7	9.0	10.2	11.5	12.8	14.1	15.4	16.6	17.9	19.2
13 00	0.0	1.3	2.6	3.9	5.2	6.5	7.8	9.1	10.4	11.7	13.0	14.3	15.6	16.9	18.2	19.5
12	0.0	1.3	2.6	4.0	5.3	6.6	7.9	9.2	10.6	11.9	13.2	14.5	15.8	17.2	18.5	19.8
24	0.0	1.3	2.7	4.0	5.4	6.7	8.0	9.4	10.7	12.1	13.4	14.7	16.1	17.4	18.8	20.1
36	0.0	1.4	2.7	4.1	5.4	6.8	8.2	9.5	10.9	12.2	13.6	15.0	16.3	17.7	19.0	20.4
48	0.0	1.4	2.8	4.1	5.5	6.9	8.3	9.7	11.0	12.4	13.8	15.2	16.6	17.9	19.3	20.7
14 00	0.0	1.4	2.8	4.2	5.6	7.0	8.4	9.8	11.2	12.6	14.0	15.4	16.8	18.2	19.6	21.0
12	0.0	1.4	2.8	4.3	5.7	7.1	8.5	9.9	11.4	12.8	14.2	15.6	17.0	18.5	19.9	21.3
24	0.0	1.4	2.9	4.3	5.8	7.2	8.6	10.1	11.5	13.0	14.4	15.8	17.3	18.7	20.2	21.6
36	0.0	1.5	2.9	4.4	5.8	7.3	8.8	10.2	11.7	13.1	14.6	16.1	17.5	19.0	20.4	21.9
48	0.0	1.5	3.0	4.4	5.9	7.4	8.9	10.4	11.8	13.3	14.8	16.3	17.8	19.2	20.7	22.2
15 00	0.0	1.5	3.0	4.5	6.0	7.5	9.0	10.5	12.0	13.5	15.0	16.5	18.0	19.5	21.0	22.5
12	0.0	1.5	3.0	4.6	6.1	7.6	9.1	10.6	12.2	13.7	15.2	16.7	18.2	19.8	21.3	22.8
24	0.0	1.5	3.1	4.6	6.2	7.7	9.2	10.8	12.3	13.9	15.4	16.9	18.5	20.0	21.6	23.1
36	0.0	1.6	3.1	4.7	6.2	7.8	9.4	10.9	12.5	14.0	15.6	17.2	18.7	20.3	21.8	23.4
48	0.0	1.6	3.2	4.7	6.3	7.9	9.5	11.1	12.6	14.2	15.8	17.4	19.0	20.5	22.1	23.7
16 00	0.0	1.6	3.2	4.8	6.4	8.0	9.6	11.2	12.8	14.4	16.0	17.6	19.2	20.8	22.4	24.0
12	0.0	1.6	3.2	4.9	6.5	8.1	9.7	11.3	13.0	14.6	16.2	17.8	19.4	21.1	22.7	24.3
24	0.0	1.6	3.3	4.9	6.6	8.2	9.8	11.5	13.1	14.8	16.4	18.0	19.7	21.3	23.0	24.6
36	0.0	1.7	3.3	5.0	6.6	8.3	10.0	11.6	13.3	14.9	16.6	18.3	19.9	21.6	23.2	24.9
48	0.0	1.7	3.4	5.0	6.7	8.4	10.1	11.8	13.4	15.1	16.8	18.5	20.2	21.8	23.5	25.2
17 00	0.0	1.7	3.4	5.1	6.8	8.5	10.2	11.9	13.6	15.3	17.0	18.7	20.4	22.1	23.8	25.5
12	0.0	1.7	3.4	5.2	6.9	8.6	10.3	12.0	13.8	15.5	17.2	18.9	20.6	22.4	24.1	25.8
24	0.0	1.7	3.5	5.2	7.0	8.7	10.4	12.2	13.9	15.7	17.4	19.1	20.9	22.6	24.4	26.1
36	0.0	1.8	3.5	5.3	7.0	8.8	10.6	12.3	14.1	15.8	17.6	19.4	21.1	22.9	24.6	26.4
48	0.0	1.8	3.6	5.3	7.1	8.9	10.7	12.5	14.2	16.0	17.8	19.6	21.4	23.1	24.9	26.7
18 00	0.0	1.8	3.6	5.4	7.2	9.0	10.8	12.6	14.4	16.2	18.0	19.8	21.6	23.4	25.2	27.0
12	0.0	1.8	3.6	5.5	7.3	9.1	10.9	12.7	14.6	16.4	18.2	20.0	21.8	23.7	25.5	27.3
24	0.0	1.8	3.7	5.5	7.4	9.2	11.0	12.9	14.7	16.6	18.4	20.2	22.1	23.9	25.8	27.6
36	0.0	1.9	3.7	5.6	7.4	9.3	11.2	13.0	14.9	16.7	18.6	20.5	22.3	24.2	26.0	27.9
48	0.0	1.9	3.8	5.6	7.5	9.4	11.3	13.2	15.0	16.9	18.8	20.7	22.6	24.4	26.3	28.2
19 00	0.0	1.9	3.8	5.7	7.6	9.5	11.4	13.3	15.2	17.1	19.0	20.9	22.8	24.7	26.6	28.5
12	0.0	1.9	3.8	5.8	7.7	9.6	11.5	13.4	15.4	17.3	19.2	21.1	23.0	25.0	26.9	28.8
24	0.0	1.9	3.9	5.8	7.8	9.7	11.6	13.6	15.5	17.5	19.4	21.3	23.3	25.2	27.2	29.1
36	0.0	2.0	3.9	5.9	7.8	9.8	11.8	13.7	15.7	17.6	19.6	21.6	23.5	25.5	27.4	29.4
48	0.0	2.0	4.0	5.9	7.9	9.9	11.9	13.9	15.8	17.8	19.8	21.8	23.8	25.7	27.7	29.7
20 00	0.0	2.0	4.0	6.0	8.0	10.0	12.0	14.0	16.0	18.0	20.0	22.0	24.0	26.0	28.0	30.0
12	0.0	2.0	4.0	6.1	8.1	10.1	12.1	14.1	16.2	18.2	20.2	22.2	24.2	26.3	28.3	30.3
24	0.0	2.0	4.1	6.1	8.2	10.2	12.2	14.3	16.3	18.4	20.4	22.4	24.5	26.5	28.6	30.6
36	0.0	2.1	4.1	6.2	8.2	10.3	12.4	14.4	16.5	18.5	20.6	22.7	24.7	26.8	28.8	30.9
48	0.0	2.1	4.2	6.2	8.3	10.4	12.5	14.6	16.6	18.7	20.8	22.9	25.0	27.0	29.1	31.2
21 00	0.0	2.1	4.2	6.3	8.4	10.5	12.6	14.7	16.8	18.9	21.0	23.1	25.2	27.3	29.4	31.5
12	0.0	2.1	4.2	6.4	8.5	10.6	12.7	14.8	17.0	19.1	21.2	23.3	25.4	27.6	29.7	31.8
24	0.0	2.1	4.3	6.4	8.6	10.7	12.8	15.0	17.1	19.3	21.4	23.5	25.7	27.8	30.0	32.1
36	0.0	2.2	4.3	6.5	8.6	10.8	13.0	15.1	17.3	19.4	21.6	23.8	25.9	28.1	30.2	32.4
48	0.0	2.2	4.4	6.5	8.7	10.9	13.1	15.3	17.4	19.6	21.8	24.0	26.2	28.3	30.5	32.7
22 00	0.0	2.2	4.4	6.6	8.8	11.0	13.2	15.4	17.6	19.8	22.0	24.2	26.4	28.6	30.8	33.0
12	0.0	2.2	4.4	6.7	8.9	11.1	13.3	15.5	17.8	20.0	22.2	24.4	26.6	28.9	31.1	33.3
24	0.0	2.2	4.5	6.7	9.0	11.2	13.4	15.7	17.9	20.2	22.4	24.6	26.9	29.1	31.4	33.6
36	0.0	2.3	4.5	6.8	9.0	11.3	13.6	15.8	18.1	20.3	22.6	24.9	27.1	29.4	31.6	33.9
48	0.0	2.3	4.6	6.8	9.1	11.4	13.7	16.0	18.2	20.5	22.8	25.1	27.4	29.6	31.9	34.2
23 00	0.0	2.3	4.6	6.9	9.2	11.5	13.8	16.1	18.4	20.7	23.0	25.3	27.6	29.9	32.2	34.5
12	0.0	2.3	4.6	7.0	9.3	11.6	13.9	16.2	18.6	20.9	23.2	25.5	27.8	30.2	32.5	34.8
24	0.0	2.3	4.7	7.0	9.4	11.7	14.0	16.4	18.7	21.1	23.4	25.7	28.1	30.4	32.8	35.1
36	0.0	2.4	4.7	7.1	9.4	11.8	14.2	16.5	18.9	21.2	23.6	26.0	28.3	30.7	33.0	35.4
48	0.0	2.4	4.8	7.1	9.5	11.9	14.3	16.7	19.0	21.4	23.8	26.2	28.6	30.9	33.3	35.7
24 00	0.0	2.4	4.8	7.2	9.6	12.0	14.4	16.8	19.2	21.6	24.0	26.4	28.8	31.2	33.6	36.0

MOON GHA CORRECTION TABLE (HOURS)
ALWAYS ADD

	Var/hr	14°20'	14°20.5'	14°21'	14°21.5'	14°22'	14°22.5'	14°23'	14°23.5'	14°24'	14°24.5'	14°25'	14°25.5'
		° '	° '	° '	° '	° '	° '	° '	° '	° '	° '	° '	° '
HOURS	1	14 20.0	14 20.5	14 21.0	14 21.5	14 22.0	14 22.5	14 23.0	14 23.5	14 24.0	14 24.5	14 25.0	14 25.5
	2	28 40.0	28 41.0	28 42.0	28 43.0	28 44.0	28 45.0	28 46.0	28 47.0	28 48.0	28 49.0	28 50.0	28 51.0
	3	43 00.0	43 01.5	43 03.0	43 04.5	43 06.0	43 07.5	43 09.0	43 10.5	43 12.0	43 13.5	43 15.0	43 16.5
	4	57 20.0	57 22.0	57 24.0	57 26.0	57 28.0	57 30.0	57 32.0	57 34.0	57 36.0	57 38.0	57 40.0	57 42.0
	5	71 40.0	71 42.5	71 45.0	71 47.5	71 50.0	71 52.5	71 55.0	71 57.5	72 00.0	72 02.5	72 05.0	72 07.5

	Var/hr	14°26'	14°26.5'	14°27'	14°27.5'	14°28'	14°28.5'	14°29'	14°29.5'	14°30'	14°30.5'	14°31'	14°31.5'
HOURS	1	14 26.0	14 26.5	14 27.0	14 27.5	14 28.0	14'28.5	14 29.0	14 29.5	14 30.0	14 30.5	14 31.0	14 31.5
	2	28 52.0	28 53.0	28 54.0	28 55.0	28 56.0	28 57.0	28 58.0	28 59.0	29 00.0	29 01.0	29 02.0	29 03.0
	3	43 18.0	43 19.5	43 21.0	43 22.5	43 24.0	43 25.5	43 27.0	43 28.5	43 30.0	43 31.5	43 33.0	43 34.5
	4	57 44.0	57 46.0	57 48.0	57 50.0	57 52.0	57 54.0	57 56.0	57 58.0	58 00.0	58 02.0	58 04.0	58 06.0
	5	72 10.0	72 12.5	72 15.0	72 17.5	72 20.0	72 22.5	72 25.0	72 27.5	72 30.0	72 32.5	72 35.0	72 37.5

	Var/hr	14°32'	14°32.5'	14°33'	14°33.5'	14°34'	14°34.5'	14°35'	14°35.5'	14°36'	14°36.5'	14°37'	14°37.5'
HOURS	1	14 32.0	14 32.5	14 33.0	14 33.5	14 34.0	14 34.5	14 35.0	14 35.5	14 36.0	14 36.5	14 37.0	14 37.5
	2	29 04.0	29 05.0	29 06.0	29 07.0	29 08.0	29 09.0	29 10.0	29 11.0	29 12.0	29 13.0	29 14.0	29 15.0
	3	43 36.0	43 37.5	43 39.0	43 40.5	43 42.0	43 43.5	43 45.0	43 46.5	43 48.0	43 49.5	43 51.0	43 52.5
	4	58 08.0	58 10.0	58 12.0	58 14.0	58 16.0	58 18.0	58 20.0	58 22.0	58 24.0	58 26.0	58 28.0	58 30.0
	5	72 40.0	72 42.5	72 45.0	72 47.5	72 50.0	72 52.5	72 55.0	72 57.5	73 00.0	73 02.5	73 05.0	73 07.5

NOTE: The correction for Moon GHA is taken in three parts. Using the figure for variation per hour given in the monthly tables, you get the correction for hours from the table above and the correction for minutes from the tables on the next two pages. Finally, take the correction for seconds from the right-hand side of whichever minutes table you used.

NOTES ON PHASES OF THE MOON

What follows below is approximate. The season of the year, the latitude of the observer and the declination of the Moon all have an influence.

The Moon's phases are shown on the third monthly pages. It is useful to be aware of them, not least because they will indicate if the Moon is likely to be available for sights. Certainly the Moon should normally be used, if visible at the right altitude, for morning and evening Stars; and don't overlook its value during the day.

1 Technically the 'New Moon' cannot be seen; it rises and sets with the Sun. As it begins its monthly journey around the Earth it will be seen as a conventional 'New Moon', setting at dusk and not yet useful to us.

2 By the time the Moon reaches its first quarter (Half Moon) it will be rising at midday and setting at midnight - very useful for afternoon sights and evening Stars.

3 Nearing Full Moon it will be rising and setting ever later until, at Full Moon, it rises at dusk and sets at dawn; it is then often too low in the sky for morning or evening Stars.

4 As the Moon approaches its last quarter (Half Moon) it will start setting after dawn and therefore be available for morning Stars and for daylight sights in the forenoon. At its last quarter it will rise at midnight and set at midday. Soon, however, it will be rising too late to be high enough for morning Stars and in two or three days will start its cycle again as the New Moon.

51

MOON GHA CORRECTION TABLE (MINUTES)

Var/hr	14°20'	14°21'	14°22'	14°23'	14°24'	14°25'	14°26'	14°27'	14°28'	Diff 1'	Sec	Corr
	° '	° '	° '	° '	° '	° '	° '	° '	° '	'	'	'
0	0 00.0	0 00.0	0 00.0	0 00.0	0 00.0	0 00.0	0 00.0	0 00.0	0 00.0	0.0	0	0.0
1	0 14.3	0 14.4	0 14.4	0 14.4	0 14.4	0 14.4	0 14.4	0 14.4	0 14.5	0.0	1	0.2
2	0 28.7	0 28.7	0 28.7	0 28.8	0 28.8	0 28.8	0 28.9	0 28.9	0 28.9	0.0	2	0.5
3	0 43.0	0 43.0	0 43.1	0 43.2	0 43.2	0 43.2	0 43.3	0 43.4	0 43.4	0.0	3	0.7
4	0 57.3	0 57.4	0 57.5	0 57.5	0 57.6	0 57.7	0 57.7	0 57.8	0 57.9	0.1	4	1.0
5	1 11.7	1 11.8	1 11.8	1 11.9	1 12.0	1 12.1	1 12.2	1 12.2	1 12.3	0.1	5	1.2
6	1 26.0	1 26.1	1 26.2	1 26.3	1 26.4	1 26.5	1 26.6	1 26.7	1 26.8	0.1	6	1.4
7	1 40.3	1 40.4	1 40.6	1 40.7	1 40.8	1 40.9	1 41.2	1 41.2	1 41.3	0.1	7	1.7
8	1 54.7	1 54.8	1 54.9	1 55.1	1 55.2	1 55.3	1 55.5	1 55.6	1 55.7	0.1	8	1.9
9	2 09.0	2 09.2	2 09.3	2 09.4	2 09.6	2 09.8	2 09.9	2 10.0	2 10.2	0.2	9	2.2
10	2 23.3	2 23.5	2 23.7	2 23.8	2 24.0	2 24.2	2 24.3	2 24.5	2 24.7	0.2	10	2.4
11	2 37.7	2 37.8	2 38.0	2 38.2	2 38.4	2 38.6	2 38.8	2 39.0	2 39.1	0.2	11	2.6
12	2 52.0	2 52.2	2 52.4	2 52.6	2 52.8	2 53.0	2 53.2	2 53.4	2 53.6	0.2	12	2.9
13	3 06.3	3 06.6	3 06.8	3 07.0	3 07.2	3 07.4	3 07.6	3 07.8	3 08.1	0.2	13	3.1
14	3 20.7	3 20.9	3 21.1	3 21.4	3 21.6	3 21.8	3 22.1	3 22.3	3 22.5	0.2	14	3.4
15	3 35.0	3 35.2	3 35.5	3 35.8	3 36.0	3 36.2	3 36.5	3 36.8	3 37.0	0.2	15	3.6
16	3 49.3	3 49.6	3 49.9	3 50.1	3 50.4	3 50.7	3 50.9	3 51.2	3 51.5	0.3	16	3.8
17	4 03.7	4 04.0	4 04.2	4 04.5	4 04.8	4 05.1	4 05.4	4 05.6	4 05.9	0.3	17	4.1
18	4 18.0	4 18.3	4 18.6	4 18.9	4 19.2	4 19.5	4 19.8	4 20.1	4 20.4	0.3	18	4.3
19	4 32.3	4 32.6	4 33.0	4 33.3	4 33.6	4 33.9	4 34.2	4 34.6	4 34.9	0.3	19	4.6
20	4 46.7	4 47.0	4 47.3	4 47.7	4 48.0	4 48.3	4 48.7	4 49.0	4 49.3	0.3	20	4.8
21	5 01.0	5 01.4	5 01.7	5 02.0	5 02.4	5 02.8	5 03.1	5 03.4	5 03.8	0.4	21	5.0
22	5 15.3	5 15.7	5 16.1	5 16.4	5 16.8	5 17.2	5 17.5	5 17.9	5 18.3	0.4	22	5.3
23	5 29.7	5 30.0	5 30.4	5 30.8	5 31.2	5 31.6	5 32.0	5 32.4	5 32.7	0.4	23	5.5
24	5 44.0	5 44.4	5 44.8	5 45.2	5 45.6	5 46.0	5 46.4	5 46.8	5 47.2	0.4	24	5.8
25	5 58.3	5 58.8	5 59.2	5 59.6	6 00.0	6 00.4	6 00.8	6 01.2	6 01.7	0.4	25	6.0
26	6 12.7	6 13.1	6 13.5	6 14.0	6 14.4	6 14.8	6 15.3	6 15.7	5 16.1	0.4	26	6.2
27	6 27.0	6 27.4	6 27.9	6 28.4	6 28.8	6 29.2	6 29.7	6 30.2	6 30.6	0.4	27	6.5
28	6 41.3	6 41.8	6 42.3	6 42.7	6 43.2	6 43.7	6 44.1	6 44.6	6 45.1	0.5	28	6.7
29	6 55.7	6 56.2	6 56.6	6 57.1	6 57.6	6 58.1	6 58.6	6 59.0	6 59.5	0.5	29	7.0
30	7 10.0	7 10.5	7 11.0	7 11.5	7 12.0	7 12.5	7 13.0	7 13.5	7 14.0	0.5	30	7.2
31	7 24.3	7 24.8	7 25.4	7 25.9	7 26.4	7 26.9	7 27.4	7 28.0	7 28.5	0.5	31	7.4
32	7 38.7	7 39.2	7 39.7	7 40.3	7 40.8	7 41.3	7 41.9	7 42.4	7 42.9	0.5	32	7.7
33	7 53.0	7 53.6	7 54.1	7 54.6	7 55.2	7 55.8	7 56.3	7 56.8	7 57.4	0.6	33	7.9
34	8 07.3	8 07.9	8 08.5	8 09.0	8 09.6	8 10.2	8 10.7	8 11.3	8 11.9	0.6	34	8.2
35	8 21.7	8 22.2	8 22.8	8 23.4	8 24.0	8 24.6	8 25.2	8 25.8	8 26.3	0.6	35	8.4
36	8 36.0	8 36.6	8 37.2	8 37.8	8 38.4	8 39.0	8 39.6	8 40.2	8 40.8	0.6	36	8.6
37	8 50.3	8 51.0	8 51.6	8 52.2	8 52.8	8 53.4	8 54.0	8 54.6	8 55.3	0.6	37	8.9
38	9 04.7	9 05.3	9 05.9	9 06.6	9 07.2	9 07.8	9 08.5	9 09.1	9 09.7	0.6	38	9.1
39	9 19.0	9 19.6	9 20.3	9 21.0	9 21.6	9 22.2	9 22.9	9 23.6	9 24.2	0.6	39	9.4
40	9 33.3	9 34.0	9 34.7	9 35.3	9 36.0	9 36.7	9 37.3	9 38.0	9 38.7	0.7	40	9.6
41	9 47.7	9 48.4	9 49.0	9 49.7	9 50.4	9 51.1	9 51.8	9 52.4	9 53.1	0.7	41	9.8
42	10 02.0	10 02.7	10 03.4	10 04.1	10 04.8	10 05.5	10 06.2	10 06.9	10 07.6	0.7	42	10.1
43	10 16.3	10 17.0	10 17.8	10 18.5	10 19.2	10 19.9	10 20.6	10 21.4	10 22.1	0.7	43	10.3
44	10 30.7	10 31.4	10 32.1	10 32.9	10 33.6	10 34.3	10 35.1	10 35.8	10 36.5	0.7	44	10.6
45	10 45.0	10 45.8	10 46.5	10 47.2	10 48.0	10 48.8	10 49.5	10 50.2	10 51.0	0.8	45	10.8
46	10 59.3	11 00.1	11 00.9	11 01.6	11 02.4	11 03.2	11 03.9	11 04.7	11 05.5	0.8	46	11.0
47	11 13.7	11 14.4	11 15.2	11 16.0	11 16.8	11 17.6	11 18.4	11 19.2	11 19.9	0.8	47	11.3
48	11 28.0	11 28.8	11 29.6	11 30.4	11 31.2	11 32.0	11 32.8	11 33.6	11 34.4	0.8	48	11.5
49	11 42.3	11 43.2	11 44.0	11 44.8	11 45.6	11 46.4	11 47.2	11 48.0	11 48.9	0.8	49	11.8
50	11 56.7	11 57.5	11 58.3	11 59.2	12 00.0	12 00.8	12 01.7	12 02.5	12 03.3	0.8	50	12.0
51	12 11.0	12 11.8	12 12.7	12 13.6	12 14.4	12 15.2	12 16.1	12 17.0	12 17.8	0.8	51	12.2
52	12 25.3	12 26.2	12 27.1	12 27.9	12 28.8	12 29.7	12 30.5	12 31.4	12 32.3	0.9	52	12.5
53	12 39.7	12 40.6	12 41.4	12 42.3	12 43.2	12 44.1	12 45.0	12 45.8	12 46.7	0.9	53	12.7
54	12 54.0	12 54.9	12 55.8	12 56.7	12 57.6	12 58.5	12 59.4	13 00.3	13 01.2	0.9	54	13.0
55	13 08.3	13 09.2	13 10.2	13 11.1	13 12.0	13 12.9	13 13.8	13 14.8	13 15.7	0.9	55	13.2
56	13 22.7	13 23.6	13 24.5	13 25.5	13 26.4	13 27.3	13 28.3	13 29.2	13 30.1	0.9	56	13.4
57	13 37.0	13 38.0	13 38.9	13 39.8	13 40.8	13 41.8	13 42.7	13 43.6	13 44.6	1.0	57	13.7
58	13 51.3	13 52.3	13 53.3	13 54.2	13 55.2	13 56.2	13 57.1	13 58.1	13 59.1	1.0	58	13.9
59	14 05.7	14 06.6	14 07.6	14 08.6	14 09.6	14 10.6	14 11.6	14 12.6	14 13.5	1.0	59	14.1
60	14 20.0	14 21.0	14 22.0	14 23.0	14 24.0	14 25.0	14 26.0	14 27.0	14 28.0	1.0	60	14.4

MOON GHA CORRECTION TABLE (MINUTES)

Var/hr	14°290'	14°30'	14°31'	14°32'	14°33'	14°34'	14°35'	14°36'	14°37'	Diff 1'	Sec	Corr
	° '	° '	° '	° '	° '	° '	° '	° '	° '	'	'	'
0	0 00.0	0 00.0	0 00.0	0 00.0	0 00.0	0 00.0	0 00.0	0 00.0	0 00.0	0.0	0	0.0
1	0 14.5	0 14.5	0 14.5	0 14.6	0 14.6	0 14.6	0 14.6	0 14.6	0 14.6	0.0	1	0.2
2	0 29.0	0 29.0	0 29.0	0 29.1	0 29.1	0 29.1	0 29.2	0 29.2	0 29.2	0.0	2	0.5
3	0 43.4	0 43.5	0 43.6	0 43.6	0 43.6	0 43.7	0 43.8	0 43.8	0 43.8	0.0	3	0.7
4	0 57.9	0 58.0	0 58.1	0 58.1	0 58.2	0 58.3	0 58.3	0 58.4	0 58.5	0.1	4	1.0
5	1 12.4	1 12.5	1 12.6	1 12.7	1 12.8	1 12.8	1 12.9	1 13.0	1 13.1	0.1	5	1.2
6	1 26.9	1 27.0	1 27.1	1 27.2	1 27.3	1 27.4	1 27.5	1 27.6	1 27.7	0.1	6	1.5
7	1 41.4	1 41.5	1 41.6	1 41.7	1 41.8	1 42.0	1 42.1	1 42.2	1 42.3	0.1	7	1.7
8	1 55.9	1 56.0	1 56.1	1 56.3	1 56.4	1 56.5	1 56.7	1 56.8	1 56.9	0.1	8	1.9
9	2 10.4	2 10.5	2 10.6	2 10.8	2 11.0	2 11.1	2 11.2	2 11.4	2 11.6	0.2	9	2.2
10	2 24.8	2 25.0	2 25.2	2 25.2	2 25.5	2 25.7	2 25.8	2 26.0	2 26.2	0.2	10	2.4
11	2 39.3	2 39.5	2 39.7	2 39.9	2 40.0	2 40.2	2 40.4	2 40.6	2 40.8	0.2	11	2.7
12	2 53.8	2 54.0	2 54.2	2 54.4	2 54.6	2 54.8	2 55.0	2 55.2	2 55.4	0.2	12	2.9
13	3 08.3	3 08.5	3 08.7	3 08.9	3 09.2	3 09.4	3 09.6	3 09.8	3 10.0	0.2	13	3.2
14	3 22.8	3 23.0	3 23.2	3 23.5	3 23.7	3 23.9	3 24.2	3 24.4	3 24.6	0.2	14	3.4
15	3 37.2	3 37.5	3 37.8	3 38.0	3 38.2	3 38.5	3 38.8	3 39.0	3 39.2	0.2	15	3.6
16	3 51.7	3 52.0	3 52.3	3 52.5	3 52.8	3 53.1	3 53.3	3 53.6	3 53.9	0.3	16	3.9
17	4 06.2	4 06.5	4 06.8	4 07.1	4 07.4	4 07.6	4 07.9	4 08.2	4 08.5	0.3	17	4.1
18	4 20.7	4 21.0	4 21.3	4 21.6	4 21.9	4 22.2	4 22.5	4 22.8	4 23.1	0.3	18	4.4
19	4 35.2	4 35.5	4 35.8	4 36.1	4 36.4	4 36.8	4 37.1	4 37.4	4 37.7	0.3	19	4.6
20	4 49.7	4 50.0	4 50.3	4 50.7	4 51.0	4 51.3	4 51.7	4 52.0	4 52.3	0.3	20	4.9
21	5 04.2	5 04.5	5 04.8	5 05.2	5 05.6	5 05.9	5 06.2	5 06.6	5 07.0	0.4	21	5.1
22	5 18.6	5 19.0	5 19.4	5 19.7	5 20.1	5 20.5	5 20.8	5 21.2	5 21.6	0.4	22	5.3
23	5 33.1	5 33.5	5 33.9	5 34.3	5 34.6	5 35.0	5 35.4	5 35.8	5 36.2	0.4	23	5.6
24	5 47.6	5 48.0	5 48.4	5 48.8	5 49.2	5 49.6	5 50.0	5 50.4	5 50.8	0.4	24	5.8
25	6 02.1	6 02.5	6 02.9	6 03.3	6 03.8	6 04.2	6 04.6	6 05.0	6 05.4	0.4	25	6.1
26	6 16.6	6 17.0	6 17.4	6 17.9	6 18.3	6 18.7	6 19.2	6 19.6	6 20.0	0.4	26	6.3
27	6 31.0	6 31.5	6 32.0	6 32.4	6 32.8	6 33.3	6 33.8	6 34.2	6 34.6	0.4	27	6.5
28	6 45.5	6 46.0	6 46.5	6 46.9	6 47.4	6 47.9	6 4&3	6 48.8	6 49.3	0.5	28	6.8
29	7 00.0	7 00.5	7 01.0	7 01.5	7 02.0	7 02.4	7 02.9	7 03.4	7 03.9	0.5	29	7.0
30	7 14.5	7 15.0	7 15.5	7 16.0	7 16.5	7 17.0	7 17.5	7 18.0	7 18.5	0.5	30	7.3
31	7 29.0	7 29.5	7 30.0	7 30.5	7 31.0	7 31.6	7 32.1	7 32.6	7 33.1	0.5	31	7.5
32	7 43.5	7 44.0	7 44.5	7 45.1	7 45.6	7 46.1	7 46.7	7 47.2	7 47.7	0.5	32	7.8
33	7 58.0	7 58.5	7 59.0	7 59.6	8 00.2	8 00.7	8 01.2	8 01.8	8 02.4	0.6	33	8.0
34	8 12.4	8 13.0	8 13.6	8 14.1	8 14.7	8 15.3	8 15.8	8 16.4	8 17.0	0.6	34	8.2
35	8 26.9	8 27.5	8 28.1	8 28.7	8 29.2	8 29.8	8 30.4	8 31.0	8 31.6	0.6	35	8.5
36	8 41.4	8 42.0	8 42.6	8 43.2	8 43.8	8 44.4	8 45.0	8 45.6	8 46.2	0.6	36	8.7
37	8 55.9	8 56.5	8 57.1	8 57.7	8 58.4	8 59.0	8 59.6	9 00.2	9 00.8	0.6	37	9.0
38	9 10.4	9 11.0	9 11.6	9 12.3	9 12.9	9 13.5	9 14.2	9 14.8	9 15.4	0.6	38	9.2
39	9 24.8	9 25.5	9 26.2	9 26.8	9 27.4	9 28.1	9 28.8	9 29.4	9 30.0	0.6	39	9.5
40	9 39.3	9 40.0	9 40.7	9 41.3	9 42.0	9 42.7	9 43.3	9 44.0	9 44.7	0.7	40	9.7
41	9 53.8	9 54.5	9 55.2	9 55.9	9 56.6	9 57.2	9 57.9	9 58.6	9 59.3	0.7	41	9.9
42	10 08.3	10 09.0	10 09.7	10 10.4	10 11.1	10 11.8	10 12.5	10 13.2	10 13.9	0.7	42	10.2
43	10 22.8	10 23.5	10 24.2	10 24.9	10 25.6	10 26.4	10 27.1	10 27.8	10 28.5	0.7	43	10.4
44	10 37.3	10 38.0	10 38.7	10 39.5	10 40.2	10 40.9	10 41.7	10 42.2	10 43.1	0.7	44	10.7
45	10 51.8	10 52.5	10 53.2	10 54.0	10 54.8	10 55.5	10 56.2	10 57.0	10 57.8	0.8	45	10.9
46	11 06.2	11 07.0	11 07.8	11 08.5	11 09.3	11 10.1	11 10.8	11 11.6	11 12.4	0.8	46	11.2
47	11 20.7	11 21.5	11 22.3	11 23.1	11 23.8	11 24.6	11 25.4	11 26.2	11 27.0	0.8	47	11.4
48	11 35.2	11 36.0	11 36.8	11 37.6	11 38.4	11 39.2	11 40.0	11 40.8	11 41.6	0.8	48	11.6
49	11 49.7	11 50.5	11 51.3	11 52.1	11 53.0	11 53.8	11 54.6	11 55.4	11 56.2	0.8	49	11.9
50	12 04.2	12 05.0	12 05.8	12 06.7	12 07.5	12 08.3	12 09.2	12 10.0	12 10.8	0.8	50	12.1
51	12 18.6	12 19.5	12 20.4	12 21.2	12 22.0	12 22.9	12 23.8	12 24.6	12 25.4	0.8	51	12.4
52	12 33.1	12 34.0	12 34.9	12 35.7	12 36.6	12 37.5	12 38.3	12 39.2	12 40.1	0.9	52	12.6
53	12 47.6	12 48.5	12 49.4	12 50.3	12 51.2	12 52.0	12 52.9	12 53.8	12 54.7	0.9	53	12.9
54	13 02.1	13 03.0	13 03.9	13 04.8	13 05.7	13 06.6	13 07.5	13 08.4	13 09.3	0.9	54	13.1
55	13 16.6	13 17.5	13 18.4	13 19.3	13 20.2	13 21.2	13 22.1	13 23.0	13 23.9	0.9	55	13.3
56	13 31.1	13 32.0	13 32.9	13 33.9	13 34.8	13 35.7	13 36.7	13 37.6	13 38.5	0.9	56	13.6
57	13 45.6	13 46.5	13 47.4	13 48.4	13 49.4	13 50.3	13 51.2	13 52.2	13 53.2	1.0	57	13.9
58	14 00.0	14 01.0	14 02.0	14 02.9	14 03.9	14 04.9	14 05.8	14 06.8	14 07.8	1.0	58	14.1
59	14 14.5	14 15.5	14 16.5	14 17.5	14 18.4	14 19.4	14 20.4	14 21.4	14 22.4	1.0	59	14.3
60	14 29.0	14 30.0	14 31.0	14 32.0	14 33.0	14 34.0	14 35.0	14 36.0	14 37.0	1.0	60	14.6

MOON DECLINATION CORRECTION TABLE

Var/hr (MINUTES)	0.0'	1.0'	2.0'	3.0'	4.0'	5.0'	6.0'	7.0'	8.0'	9.0'	10.0'	11.0'	12.0'	13.0'	14.0'	15.0'	16.0'	17.0'	18.0'
0	0.0	0.0	0.0	0.0	0.0	0.0	0.0	0.0	0.0	0.0	0.0	0.0	0.0	0.0	0.0	0.0	0.0	0.0	0.0
1	0.0	0.0	0.0	0.0	0.1	0.1	0.1	0.1	0.1	0.2	0.2	0.2	0.2	0.2	0.2	0.2	0.3	0.3	0.3
2	0.0	0.0	0.1	0.1	0.1	0.2	0.2	0.2	0.3	0.3	0.3	0.4	0.4	0.4	0.5	0.5	0.5	0.6	0.6
3	0.0	0.0	0.1	0.2	0.2	0.2	0.3	0.4	0.4	0.4	0.5	0.6	0.6	0.6	0.7	0.8	0.8	0.8	0.9
4	0.0	0.1	0.1	0.2	0.3	0.3	0.4	0.5	0.5	0.6	0.7	0.7	0.8	0.9	0.9	1.0	1.1	1.1	1.2
5	0.0	0.1	0.2	0.2	0.3	0.4	0.5	0.6	0.7	0.8	0.8	0.9	1.0	1.1	1.2	1.2	1.3	1.4	1.5
6	0.0	0.1	0.2	0.3	0.4	0.5	0.6	0.7	0.8	0.9	1.0	1.1	1.2	1.3	1.4	1.5	1.6	1.7	1.8
7	0.0	0.1	0.2	0.4	0.5	0.6	0.7	0.8	0.9	1.0	1.2	1.3	1.4	1.5	1.6	1.8	1.9	2.0	2.1
8	0.0	0.1	0.3	0.4	0.5	0.7	0.8	0.9	1.1	1.2	1.3	1.5	1.6	1.7	1.9	2.0	2.1	2.3	2.4
9	0.0	0.2	0.3	0.4	0.6	0.8	0.9	1.0	1.2	1.4	1.5	1.6	1.8	2.0	2.1	2.2	2.4	2.6	2.7
10	0.0	0.2	0.3	0.5	0.7	0.8	1.0	1.2	1.3	1.5	1.7	1.8	2.0	2.2	2.3	2.5	2.7	2.8	3.0
11	0.0	0.2	0.4	0.6	0.7	0.9	1.1	1.3	1.5	1.6	1.8	2.0	2.2	2.4	2.6	2.8	2.9	3.1	3.3
12	0.0	0.2	0.4	0.6	0.8	1.0	1.2	1.4	1.6	1.8	2.0	2.2	2.4	2.6	2.8	3.0	3.2	3.4	3.6
13	0.0	0.2	0.4	0.6	0.9	1.1	1.3	1.5	1.7	2.0	2.2	2.4	2.6	2.8	3.0	3.2	3.5	3.7	3.9
14	0.0	0.2	0.5	0.7	0.9	1.2	1.4	1.6	1.9	2.1	2.3	2.6	2.8	3.0	3.3	3.5	3.7	4.0	4.2
15	0.0	0.2	0.5	0.8	1.0	1.2	1.5	1.8	2.0	2.2	2.5	2.8	3.0	3.2	3.5	3.8	4.0	4.2	4.5
16	0.0	0.3	0.5	0.8	1.1	1.3	1.6	1.9	2.1	2.4	2.7	2.9	3.2	3.5	3.7	4.0	4.3	4.5	4.8
17	0.0	0.3	0.6	0.8	1.1	1.4	1.7	2.0	2.3	2.6	2.8	3.1	3.4	3.7	4.0	4.2	4.5	4.8	5.1
18	0.0	0.3	0.6	0.9	1.2	1.5	1.8	2.1	2.4	2.7	3.0	3.3	3.6	3.9	4.2	4.5	4.8	5.1	5.4
19	0.0	0.3	0.6	1.0	1.3	1.6	1.9	2.2	2.5	2.8	3.2	3.5	3.8	4.1	4.4	4.8	5.1	5.4	5.7
20	0.0	0.3	0.7	1.0	1.3	1.7	2.0	2.3	2.7	3.0	3.3	3.7	4.0	4.3	4.7	5.0	5.3	5.7	6.0
21	0.0	0.4	0.7	1.0	1.4	1.8	2.1	2.4	2.8	3.2	3.5	3.8	4.2	4.6	4.9	5.2	5.6	6.0	6.3
22	0.0	0.4	0.7	1.1	1.5	1.8	2.2	2.6	2.9	3.3	3.7	4.0	4.4	4.8	5.1	5.5	5.9	6.2	6.6
23	0.0	0.4	0.8	1.2	1.5	1.9	2.3	2.7	3.1	3.4	3.8	4.2	4.6	5.0	5.4	5.8	6.1	6.5	6.9
24	0.0	0.4	0.8	1.2	1.6	2.0	2.4	2.8	3.2	3.6	4.0	4.4	4.8	5.2	5.6	6.0	6.4	6.8	7.2
25	0.0	0.4	0.8	1.2	1.7	2.1	2.5	2.9	3.3	3.8	4.2	4.6	5.0	5.4	5.8	6.2	6.7	7.1	7.5
26	0.0	0.4	0.9	1.3	1.7	2.2	2.6	3.0	3.5	3.9	4.3	4.8	5.2	5.6	6.1	6.5	6.9	7.4	7.8
27	0.0	0.4	0.9	1.3	1.8	2.2	2.7	3.2	3.6	4.0	4.5	5.0	5.4	5.8	6.3	6.8	7.2	7.6	8.1
28	0.0	0.5	0.9	1.4	1.9	2.3	2.8	3.3	3.7	4.2	4.7	5.1	5.6	6.1	6.5	7.0	7.5	7.9	8.4
29	0.0	0.5	1.0	1.4	1.9	2.4	2.9	3.4	3.9	4.4	4.8	5.3	5.8	6.3	6.8	7.2	7.7	8.2	8.7
30	0.0	0.5	1.0	1.5	2.0	2.5	3.0	3.5	4.0	4.5	5.0	5.5	6.0	6.5	7.0	7.5	8.0	8.5	9.0
31	0.0	0.5	1.0	1.6	2.1	2.6	3.1	3.6	4.1	4.6	5.2	5.7	6.2	6.7	7.2	7.8	8.3	8.8	9.3
32	0.0	0.5	1.1	1.6	2.1	2.7	3.2	3.7	4.3	4.8	5.3	5.9	6.4	6.9	7.5	8.0	8.5	9.1	9.6
33	0.0	0.6	1.1	1.6	2.2	2.8	3.3	3.8	4.4	5.0	5.5	6.0	6.6	7.2	7.7	8.2	8.8	9.4	9.9
34	0.0	0.6	1.1	1.7	2.3	2.8	3.4	4.0	4.5	5.1	5.7	6.2	6.8	7.4	7.9	8.5	9.1	9.6	10.2
35	0.0	0.6	1.2	1.8	2.3	2.9	3.5	4.1	4.7	5.2	5.8	6.4	7.0	7.6	8.2	8.8	9.3	9.9	10.5
36	0.0	0.6	1.2	1.8	2.4	3.0	3.6	4.2	4.8	5.4	6.0	6.6	7.2	7.8	8.4	9.0	9.6	10.2	10.8
37	0.0	0.6	1.2	1.8	2.5	3.1	3.7	4.3	4.9	5.6	6.2	6.8	7.4	8.0	8.6	9.2	9.9	10.5	11.1
38	0.0	0.6	1.3	1.9	2.5	3.2	3.8	4.4	5.1	5.7	6.3	7.0	7.6	8.2	8.9	9.5	10.1	10.8	11.4
39	0.0	0.6	1.3	2.0	2.6	3.2	3.9	4.6	5.2	5.8	6.5	7.2	7.8	8.4	9.1	9.8	10.4	11.0	11.7
40	0.0	0.7	1.3	2.0	2.7	3.3	4.0	4.7	5.3	6.0	6.7	7.3	8.0	8.7	9.3	10.0	10.7	11.3	12.0
41	0.0	0.7	1.4	2.0	2.7	3.4	4.1	4.8	5.5	6.2	6.8	7.5	8.2	8.9	9.6	10.2	10.9	11.6	12.3
42	0.0	0.7	1.4	2.1	2.8	3.5	4.2	4.9	5.6	6.3	7.0	7.7	8.4	9.1	9.8	10.5	11.2	11.9	12.6
43	0.0	0.7	1.4	2.2	2.9	3.6	4.3	5.0	5.7	6.4	7.2	7.9	8.6	9.3	10.0	10.8	11.5	12.2	12.9
44	0.0	0.7	1.5	2.2	2.9	3.7	4.4	5.1	5.9	6.6	7.3	8.1	8.8	9.5	10.3	11.0	11.7	12.5	13.2
45	0.0	0.7	1.5	2.2	3.0	3.8	4.5	5.2	6.0	6.8	7.5	8.2	9.0	9.8	10.5	11.2	12.0	12.8	13.5
46	0.0	0.8	1.5	2.3	3.1	3.8	4.6	5.4	6.1	6.9	7.7	8.4	9.2	10.0	10.7	11.5	12.3	13.0	13.8
47	0.0	0.8	1.6	2.4	3.1	3.9	4.7	5.5	6.3	7.0	7.8	8.6	9.4	10.2	11.0	11.8	12.5	13.3	14.1
48	0.0	0.8	1.6	2.4	3.2	4.0	4.8	5.6	6.4	7.2	8.0	8.8	9.6	10.4	11.2	12.0	12.8	13.6	14.4
49	0.0	0.8	1.6	2.4	3.3	4.1	4.9	5.7	6.5	7.4	8.2	9.0	9.8	10.6	11.4	12.2	13.1	13.9	14.7
50	0.0	0.8	1.7	2.5	3.3	4.2	5.0	5.8	6.7	7.5	8.3	9.2	10.0	10.8	11.7	12.5	13.3	14.2	15.0
51	0.0	0.8	1.7	2.6	3.4	4.2	5.1	6.0	6.8	7.6	8.5	9.4	10.2	11.0	11.9	12.8	13.6	14.4	15.3
52	0.0	0.9	1.7	2.6	3.5	4.3	5.2	6.1	6.9	7.8	8.7	9.5	10.4	11.3	12.1	13.0	13.9	14.7	15.6
53	0.0	0.9	1.8	2.6	3.5	4.4	5.3	6.2	7.1	8.0	8.8	9.7	10.6	11.5	12.4	13.2	14.1	15.0	15.9
54	0.0	0.9	1.8	2.7	3.6	4.5	5.4	6.3	7.2	8.1	9.0	9.9	10.8	11.7	12.6	13.5	14.4	15.3	16.2
55	0.0	0.9	1.8	2.8	3.7	4.6	5.5	6.4	7.3	8.2	9.2	10.1	11.0	11.9	12.8	13.8	14.7	15.6	16.5
56	0.0	0.9	1.9	2.8	3.7	4.7	5.6	6.5	7.5	8.4	9.3	10.3	11.2	12.1	13.1	14.0	14.9	15.9	16.8
57	0.0	1.0	1.9	2.8	3.8	4.8	5.7	6.6	7.6	8.6	9.5	10.4	11.4	12.4	13.3	14.2	15.2	16.2	17.1
58	0.0	1.0	1.9	2.9	3.9	4.8	5.8	6.8	7.7	8.7	9.7	10.6	11.6	12.6	13.5	14.5	15.5	16.4	17.4
59	0.0	1.0	2.0	3.0	3.9	4.9	5.9	6.9	7.9	8.8	9.8	10.8	11.8	12.8	13.8	14.8	15.7	16.7	17.7
60	0.0	1.0	2.0	3.0	4.0	5.0	6.0	7.0	8.0	9.0	10.0	11.0	12.0	13.0	14.0	15.0	16.0	17.0	18.0

ARC TO TIME CONVERSION TABLE

Arc °	Time h m	Arc °	Time h m	Arc °	Time h m	Arc °	Time h m	Arc °	Time h m	Arc °	Time h m	Arc '	Time m s	Arc " '	Time s
0	0 00	**60**	4 00	**120**	8 00	**180**	12 00	**240**	16 00	**300**	20 00	**0**	0 00	**0 = 0.0**	0.00
1	0 04	**61**	4 04	**121**	8 04	**181**	12 04	**241**	16 04	**301**	20 04	**1**	0 04	**1**	0.07
2	0 08	**62**	4 08	**122**	8 08	**182**	12 08	**242**	16 08	**302**	20 08	**2**	0 08	**2**	0.13
3	0 12	**63**	4 12	**123**	8 12	**183**	12 12	**243**	16 12	**303**	20 12	**3**	0 12	**3**	0.20
4	0 16	**64**	4 16	**124**	8 16	**184**	12 16	**244**	16 16	**304**	20 16	**4**	0 16	**4**	0.27
5	0 20	**65**	4 20	**125**	8 20	**185**	12 20	**245**	16 20	**305**	20 20	**5**	0 20	**5**	0.33
6	0 24	**66**	4 24	**126**	8 24	**186**	12 24	**246**	16 24	**306**	20 24	**6**	0 24	**6 = 0.1**	0.40
7	0 28	**67**	4 28	**127**	8 28	**187**	12 28	**247**	16 28	**307**	20 28	**7**	0 28	**7**	0.47
8	0 32	**68**	4 32	**128**	8 32	**188**	12 32	**248**	16 32	**308**	20 32	**8**	0 32	**8**	0.53
9	0 36	**69**	4 36	**129**	8 36	**189**	12 36	**249**	16 36	**309**	20 36	**9**	0 36	**9**	0.60
10	0 40	**70**	4 40	**130**	8 40	**190**	12 40	**250**	16 40	**310**	20 40	**10**	0 40	**10**	0.67
11	0 44	**71**	4 44	**131**	8 44	**191**	12 44	**251**	16 44	**311**	20 44	**11**	0 44	**11**	0.73
12	0 48	**72**	4 48	**132**	8 48	**192**	12 48	**252**	16 48	**312**	20 48	**12**	0 48	**12 = 0.2**	0.80
13	0 52	**73**	4 52	**133**	8 52	**193**	12 52	**253**	16 52	**313**	20 52	**13**	0 52	**13**	0.87
14	0 56	**74**	4 56	**134**	8 56	**194**	12 56	**254**	16 56	**314**	20 56	**14**	0 56	**14**	0.93
15	1 00	**75**	5 00	**135**	9 00	**195**	13 00	**255**	17 00	**315**	21 00	**15**	1 00	**15**	1.00
16	1 04	**76**	5 04	**136**	9 04	**196**	13 04	**256**	17 04	**316**	21 04	**16**	1 04	**16**	1.07
17	1 08	**77**	5 08	**137**	9 08	**197**	13 08	**257**	17 08	**317**	21 08	**17**	1 08	**17**	1.13
18	1 12	**78**	5 12	**138**	9 12	**198**	13 12	**258**	17 12	**318**	21 12	**18**	1 12	**18 = 0.3**	1.20
19	1 16	**79**	5 16	**139**	9 16	**199**	13 16	**259**	17 16	**319**	21 16	**19**	1 16	**19**	1.27
20	1 20	**80**	5 20	**140**	9 20	**200**	13 20	**260**	17 20	**320**	21 20	**20**	1 20	**20**	1.33
21	1 24	**81**	5 24	**141**	9 24	**201**	13 24	**261**	17 24	**321**	21 24	**21**	1 24	**21**	1.40
22	1 28	**82**	5 28	**142**	9 28	**202**	13 28	**262**	17 28	**322**	21 28	**22**	1 28	**22**	1.47
23	1 32	**83**	5 32	**143**	9 32	**203**	13 32	**263**	17 32	**323**	21 32	**23**	1 32	**23**	1.53
24	1 36	**84**	5 36	**144**	9 36	**204**	13 36	**264**	17 36	**324**	21 36	**24**	1 36	**24 = 0.4**	1.60
25	1 40	**85**	5 40	**145**	9 40	**205**	13 40	**265**	17 40	**325**	21 40	**25**	1 40	**25**	1.67
26	1 44	**86**	5 44	**146**	9 44	**206**	13 44	**266**	17 44	**326**	21 44	**26**	1 44	**26**	1.73
27	1 48	**87**	5 48	**147**	9 48	**207**	13 48	**267**	17 48	**327**	21 48	**27**	1 48	**27**	1.80
28	1 52	**88**	5 52	**148**	9 52	**208**	13 52	**268**	17 52	**328**	21 52	**28**	1 52	**28**	1.87
29	1 56	**89**	5 56	**149**	9 56	**209**	13 56	**269**	17 56	**329**	21 56	**29**	1 56	**29**	1.93
30	2 00	**90**	6 00	**150**	10 00	**210**	14 00	**270**	18 00	**330**	22 00	**30**	2 00	**30 = 0.5**	2.00
31	2 04	**91**	6 04	**151**	10 04	**211**	14 04	**271**	18 04	**331**	22 04	**31**	2 04	**31**	2.07
32	2 08	**92**	6 08	**152**	10 08	**212**	14 08	**272**	18 08	**332**	22 08	**32**	2 08	**32**	2.13
33	2 12	**93**	6 12	**153**	10 12	**213**	14 12	**273**	18 12	**333**	22 12	**33**	2 12	**33**	2.20
34	2 16	**94**	6 16	**154**	10 16	**214**	14 16	**274**	18 16	**334**	22 16	**34**	2 16	**34**	2.27
35	2 20	**95**	6 20	**155**	10 20	**215**	14 20	**275**	18 20	**335**	22 20	**35**	2 20	**35**	2.33
36	2 24	**96**	6 24	**156**	10 24	**216**	14 24	**276**	18 24	**336**	22 24	**36**	2 24	**36 = 0.6**	2.40
37	2 28	**97**	6 28	**157**	10 28	**217**	14 28	**277**	18 28	**337**	22 28	**37**	2 28	**37**	2.47
38	2 32	**98**	6 32	**158**	10 32	**218**	14 32	**278**	18 32	**338**	22 32	**38**	2 32	**38**	2.53
39	2 36	**99**	6 36	**159**	10 36	**219**	14 36	**279**	18 36	**339**	22 36	**39**	2 36	**39**	2.60
40	2 40	**100**	6 40	**160**	10 40	**220**	14 40	**280**	18 40	**340**	22 40	**40**	2 40	**40**	2.67
41	2 44	**101**	6 44	**161**	10 44	**221**	14 44	**281**	18 44	**341**	22 44	**41**	2 44	**41**	2.73
42	2 48	**102**	6 48	**162**	10 48	**222**	14 48	**282**	18 48	**342**	22 48	**42**	2 48	**42 = 0.7**	2.80
43	2 52	**103**	6 52	**163**	10 52	**223**	14 52	**283**	18 52	**343**	22 52	**43**	2 52	**43**	2.87
44	2 56	**104**	6 56	**164**	10 56	**224**	14 56	**284**	18 56	**344**	22 56	**44**	2 56	**44**	2.93
45	3 00	**105**	7 00	**165**	11 00	**225**	15 00	**285**	19 00	**345**	23 00	**45**	3 00	**45**	3.00
46	3 04	**106**	7 04	**166**	11 04	**226**	15 04	**286**	19 04	**346**	23 04	**46**	3 04	**46**	3.07
47	3 08	**107**	7 08	**167**	11 08	**227**	15 08	**287**	19 08	**347**	23 08	**47**	3 08	**47**	3.13
48	3 12	**108**	7 12	**168**	11 12	**228**	15.12	**288**	19 12	**348**	23 12	**48**	3 12	**48 = 0.8**	3.20
49	3 16	**109**	7 16	**169**	11 16	**229**	15 16	**289**	19 16	**349**	23 16	**49**	3 16	**49**	3.27
50	3 20	**110**	7 20	**170**	11 20	**230**	15 20	**290**	19 20	**350**	23 20	**50**	3 20	**50**	3.33
51	3 24	**111**	7 24	**171**	11 24	**231**	15 24	**291**	19 24	**351**	23 24	**51**	3 24	**51**	3.40
52	3 28	**112**	7 28	**172**	11 28	**232**	15 28	**292**	19 28	**353**	23 28	**52**	3 28	**52**	3.47
53	3 32	**113**	7 32	**173**	11 32	**233**	15 32	**293**	19 32	**353**	23 32	**53**	3 32	**53**	3.53
54	3 36	**114**	7 36	**174**	11 36	**234**	15 36	**294**	19 36	**354**	23 36	**54**	3 36	**54 = 0.9**	3.60
55	3 40	**115**	7 40	**175**	11 40	**235**	15 40	**295**	19 40	**355**	23 40	**55**	3 40	**55**	3.67
56	3 44	**116**	7 44	**176**	11 44	**236**	15 44	**296**	19 44	**356**	23 44	**56**	3 44	**56**	3.73
57	3 48	**117**	7 48	**177**	11 48	**237**	15 48	**297**	19 48	**357**	23 48	**57**	3 48	**57**	3.80
58	3 52	**118**	7 52	**178**	11 52	**238**	15 52	**298**	19 52	**358**	23 52	**58**	3 52	**58**	3.87
59	3 56	**119**	7 56	**179**	11 56	**239**	15 56	**299**	19 56	**359**	23 56	**59**	3 56	**59**	3.93
60	4 00	**120**	8 00	**180**	12 00	**240**	16 00	**300**	20 00	**360**	24 00	**60**	4 00	**60 = 1.0**	4.00

53

ALPHABETICAL INDEX OF PRINCIPAL STARS
WITH THEIR APPROXIMATE PLACES – 2012

Proper Name		Constellation Name	Mag	RA	Dec	SHA	No
Acamar	θ	**Eridani**	**3.2**	2 59	S 40	315	**8**
Achernar	α	**Eridan**	**0.5**	1 38	S 57	335	**5**
Acrux	α	**Crucis**	**1.3**	12 27	S 63	173	**32**
Adhara	ε	**Canis Majoris**	**1.5**	6 59	S 29	255	**20**
Aldebaran	α	**Tauri**	**0.9**	4 37	N 17	291	**11**
Alioth	ε	**Ursae Majoris**	**1.8**	12 55	N 56	166	**35**
Alkaid	η	**Ursae Majoris**	**1.9**	13 48	N 49	153	**37**
Al Na'ir	α	**Gruis**	**1.7**	22 09	S 47	28	**58**
Alnilam	ε	**Orionis**	**1.7**	5 37	S 1	276	**16**
Alphard	α	**Hydrae**	**2.0**	9 28	S 9	218	**27**
Alphecca	α	**Coronae Bor**	**2.2**	15 35	N 27	126	**44**
Alpheratz	α	**Andromedae**	**2.1**	0 09	N 29	358	**1**
Altair	α	**Aquilae**	**0.8**	19 51	N 9	62	**54**
Ankaa	α	**Phoenicis**	**2.4**	0 27	S 42	353	**2**
Antares	α	**Scorpii**	**1.0**	16 30	S 26	112	**45**
Arcturus	α	**Bootis**	**0.0**	14 16	N 19	146	**40**
Atria	α	**Triang Aust**	**1.9**	16 50	S 69	107	**46**
Avior	ε	**Carinae**	**1.9**	8 23	S 60	234	**24**
Bellatrix	γ	**Orionis**	**1.6**	5 26	N 6	279	**14**
Betelgeuse	α	**Orionis**	**0.1–1.2**	5 56	N 7	271	**17**
Canopus	α	**Carinae**	**–0.7**	6 24	S 53	264	**18**
Capella	α	**Aurigae**	**0.1**	5 18	N 46	281	**13**
Castor	α	**Geminorum**	**1.6**	7 35	N 32	246	**21**
Deneb	α	**Cygni**	**1.3**	20 42	N 45	50	**56**
Denebola	β	**Leonis**	**2.1**	11 50	N 15	183	**30**
Diphda	β	**Ceti**	**2.0**	0 44	S 18	349	**4**
Dubhe	α	**Ursae Majoris**	**1.8**	11 04	N 62	194	**29**
Elnath	β	**Tauri**	**1.7**	5 27	N 29	278	**15**
Eltanin	γ	**Draconis**	**2.2**	17 57	N 51	91	**50**
Enif	ε	**Pegasi**	**2.4**	21 45	N 10	34	**57**
Fomalhaut	α	**Piscis Aust**	**1.2**	22 58	S 30	15	**59**
Gacrux	γ	**Crucis**	**1.6**	12 32	S 57	172	**33**
Gienah	γ	**Corvi**	**2.6**	12 16	S 18	176	**31**
Hadar	β	**Centauri**	**0.6**	14 05	S 60	149	**38**
Hamal	α	**Arietis**	**2.0**	2 08	N 24	328	**7**
Kaus Aust.	ε	**Sagittarii**	**1.9**	18 25	S 34	84	**51**
Kochab	β	**Ursae Minoris**	**2.1**	14 51	N 74	137	**43**
Markab	α	**Pegasi**	**2.5**	23 05	N 15	14	**60**
Menkar	α	**Ceti**	**2.5**	3 03	N 4	314	**9**
Menkent	θ	**Centauri**	**2.1**	14 07	S 36	148	**39**
Miaplacidus	β	**Carinae**	**1.7**	9 13	S 70	222	**26**
Mimosa	β	**Crucis**	**1.3**	12 48	S 60	168	**34**
Mirfak	α	**Persei**	**1.8**	3 25	N 50	309	**10**
Nunki	σ	**Sagittarii**	**2.0**	18 56	S 26	76	**53**
Peacock	α	**Pavonis**	**1.9**	20 27	S 57	53	**55**
POLARIS	α	**Urase Minoris**	**2.0**	2 46	N 89	318	**6**
Pollux	β	**Geminorum**	**1.1**	7 46	N 28	243	**23**
Procyon	α	**Canis Minoris**	**0.4**	7 40	N 5	245	**22**
Rasalhague	α	**Ophiuchi**	**2.1**	17 36	N 13	96	**49**
Regulus	α	**Leonis**	**1.4**	10 09	N 12	208	**28**
Rigel	β	**Orionis**	**0.1**	5 15	S 8	281	**12**
Rigil Kent.	α	**Centauri**	**–0.3**	14 40	S 61	140	**41**
Sabik	η	**Ophiuchi**	**2.4**	17 11	S 16	102	**47**
Schedar	α	**Cassiopeiae**	**2.2**	0 41	N 57	350	**3**
Shaula	γ	**Scorpii**	**1.6**	17 34	S 37	96	**48**
Sirius	α	**Canis Majoris**	**–1.5**	6 46	S 17	259	**19**
Spica	α	**Virginis**	**1.0**	13 26	S 11	159	**36**
Suhail	γ	**Velorum**	**2.2**	9 08	S 43	223	**25**
Vega	α	**Lyrae**	**0.0**	18 37	N 39	81	**52**
Zuben'ubi	α	**Librae**	**2.8**	14 52	S 16	137	**42**

The last column refers to the number given to the star in this almanac.The star's exact position may be found according to this number on the monthly pages.

MOON MERIDIAN PASSAGE CORRECTION TABLE

	DAILY DIFFERENCE OF MERIDIAN PASSAGE											
Long.	**39**	**42**	**45**	**48**	**51**	**54**	**57**	**60**	**63**	**66**	**69**	**Long.**
	min	min	min	min	min	min	min	min	min	min	min	
0°	0	0	0	0	0	0	0	0	0	0	0	**0°**
10°	1	1	1	1	1	1	2	2	2	2	2	**10°**
20°	2	2	2	3	3	3	3	3	3	4	4	**20°**
30°	3	3	4	4	4	4	5	5	5	5	6	**30°**
40°	4	5	5	5	6	6	6	7	7	7	8	**40°**
50°	5	6	6	7	7	7	8	8	9	9	10	**50°**
60°	6	7	7	8	8	9	9	10	10	11	11	**60°**
70°	8	8	9	9	10	10	11	12	12	13	13	**70°**
80°	9	9	10	11	11	12	13	13	14	15	15	**80°**
90°	10	10	11	12	13	13	14	15	16	16	17	**90°**
100°	11	12	12	13	14	15	16	17	17	18	19	**100°**
110°	12	13	14	15	16	16	17	18	19	20	21	**110°**
120°	13	14	15	16	17	18	19	20	21	22	23	**120°**
130°	14	15	16	17	18	19	21	22	23	24	25	**130°**
140°	15	16	17	19	20	21	22	23	24	26	27	**140°**
150°	16	17	19	20	21	22	24	25	26	27	29	**150°**
160°	17	19	20	21	23	24	25	27	28	29	31	**160°**
170°	18	20	21	23	24	25	27	28	30	31	33	**170°**
180°	19	21	22	24	25	27	28	30	31	33	34	**180°**

NOTE: Apply correction in **minutes** to the time of meridian passage given in the monthly tables. **Add** the correction if longitude is West; **subtract** if longitude is East.

′	0° Log	0° Nat	1° Log	1° Nat	2° Log	2° Nat	3° Log	3° Nat	4° Log	4° Nat	5° Log	5° Nat	6° Log	6° Nat	′
0	∞	0.0000	1827	0002	6.7847	0006	1369	0014	7.3867	0024	5804	0038	7.7386	0.0055	60
1	2.6264	0.0000	1971	0002	6.7919	0006	1417	0014	7.3903	0025	5833	0038	7.7410	0.0055	59
2	3.2285	0.0000	2112	0002	6.7991	0006	1465	0014	7.3939	0025	5862	0039	7.7434	0.0055	58
3	3.5807	0.0000	2251	0002	6.8062	0006	1512	0014	7.3975	0025	5890	0039	7.7458	0.0056	57
4	3.8305	0.0000	2388	0002	6.8132	0007	1560	0014	7.4010	0025	5919	0039	7.7482	0.0056	56
5	4.0244	0.0000	2522	0002	6.8202	0007	1607	0014	7.4046	0025	5947	0039	7.7506	0.0056	55
6	4.1827	0.0000	2655	0002	6.8271	0007	1653	0015	7.4081	0026	5976	0040	7.7530	0.0057	54
7	4.3166	0.0000	2786	0002	6.8340	0007	1700	0015	7.4116	0026	6004	0040	7.7553	0.0057	53
8	4.4326	0.0000	2914	0002	6.8408	0007	1746	0015	7.4151	0026	6032	0040	7.7577	0.0057	52
9	4.5349	0.0000	3041	0002	6.8476	0007	1792	0015	7.4186	0026	6060	0040	7.7601	0.0058	51
10	4.6264	0.0000	3166	0002	6.8543	0007	1838	0015	7.4221	0026	6089	0041	7.7624	0.0058	50
11	4.7092	0.0000	3289	0002	6.8609	0007	1884	0015	7.4256	0027	6116	0041	7.7647	0.0058	49
12	4.7848	0.0000	3411	0002	6.8675	0007	1929	0016	7.4290	0027	6144	0041	7.7671	0.0058	48
13	4.8543	0.0000	3531	0002	6.8741	0007	1974	0016	7.4325	0027	6172	0041	7.7694	0.0059	47
14	4.9187	0.0000	3649	0002	6.8806	0008	2019	0016	7.4359	0027	6200	0042	7.7717	0.0059	46
15	4.9786	0.0000	3765	0002	6.8870	0008	2064	0016	7.4393	0027	6227	0042	7.7741	0.0059	45
16	5.0347	0.0000	3880	0002	6.8934	0008	2108	0016	7.4427	0028	6255	0042	7.7764	0.0060	44
17	5.0873	0.0000	3994	0003	6.8998	0008	2152	0016	7.4461	0028	6282	0042	7.7787	0.0060	43
18	5.1370	0.0000	4106	0003	6.9061	0008	2196	0017	7.4495	0028	6310	0043	7.7810	0.0060	42
19	5.1839	0.0000	4217	0003	6.9124	0008	2240	0017	7.4528	0028	6337	0043	7.7833	0.0061	41
20	5.2285	0.0000	4326	0003	6.9186	0008	2284	0017	7.4562	0029	6364	0043	7.7855	0.0061	40
21	5.2709	0.0000	4434	0003	6.9248	0008	2327	0017	7.4595	0029	6391	0044	7.7878	0.0061	39
22	5.3113	0.0000	4540	0003	6.9309	0009	2370	0017	7.4628	0029	6418	0044	7.7901	0.0062	38
23	5.3499	0.0000	4646	0003	6.9370	0009	2413	0017	7.4661	0029	6445	0044	7.7924	0.0062	37
24	5.3868	0.0000	4750	0003	6.9431	0009	2456	0018	7.4694	0029	6472	0044	7.7946	0.0062	36
25	5.4223	0.0000	4852	0003	6.9491	0009	2498	0018	7.4727	0030	6499	0045	7.7969	0.0063	35
26	5.4564	0.0000	4954	0003	6.9551	0009	2540	0018	7.4760	0030	6525	0045	7.7991	0.0063	34
27	5.4891	0.0000	5054	0003	6.9610	0009	2582	0018	7.4792	0030	6552	0045	7.8014	0.0063	33
28	5.5207	0.0000	5154	0003	6.9669	0009	2624	0018	7.4825	0030	6578	0045	7.8036	0.0064	32
29	5.5512	0.0000	5252	0003	6.9727	0009	2666	0018	7.4857	0031	6605	0046	7.8059	0.0064	31
30	5.5807	0.0000	5349	0003	6.9785	0010	2707	0019	7.4889	0031	6631	0046	7.8081	0.0064	30
31	5.6091	0.0000	5445	0004	6.9843	0010	2749	0019	7.4921	0031	6657	0046	7.8103	0.0065	29
32	5.6367	0.0000	5540	0004	6.9900	0010	2790	0019	7.4953	0031	6684	0047	7.8125	0.0065	28
33	5.6634	0.0000	5634	0004	6.9957	0010	2830	0019	7.4985	0032	6710	0047	7.8147	0.0065	27
34	5.6894	0.0000	5727	0004	7.0014	0010	2871	0019	7.5017	0032	6736	0047	7.8169	0.0066	26
35	5.7146	0.0001	5818	0004	7.0070	0010	2912	0020	7.5049	0032	6762	0047	7.8191	0.0066	25
36	5.7390	0.0001	5909	0004	7.0126	0010	2952	0020	7.5080	0032	6788	0048	7.8213	0.0066	24
37	5.7628	0.0001	5999	0004	7.0181	0010	2992	0020	7.5111	0032	6813	0048	7.8235	0.0067	23
38	5.7860	0.0001	6088	0004	7.0237	0011	3032	0020	7.5143	0033	6839	0048	7.8257	0.0067	22
39	5.8085	0.0001	6177	0004	7.0291	0011	3072	0020	7.5174	0033	6865	0049	7.8279	0.0067	21
40	5.8305	0.0001	6264	0004	7.0346	0011	3111	0020	7.5205	0033	6890	0049	7.8301	0.0068	20
41	5.8520	0.0001	6350	0004	7.0400	0011	3151	0021	7.5236	0033	6916	0049	7.8322	0.0068	19
42	5.8729	0.0001	6436	0004	7.0454	0011	3190	0021	7.5267	0034	6941	0049	7.8344	0.0068	18
43	5.8934	0.0001	6521	0004	7.0507	0011	3229	0021	7.5297	0034	6967	0050	7.8365	0.0069	17
44	5.9133	0.0001	6605	0005	7.0560	0011	3268	0021	7.5328	0034	6992	0050	7.8387	0.0069	16
45	5.9328	0.0001	6688	0005	7.0613	0012	3306	0021	7.5359	0034	7017	0050	7.8408	0.0069	15
46	5.9519	0.0001	6770	0005	7.0666	0012	3345	0022	7.5389	0035	7042	0051	7.8430	0.0070	14
47	5.9706	0.0001	6852	0005	7.0718	0012	3383	0022	7.5419	0035	7067	0051	7.8451	0.0070	13
48	5.9889	0.0001	6932	0005	7.0770	0012	3421	0022	7.5450	0035	7092	0051	7.8472	0.0070	12
49	6.0068	0.0001	7012	0005	7.0821	0012	3459	0022	7.5480	0035	7117	0051	7.8494	0.0071	11
50	6.0244	0.0001	7092	0005	7.0872	0012	3497	0022	7.5510	0036	7142	0052	7.8515	0.0071	10
51	6.0416	0.0001	7170	0005	7.0923	0012	3535	0023	7.5539	0036	7167	0052	7.8536	0.0071	9
52	6.0584	0.0001	7248	0005	7.0974	0013	3572	0023	7.5569	0036	7191	0052	7.8557	0.0072	8
53	6.0750	0.0001	7325	0005	7.1024	0013	3610	0023	7.5599	0036	7216	0053	7.8578	0.0072	7
54	6.0912	0.0001	7402	0005	7.1074	0013	3647	0023	7.5629	0037	7240	0053	7.8599	0.0072	6
55	6.1071	0.0001	7478	0006	7.1124	0013	3684	0023	7.5658	0037	7265	0053	7.8620	0.0073	5
56	6.1228	0.0001	7553	0006	7.1174	0013	3721	0024	7.5687	0037	7289	0054	7.8641	0.0073	4
57	6.1382	0.0001	7628	0006	7.1223	0013	3757	0024	7.5717	0037	7314	0054	7.8662	0.0073	3
58	6.1533	0.0001	7701	0006	7.1272	0013	3794	0024	7.5746	0038	7338	0054	7.8682	0.0074	2
59	6.1681	0.0001	7775	0006	7.1320	0014	3830	0024	7.5775	0038	7362	0054	7.8703	0.0074	1
60	6.1827	0.0002	7847	0006	7.1369	0014	3867	0024	7.5804	0038	7386	0055	7.8724	0.0075	0
′	Log 359°	Nat	Log 358°	Nat	Log 357°	Nat	Log 356°	Nat	Log 355°	Nat	Log 354°	Nat	Log 353°	Nat	′

NOTE: For compactness, the leading digit is not displayed in every column. Scan other columns for necessary digit.

VERSINES

′	7° Log	7° Nat	8° Log	8° Nat	9° Log	9° Nat	10° Log	10° Nat	11° Log	11° Nat	12° Log	12° Nat	13° Log	13° Nat	′
0	7.8724	0.0075	9882	0097	8.0903	0123	1816	0152	8.2642	0184	3395	0219	8.4087	0.0256	60
1	7.8744	0.0075	9900	0098	8.0919	0124	1831	0152	8.2655	0184	3407	0219	8.4099	0.0257	59
2	7.8765	0.0075	9918	0098	8.0935	0124	1845	0153	8.2668	0185	3419	0220	8.4110	0.0258	58
3	7.8786	0.0076	9936	0099	8.0951	0124	1859	0153	8.2681	0185	3431	0220	8.4121	0.0258	57
4	7.8806	0.0076	9954	0099	8.0967	0125	1874	0154	8.2694	0186	3443	0221	0.4132	0.0259	56
5	7.8826	0.0076	9972	0099	8.0983	0125	1888	0154	8.2707	0187	3455	0222	8.4143	0.0260	55
6	7.8847	0.0077	9990	0100	8.0999	0126	1902	0155	8.2720	0187	3467	0222	8.4154	0.0260	54
7	7.8867	0.0077	0008	0100	8.1015	0126	1917	0155	8.2733	0188	3479	0223	8.4165	0.0261	53
8	7.8887	0.0077	0025	0101	8.1031	0127	1931	0156	8.2746	0188	3491	0223	8.4176	0.0262	52
9	7.8908	0.0078	0043	0101	8.1046	0127	1945	0157	8.2759	0189	3502	0224	8.4187	0.0262	51
10	7.8928	0.0078	0061	0101	8.1062	0128	1959	0157	8.2772	0189	3514	0225	8.4198	0.0263	50
11	7.8948	0.0078	0078	0102	8.1078	0128	1974	0158	8.2785	0190	3526	0225	8.4209	0.0264	49
12	7.8968	0.0079	0096	0102	8.1094	0129	1988	0158	8.2798	0190	3538	0226	8.4220	0.0264	48
13	7.8988	0.0079	0114	0103	8.1109	0129	2002	0159	8.2811	0191	3550	0226	8.4230	0.0265	47
14	7.9008	0.0080	0131	0103	8.1125	0130	2016	0159	8.2824	0192	3562	0227	8.4241	0.0266	46
15	7.9028	0.0080	0149	0103	8.1141	0130	2030	0160	8.2836	0192	3573	0228	8.4252	0.0266	45
16	7.9048	0.0080	0166	0104	8.1156	0131	2044	0160	8.2849	0193	3585	0228	8.4263	0.0267	44
17	7.9068	0.0081	0184	0104	8.1172	0131	2058	0161	8.2862	0193	3597	0229	8.4274	0.0268	43
18	7.9088	0.0081	0201	0105	8.1187	0131	2072	0161	8.2875	0194	3609	0230	8.4285	0.0268	42
19	7.9108	0.0081	0219	0105	8.1203	0132	2086	0162	8.2887	0194	3620	0230	8.4296	0.0269	41
20	7.9127	0.0082	0236	0106	8.1218	0132	2100	0162	8.2900	0195	3632	0231	8.4306	0.0270	40
21	7.9147	0.0082	0253	0106	8.1234	0133	2114	0163	8.2913	0196	3644	0231	8.4317	0.0270	39
22	7.9167	0.0083	0271	0106	8.1249	0133	2128	0163	8.2926	0196	3655	0232	8.4328	0.0271	38
23	7.9186	0.0083	0288	0107	8.1265	0134	2142	0164	8.2938	0197	3667	0233	8.4339	0.0272	37
24	7.9206	0.0083	0305	0107	8.1280	0134	2156	0164	8.2951	0197	3679	0233	8.4350	0.0272	36
25	7.9225	0.0084	0322	0108	8.1295	0135	2170	0165	8.2964	0198	3690	0234	8.4360	0.0273	35
26	7.9245	0.0084	0339	0108	8.1311	0135	2184	0165	8.2976	0198	3702	0235	8.4371	0.0274	34
27	7.9264	0.0084	0357	0109	8.1326	0136	2198	0166	8.2989	0199	3714	0235	8.4382	0.0274	33
28	7.9284	0.0085	0374	0109	8.1341	0136	2211	0166	8.3001	0200	3725	0236	8.4392	0.0275	32
29	7.9303	0.0085	0391	0109	8.1357	0137	2225	0167	8.3014	0200	3737	0236	8.4403	0.0276	31
30	7.9322	0.0086	0408	0110	8.1372	0137	2239	0167	8.3027	0201	3748	0237	8.4414	0.0276	30
31	7.9342	0.0086	0425	0110	8.1387	0138	2253	0168	8.3039	0201	3760	0238	8.4424	0.0277	29
32	7.9361	0.0086	0442	0111	8.1402	0138	2266	0169	8.3052	0202	3771	0238	8.4435	0.0278	28
33	7.9380	0.0087	0459	0111	8.1417	0139	2280	0169	8.3064	0202	3783	0239	8.4446	0.0278	27
34	7.9399	0.0087	0475	0112	8.1432	0139	2294	0170	8.3077	0203	3794	0240	8.4456	0.0279	26
35	7.9418	0.0087	0492	0112	8.1447	0140	2307	0170	8.3089	0204	3806	0240	8.4467	0.0280	25
36	7.9437	0.0088	0509	0112	8.1463	0140	2321	0171	8.3102	0204	3817	0241	8.4478	0.0280	24
37	7.9456	0.0088	0526	0113	8.1478	0141	2335	0171	8.3114	0205	3829	0241	8.4488	0.0281	23
38	7.9475	0.0089	0543	0113	8.1493	0141	2348	0172	8.3126	0205	3840	0242	8.4499	0.0282	22
39	7.9494	0.0089	0559	0114	8.1508	0141	2362	0172	8.3139	0206	3851	0243	8.4509	0.0282	21
40	7.9513	0.0089	0576	0114	8.1522	0142	2375	0173	8.3151	0207	3863	0243	8.4520	0.0283	20
41	7.9532	0.0090	0593	0115	8.1537	0142	2389	0173	8.3164	0207	3874	0244	8.4530	0.0284	19
42	7.9551	0.0090	0609	0115	8.1552	0143	2402	0174	8.3176	0208	3886	0245	8.4541	0.0285	18
43	7.9569	0.0091	0626	0116	8.1567	0143	2416	0174	8.3188	0208	3897	0245	8.4551	0.0285	17
44	7.9588	0.0091	0642	0116	8.1582	0144	2429	0175	8.3200	0209	3908	0246	8.4562	0.0286	16
45	7.9607	0.0091	0659	0116	8.1597	0144	2443	0175	8.3213	0210	3920	0247	8.4572	0.0287	15
46	7.9625	0.0092	0675	0117	8.1612	0145	2456	0176	8.3225	0210	3931	0247	8.4583	0.0287	14
47	7.9644	0.0092	0692	0117	8.1626	0145	2469	0177	8.3237	0211	3942	0248	8.4593	0.0288	13
48	7.9662	0.0093	0708	0118	8.1641	0146	2483	0177	8.3250	0211	3953	0249	8.4604	0.0289	12
49	7.9681	0.0093	0725	0118	8.1656	0146	2496	0178	8.3262	0212	3965	0249	8.4614	0.0289	11
50	7.9699	0.0093	0741	0119	8.1671	0147	2510	0178	8.3274	0213	3976	0250	8.4625	0.0290	10
51	7.9718	0.0094	0757	0119	8.1685	0147	2523	0179	8.3286	0213	3987	0250	8.4635	0.0291	9
52	7.9736	0.0094	0774	0120	8.1700	0148	2536	0179	8.3298	0214	3998	0251	8.4645	0.0291	8
53	7.9755	0.0095	0790	0120	8.1715	0148	2549	0180	8.3310	0214	4010	0252	8.4656	0.0292	7
54	7.9773	0.0095	0806	0120	8.1729	0149	2563	0180	8.3323	0215	4021	0252	8.4666	0.0293	6
55	7.9791	0.0095	0823	0121	8.1744	0149	2576	0181	8.3335	0216	4032	0253	8.4677	0.0294	5
56	7.9809	0.0096	0839	0121	8.1758	0150	2589	0182	8.3347	0216	4043	0254	8.4687	0.0294	4
57	7.9828	0.0096	0855	0122	8.1773	0150	2602	0182	8.3359	0217	4054	0254	8.4697	0.0295	3
58	7.9846	0.0097	0871	0122	8.1787	0151	2615	0183	8.3371	0217	4065	0255	8.4708	0.0296	2
59	7.9864	0.0097	0887	0123	8.1802	0151	2629	0183	8.3383	0218	4076	0256	8.4718	0.0296	1
60	7.9882	0.0097	0903	0123	8.1816	0152	2642	0184	8.3395	0219	4087	0256	8.4728	0.0297	0
′	Log 352°	Nat	Log 351°	Nat	Log 350°	Nat	Log 349°	Nat	Log 348°	Nat	Log 347°	Nat	Log 346°	Nat	′

NOTE: For compactness, the leading digit is not displayed in every column. Scan other columns for necessary digit.

55

VERSINES

VERSINES

'	14° Log	14° Nat	15° Log	15° Nat	16° Log	16° Nat	17° Log	17° Nat	18° Log	18° Nat	19° Log	19° Nat	20° Log	20° Nat	'
0	8.4728	0.0297	5324	0341	8.5881	0387	6404	0437	8.6897	0489	7362	0545	8.7804	0.0603	60
1	8.4738	0.0298	5334	0341	8.5890	0388	6413	0438	8.6905	0490	7370	0546	8.7811	0.0604	59
2	8.4749	0.0298	5343	0342	8.5899	0389	6421	0439	8.6913	0491	7378	0547	8.7818	0.0605	58
3	8.4759	0.0299	5353	0343	8.5908	0390	6430	0440	8.6921	0492	7385	0548	8.7825	0.0606	57
4	8.4769	0.0300	5363	0344	8.5917	0391	6438	0440	8.6929	0493	7393	0549	8.7832	0.0607	56
5	8.4779	0.0301	5372	0345	8.5926	0391	6447	0441	8.6937	0494	7400	0550	8.7839	0.0608	55
6	8.4790	0.0301	5382	0345	8.5935	0392	6455	0442	8.6945	0495	7408	0551	8.7847	0.0609	54
7	8.4800	0.0302	5391	0346	8.5944	0393	6463	0443	8.6953	0496	7415	0551	8.7854	0.0610	53
8	8.4810	0.0303	5401	0347	8.5953	0394	6472	0444	8.6961	0497	7423	0552	8.7861	0.0611	52
9	8.4820	0.0303	5410	0348	8.5962	0395	6480	0445	8.6968	0498	7430	0553	8.7868	0.0612	51
10	8.4830	0.0304	5420	0348	8.5971	0395	6488	0445	8.6976	0496	7438	0554	8.7875	0.0613	50
11	8.4841	0.0305	5429	0349	8.5980	0396	6497	0446	8.6984	0499	7445	0555	8.7882	0.0614	49
12	8.4851	0.0306	5439	0350	8.5989	0397	6505	0447	8.6992	0500	7453	0556	8.7889	0.0615	48
13	8.4861	0.0306	5448	0351	8.5997	0398	6514	0448	8.7000	0501	7460	0557	8.7896	0.0616	47
14	8.4871	0.0307	5458	0351	8.6006	0399	6522	0449	8.7008	0502	7468	0558	8.7903	0.0617	46
15	8.4881	0.0308	5467	0352	8.6015	0400	6530	0450	8.7016	0503	7475	0559	8.7910	0.0618	45
16	8.4891	0.0308	5476	0353	8.6024	0400	6539	0451	8.7024	0504	7482	0560	8.7918	0.0619	44
17	8.4901	0.0309	5486	0354	8.6033	0401	6547	0452	8.7031	0505	7490	0561	8.7925	0.0620	43
18	8.4911	0.0310	5495	0354	8.6042	0402	6555	0452	8.7039	0506	7497	0562	8.7932	0.0621	42
19	8.4921	0.0311	5505	0355	8.6051	0403	6563	0453	8.7047	0507	7505	0563	8.7939	0.0622	41
20	8.4932	0.0311	5514	0356	8.6059	0404	6572	0454	8.7055	0508	7512	0564	8.7946	0.0623	40
21	8.4942	0.0312	5523	0357	8.6068	0404	6580	0455	8.7063	0508	7520	0565	8.7963	0.0624	39
22	8.4952	0.0313	5533	0358	8.6077	0405	6588	0456	8.7071	0509	7527	0566	8.7960	0.0625	38
23	8.4962	0.0313	5542	0358	8.6086	0406	6597	0457	8.7078	0510	7534	0567	8.7967	0.0626	37
24	8.4972	0.0314	5552	0359	8.6094	0407	6605	0458	8.7086	0511	7542	0568	8.7974	0.0627	36
25	8.4982	0.0315	5561	0360	8.6103	0408	6613	0458	8.7094	0512	7549	0569	8.7981	0.0628	35
26	8.4992	0.0316	5570	0361	8.6112	0409	6621	0459	8.7102	0513	7557	0570	8.7988	0.0629	34
27	8.5002	0.0316	5579	0361	8.6121	0409	6630	0460	8.7110	0514	7564	0571	8.7995	0.0630	33
28	8.5012	0.0317	5589	0362	8.6129	0410	6638	0461	8.7117	0515	7571	0572	8.8002	0.0631	32
29	8.5021	0.0318	5598	0363	8.6138	0411	6646	0462	8.7125	0516	7579	0573	8.8009	0.0632	31
30	8.5031	0.0319	5607	0364	8.6147	0412	6654	0463	8.7133	0517	7586	0574	8.8016	0.0633	30
31	8.5041	0.0319	5617	0364	8.6156	0413	6662	0464	8.7141	0518	7593	0575	8.8023	0.0634	29
32	8.5051	0.0320	5626	0365	8.6164	0413	6671	0465	8.7148	0519	7601	0576	8.8030	0.0635	28
33	8.5061	0.0321	5635	0366	8.6173	0414	6679	0465	8.7156	0520	7608	0577	8.8037	0.0636	27
34	8.5071	0.0321	5644	0367	8.6182	0415	6687	0466	8.7164	0520	7615	0577	8.8044	0.0637	26
35	8.5081	0.0322	5654	0368	8.6190	0416	6695	0467	8.7172	0521	7623	0578	8.8051	0.0638	25
36	8.5091	0.0323	5663	0368	8.6199	0417	6703	0468	8.7179	0522	7630	0579	8.8058	0.0639	24
37	8.5101	0.0324	5672	0369	8.6208	0418	6711	0469	8.7187	0523	7637	0580	8.8065	0.0640	23
38	8.5110	0.0324	5681	0370	8.6216	0418	6720	0470	8.7195	0524	7645	0581	8.8072	0.0641	22
39	8.5120	0.0325	5691	0371	8.6225	0419	6728	0471	8.7202	0525	7652	0582	8.8079	0.0642	21
40	8.5130	0.0326	5700	0372	8.6234	0420	6736	0472	8.7210	0526	7659	0583	8.8086	0.0644	20
41	8.5140	0.0327	5709	0372	8.6242	0421	6744	0473	8.7218	0527	7666	0584	8.8092	0.0645	19
42	8.5150	0.0327	5718	0373	8.6251	0422	6752	0473	8.7225	0528	7674	0585	8.8099	0.0646	18
43	8.5160	0.0328	5727	0374	8.6259	0423	6760	0474	8.7233	0529	7681	0586	8.8106	0.0647	17
44	8.5169	0.0329	5736	0375	8.6268	0423	6768	0475	8.7241	0530	7688	0587	8.8113	0.0648	16
45	8.5179	0.0330	5745	0375	8.6277	0424	6776	0476	8.7248	0531	7696	0588	8.8120	0.0649	15
46	8.5189	0.0330	5755	0376	8.6285	0425	6785	0477	8.7256	0532	7703	0589	8.8127	0.0650	14
47	8.5199	0.0331	5764	0377	8.6294	0426	6793	0478	8.7264	0533	7710	0590	8.8134	0.0651	13
48	8.5208	0.0332	5773	0378	8.6302	0427	6801	0479	8.7271	0534	7717	0591	8.8141	0.0652	12
49	8.5218	0.0333	5782	0379	8.6311	0428	6809	0480	8.7279	0534	7725	0592	8.8148	0.0653	11
50	8.5228	0.0333	5791	0379	8.6319	0428	6817	0480	8.7287	0535	7732	0593	8.8155	0.0654	10
51	8.5237	0.0334	5800	0380	8.6328	0429	6825	0481	8.7294	0536	7739	0594	8.8161	0.0655	9
52	8.5247	0.0335	5809	0381	8.6336	0430	6833	0482	8.7302	0537	7746	0595	8.8168	0.0656	8
53	8.5257	0.0335	5818	0382	8.6345	0431	6841	0483	8.7309	0538	7753	0596	8.8175	0.0657	7
54	8.5266	0.0336	5827	0383	8.6353	0432	6849	0484	8.7317	0539	7761	0597	8.8182	0.0658	6
55	8.5276	0.0337	5836	0383	8.6362	0433	6857	0485	8.7325	0540	7768	0598	8.8189	0.0659	5
56	8.5286	0.0338	5845	0384	8.6370	0434	6865	0486	8.7332	0541	7775	0599	8.8196	0.0660	4
57	8.5295	0.0338	5854	0385	8.6379	0434	6873	0487	8.7340	0542	7782	0600	8.8202	0.0661	3
58	8.5305	0.0339	5863	0386	8.6387	0435	6881	0488	8.7347	0543	7789	0601	8.8209	0.0662	2
59	8.5315	0.0340	5872	0387	8.6396	0436	6889	0489	8.7355	0544	7797	0602	8.8216	0.0663	1
60	8.5324	0.0341	5881	0387	8.6404	0437	6897	0489	8.7362	0545	7804	0603	8.8223	0.0664	0
'	Log 345°	Nat	Log 344°	Nat	Log 343°	Nat	Log 342°	Nat	Log 341°	Nat	Log 340°	Nat	Log 339°	Nat	'

NOTE: For compactness. the leading digit is not displayed in every column. Scan other columns for necessary digit.

VERSINES

'	21° Log	21° Nat	22° Log	22° Nat	23° Log	23° Nat	24° Log	24° Nat	25° Log	25° Nat	26° Log	26° Nat	27° Log	27° Nat	'
0	8.8223	0.0664	8622	0728	8.9003	0795	9368	0865	8.9717	0937	0052	1012	9.0374	0.1090	60
1	8.8230	0.0665	8629	0729	8.9010	0796	9374	0866	8.9723	0938	0058	1013	9.0379	0.1091	59
2	8.8237	0.0666	8635	0730	8.9016	0797	9380	0867	8.9728	0939	0063	1015	9.0385	0.1093	58
3	8.8243	0.0667	8642	0731	8.9022	0798	9386	0868	8.9734	0941	0068	1016	9.0390	0.1094	57
4	8.8250	0.0668	8648	0733	8.9028	0800	9392	0869	8.9740	0942	0074	1017	9.0395	0.1095	56
5	8.8257	0.0669	8655	0734	8.9034	0801	9398	0870	8.9745	0943	0079	1018	9.0400	0.1097	55
6	8.8264	0.0670	8661	0735	8.9041	0802	9403	0872	8.9751	0944	0085	1020	9.0406	0.1098	54
7	8.8271	0.0672	8668	0736	8.9047	0803	9409	0873	8.9757	0946	0090	1021	9.0411	0.1099	53
8	8.8277	0.0673	8674	0737	8.9053	0804	9415	0874	8.9762	0947	0096	1022	9.0416	0.1101	52
9	8.8284	0.0674	8681	0738	8.9059	0805	9421	0875	8.9768	0948	0101	1024	9.0421	0.1102	51
10	8.8291	0.0675	8687	0739	8.9065	0806	9427	0876	8.9774	0949	0107	1025	9.0426	0.1103	50
11	8.8298	0.0676	8693	0740	8.9071	0808	9433	0878	8.9779	0950	0112	1026	9.0432	0.1105	49
12	8.8304	0.0677	8700	0741	8.9078	0809	9439	0879	8.9785	0952	0117	1027	9.0437	0.1106	48
13	8.8311	0.0678	8706	0742	8.9084	0810	9445	0880	8.9791	0953	0123	1029	9.0442	0.1107	47
14	8.8318	0.0679	8713	0743	8.9090	0811	9451	0881	8.9796	0954	0128	1030	9.0447	0.1108	46
15	8.8325	0.0680	8719	0745	8.9096	0812	9457	0882	8.9802	0955	0134	1031	9.0453	0.1110	45
16	8.8331	0.0681	8726	0746	8.9102	0813	9462	0884	8.9808	0957	0139	1033	9.0458	0.1111	44
17	8.8338	0.0682	8732	0747	8.9108	0814	9468	0885	8.9813	0958	0145	1034	9.0463	0.1112	43
18	8.8345	0.0683	8738	0748	8.9114	0816	9474	0886	8.9819	0959	0150	1035	9.0468	0.1114	42
19	8.8351	0.0684	8745	0749	8.9121	0817	9480	0887	8.9825	0960	0155	1036	9.0473	0.1115	41
20	8.8358	0.0685	8751	0750	8.9127	0818	9486	0888	8.9830	0962	0161	1038	9.0479	0.1116	40
21	8.8365	0.0686	8758	0751	8.9133	0819	9492	0890	8.9836	0963	0166	1039	9.0484	0.1118	39
22	8.8372	0.0687	8764	0752	8.9139	0820	9498	0891	8.9841	0964	0172	1040	9.0489	0.1119	38
23	8.8378	0.0688	8770	0753	8.9145	0821	9503	0892	8.9847	0965	0177	1042	9.0494	0.1121	37
24	8.8385	0.0689	8777	0755	8.9151	0822	9509	0893	8.9853	0967	0182	1043	9.0499	0.1122	36
25	8.8392	0.0691	8783	0756	8.9157	0824	9515	0894	8.9858	0968	0188	1044	9.0505	0.1123	35
26	8.8398	0.0692	8790	0757	8.9163	0825	9521	0896	8.9864	0969	0193	1045	9.0510	0.1125	34
27	8.8405	0.0693	8796	0758	8.9169	0826	9527	0897	8.9869	0970	0198	1047	9.0515	0.1126	33
28	8.8412	0.0694	8802	0759	8.9175	0827	9533	0898	8.9875	0972	0204	1048	9.0520	0.1127	32
29	8.8418	0.0695	8809	0760	8.9182	0828	9538	0899	8.9881	0973	0209	1049	9.0525	0.1129	31
30	8.8425	0.0696	8815	0761	8.9188	0829	9544	0900	8.9886	0974	0215	1051	9.0530	0.1130	30
31	8.8432	0.0697	8821	0762	8.9194	0831	9550	0902	8.9892	0975	0220	1052	9.0536	0.1131	29
32	8.8438	0.0698	8828	0763	8.9200	0832	9556	0903	8.9897	0977	0225	1053	9.0541	0.1133	28
33	8.8445	0.0699	8834	0765	8.9206	0833	9562	0904	8.9903	0978	0231	1055	9.0546	0.1134	27
34	8.8452	0.0700	8840	0766	8.9212	0834	9568	0905	8.9909	0979	0236	1056	9.0551	0.1135	26
35	8.8458	0.0701	8847	0767	8.9218	0835	9573	0906	8.9914	0980	0241	1057	9.0556	0.1137	25
36	8.8465	0.0702	8853	0768	8.9224	0836	9579	0908	8.9920	0982	0247	1058	9.0561	0.1138	24
37	8.8471	0.0703	8859	0769	8.9230	0838	9585	0909	8.9925	0983	0252	1060	9.0566	0.1139	23
38	8.8478	0.0704	8866	0770	8.9236	0839	9591	0910	8.9931	0984	0257	1061	9.0572	0.1141	22
39	8.8485	0.0705	8872	0771	8.9242	0840	9596	0911	8.9936	0985	0263	1062	9.0577	0.1142	21
40	8.8491	0.0707	8878	0772	8.9248	0841	9602	0912	8.9942	0987	0268	1064	9.0582	0.1143	20
41	8.8498	0.0708	8885	0773	8.9254	0842	9608	0914	8.9947	0988	0273	1065	9.0587	0.1145	19
42	8.8504	0.0709	8891	0775	8.9260	0843	9614	0915	8.9953	0989	0279	1066	9.0592	0.1146	18
43	8.8511	0.0710	8897	0776	8.9266	0845	9620	0916	8.9959	0990	0284	1068	9.0597	0.1147	17
44	8.8518	0.0711	8903	0777	8.9272	0846	9625	0917	8.9964	0992	0289	1069	9.0602	0.1149	16
45	8.8524	0.0712	8910	0778	8.9278	0847	9631	0919	8.9970	0993	0295	1070	9.0607	0.1150	15
46	8.8531	0.0713	8916	0779	8.9284	0848	9637	0920	8.9975	0994	0300	1072	9.0613	0.1151	14
47	8.8537	0.0714	8922	0780	8.9290	0849	9643	0921	8.9981	0996	0305	1073	9.0618	0.1153	13
48	8.8544	0.0715	8929	0781	8.9296	0850	9648	0922	8.9986	0997	0311	1074	9.0623	0.1154	12
49	8.8550	0.0716	8935	0782	8.9302	0852	9654	0923	8.9992	0998	0316	1075	9.0628	0.1156	11
50	8.8557	0.0717	8941	0784	8.9308	0853	9660	0925	8.9997	0999	0321	1077	9.0633	0.1157	10
51	8.8564	0.0718	8947	0785	8.9314	0854	9666	0926	9.0003	1001	0327	1078	9.0638	0.1158	9
52	8.8570	0.0719	8954	0786	8.9320	0855	9671	0927	9.0008	1002	0332	1079	9.0643	0.1160	8
53	8.8577	0.0721	8960	0787	8.9326	0856	9677	0928	9.0014	1003	0337	1081	9.0648	0.1161	7
54	8.8583	0.0722	8966	0788	8.9332	0857	9683	0930	9.0019	1004	0342	1082	9.0653	0.1162	6
55	8.8590	0.0723	8972	0789	8.9338	0859	9688	0931	9.0025	1006	0348	1083	9.0658	0.1164	5
56	8.8596	0.0724	8979	0790	8.9344	0860	9694	0932	9.0030	1007	0353	1085	9.0664	0.1165	4
57	8.8603	0.0725	8985	0792	8.9350	0861	9700	0933	9.0036	1008	0358	1086	9.0669	0.1166	3
58	8.8609	0.0726	8991	0793	8.9356	0862	9706	0934	9.0041	1010	0363	1087	9.0674	0.1168	2
59	8.8616	0.0727	8997	0794	8.9362	0863	9711	0936	9.0047	1011	0369	1089	9.0679	0.1169	1
60	8.8622	0.0728	9003	0795	8.9368	0865	9717	0937	9.0052	1012	0374	1090	9.0684	0.1171	0
'	Log 338°	Nat	Log 337°	Nat	Log 336°	Nat	Log 335°	Nat	Log 334°	Nat	Log 333°	Nat	Log 332°	Nat	'

NOTE: For compactness. the leading digit is not displayed in every column. Scan other columns for necessary digit.

VERSINES

′	28° Log	Nat	29° Log	Nat	30° Log	Nat	31° Log	Nat	32° Log	Nat	33° Log	Nat	34° Log	Nat	′
0	9.0684	0.1171	0982	1254	9.1270	1340	1548	1428	9.1817	1520	2077	1613	9.2329	0.1710	60
1	9.0689	0.1172	0987	1255	9.1275	1341	1553	1430	9.1821	1521	2081	1615	9.2333	0.1711	59
2	9.0694	0.1173	0992	1257	9.1280	1343	1557	1431	9.1826	1523	2086	1616	9.2337	0.1713	58
3	9.0699	0.1175	0997	1258	9.1284	1344	1562	1433	9.1830	1524	2090	1618	9.2341	0.1715	57
4	9.0704	0.1176	1002	1259	9.1289	1346	1566	1434	9.1835	1526	2094	1620	9.2346	0.1716	56
5	9.0709	0.1177	1007	1261	9.1294	1347	1571	1436	9.1839	1527	2098	1621	9.2350	0.1718	55
6	9.0714	0.1179	1012	1262	9.1298	1348	1576	1437	9.1843	1529	2103	1623	9.2354	0.1719	54
7	9.0719	0.1180	1016	1264	9.1303	1350	1580	1439	9.1848	1530	2107	1624	9.2358	0.1721	53
8	9.0724	0.1181	1021	1265	9.1308	1351	1585	1440	9.1852	1532	2111	1626	9.2362	0.1723	52
9	9.0729	0.1183	1026	1267	9.1313	1353	1589	1442	9.1857	1533	2115	1628	9.2366	0.1724	51
10	9.0734	0.1184	1031	1268	9.1317	1354	1594	1443	9.1861	1535	2120	1629	9.2370	0.1726	50
11	9.0739	0.1186	1036	1269	9.1322	1356	1598	1445	9.1865	1537	2124	1631	9.2374	0.1728	49
12	9.0744	0.1187	1041	1271	9.1327	1357	1603	1446	9.1870	1538	2128	1632	9.2378	0.1729	48
13	9.0749	0.1188	1046	1272	9.1331	1359	1607	1448	9.1874	1540	2132	1634	9.2383	0.1731	47
14	9.0754	0.1190	1050	1274	9.1336	1360	1612	1449	9.1879	1541	2137	1636	9.2387	0.1732	46
15	9.0759	0.1191	1055	1275	9.1341	1362	1616	1451	9.1883	1543	2141	1637	9.2391	0.1734	45
16	9.0764	0.1192	1060	1276	9.1345	1363	1621	1452	9.1887	1544	2145	1639	9.2395	0.1736	44
17	9.0769	0.1194	1065	1278	9.1350	1365	1625	1454	9.1892	1546	2149	1640	9.2399	0.1737	43
18	9.0775	0.1195	1070	1279	9.1355	1366	1630	1455	9.1896	1547	2154	1642	9.2403	0.1739	42
19	9.0780	0.1197	1075	1281	9.1359	1368	1634	1457	9.1900	1549	2158	1644	9.2407	0.1741	41
20	9.0785	0.1198	1079	1282	9.1364	1369	1639	1458	9.1905	1550	2162	1645	9.2411	0.1742	40
21	9.0790	0.1199	1084	1284	9.1369	1370	1643	1460	9.1909	1552	2166	1647	9.2415	0.1744	39
22	9.0795	0.1201	1089	1285	9.1373	1372	1648	1461	9.1913	1554	2170	1648	9.2419	0.1746	38
23	9.0800	0.1202	1094	1286	9.1378	1373	1652	1463	9.1918	1555	2175	1650	9.2423	0.1747	37
24	9.0805	0.1204	1099	1288	9.1383	1375	1657	1464	9.1922	1557	2179	1652	9.2428	0.1749	36
25	9.0810	0.1205	1104	1289	9.1387	1376	1661	1466	9.1926	1558	2183	1653	9.2432	0.1751	35
26	9.0814	0.1206	1108	1291	9.1392	1378	1666	1468	9.1931	1560	2187	1655	9.2436	0.1752	34
27	9.0819	0.1208	1113	1292	9.1397	1379	1670	1469	9.1935	1561	2191	1656	9.2440	0.1754	33
28	9.0824	0.1209	1118	1294	9.1401	1381	1675	1471	9.1939	1563	2196	1658	9.2444	0.1755	32
29	9.0829	0.1210	1123	1295	9.1406	1382	1679	1472	9.1944	1565	2200	1660	9.2448	0.1757	31
30	9.0834	0.1212	1128	1296	9.1410	1384	1684	1474	9.1948	1566	2204	1661	9.2452	0.1759	30
31	9.0839	0.1213	1132	1298	9.1415	1385	1688	1475	9.1952	1568	2208	1663	9.2456	0.1760	29
32	9.0844	0.1215	1137	1299	9.1420	1387	1693	1477	9.1957	1569	2212	1664	9.2460	0.1762	28
33	9.0849	0.1216	1142	1301	9.1424	1388	1697	1478	9.1961	1571	2217	1666	9.2464	0.1764	27
34	9.0854	0.1217	1147	1302	9.1429	1390	1702	1480	9.1965	1572	2221	1668	9.2468	0.1765	26
35	9.0859	0.1219	1151	1304	9.1434	1391	1706	1481	9.1970	1574	2225	1669	9.2472	0.1767	25
36	9.0864	0.1220	1156	1305	9.1438	1393	1711	1483	9.1974	1575	2229	1671	9.2476	0.1769	24
37	9.0869	0.1222	1161	1306	9.1443	1394	1715	1484	9.1978	1577	2233	1672	9.2480	0.1770	23
38	9.0874	0.1223	1166	1308	9.1447	1396	1720	1486	9.1983	1579	2238	1674	9.2484	0.1772	22
39	9.0879	0.1224	1171	1309	9.1452	1397	1724	1487	9.1987	1580	2242	1676	9.2489	0.1774	21
40	9.0884	0.1226	1175	1311	9.1457	1399	1728	1489	9.1991	1582	2246	1677	9.2493	0.1775	20
41	9.0889	0.1227	1180	1312	9.1461	1400	1733	1490	9.1996	1583	2250	1679	9.2497	0.1777	19
42	9.0894	0.1229	1185	1314	9.1466	1401	1737	1492	9.2000	1585	2254	1680	9.2501	0.1779	18
43	9.0899	0.1230	1190	1315	9.1470	1403	1742	1493	9.2004	1586	2258	1682	9.2505	0.1780	17
44	9.0904	0.1231	1194	1317	9.1475	1404	1746	1495	9.2009	1588	2263	1684	9.2509	0.1782	16
45	9.0909	0.1233	1199	1318	9.1480	1406	1751	1496	9.2013	1590	2267	1685	9.2513	0.1784	15
46	9.0914	0.1234	1204	1319	9.1484	1407	1755	1498	9.2017	1591	2271	1687	9.2517	0.1785	14
47	9.0919	0.1236	1209	1321	9.1489	1409	1760	1500	9.2021	1593	2275	1689	9.2521	0.1787	13
48	9.0923	0.1237	1213	1322	9.1493	1410	1764	1501	9.2026	1594	2279	1690	9.2525	0.1789	12
49	9.0928	0.1238	1218	1324	9.1498	1412	1768	1503	9.2030	1596	2283	1692	9.2529	0.1790	11
50	9.0933	0.1240	1223	1325	9.1503	1413	1773	1504	9.2034	1597	2288	1693	9.2533	0.1792	10
51	9.0938	0.1241	1228	1327	9.1507	1415	1777	1506	9.2039	1599	2292	1695	9.2537	0.1793	9
52	9.0943	0.1243	1232	1328	9.1512	1416	1782	1507	9.2043	1601	2296	1697	9.2541	0.1795	8
53	9.0948	0.1244	1237	1330	9.1516	1418	1786	1509	9.2047	1602	2300	1698	9.2545	0.1797	7
54	9.0953	0.1245	1242	1331	9.1521	1419	1791	1510	9.2052	1604	2304	1700	9.2549	0.1798	6
55	9.0958	0.1247	1247	1332	9.1525	1421	1795	1512	9.2056	1605	2308	1701	9.2553	0.1800	5
56	9.0963	0.1248	1251	1334	9.1530	1422	1799	1513	9.2060	1607	2312	1703	9.2557	0.1802	4
57	9.0968	0.1250	1256	1335	9.1535	1424	1804	1515	9.2064	1609	2317	1705	9.2561	0.1803	3
58	9.0973	0.1251	1261	1337	9.1539	1425	1808	1516	9.2069	1610	2321	1706	9.2565	0.1805	2
59	9.0977	0.1252	1266	1338	9.1544	1427	1813	1518	9.2073	1612	2325	1708	9.2569	0.1807	1
60	9.0982	0.1254	1270	1340	9.1548	1428	1817	1520	9.2077	1613	2329	1710	9.2573	0.1808	0
′	Log 331°	Nat	Log 330°	Nat	Log 329°	Nat	Log 328°	Nat	Log 327°	Nat	Log 327°	Nat	Log 325°	Nat	′

NOTE: For compactness. the leading digit is not displayed in every column. Scan other columns for necessary digit.

VERSINES

′	35° Log	Nat	36° Log	Nat	37° Log	Nat	38° Log	Nat	39° Log	Nat	40° Log	Nat	41° Log	Nat	′
0	9.2573	0.1808	2810	1910	9.3040	2014	3263	2120	9.3480	2229	3691	2340	9.3897	0.2453	60
1	9.2577	0.1810	2814	1912	9.3044	2015	3267	2122	9.3484	2230	3695	2341	9.3900	0.2455	59
2	9.2581	0.1812	2818	1913	9.3047	2017	3270	2123	9.3487	2232	3698	2343	9.3904	0.2457	58
3	9.2585	0.1813	2822	1915	9.3051	2019	3274	2125	9.3491	2234	3702	2345	9.3907	0.2459	57
4	9.2589	0.1815	2825	1917	9.3055	2021	3278	2127	9.3494	2236	3705	2347	9.3910	0.2461	56
5	9.2593	0.1817	2829	1918	9.3059	2022	3281	2129	9.3498	2238	3709	2349	9.3914	0.2462	55
6	9.2597	0.1819	2833	1920	9.3062	2024	3285	2131	9.3502	2240	3712	2351	9.3917	0.2464	54
7	9.2601	0.1820	2837	1922	9.3066	2026	3289	2132	9.3505	2241	3716	2353	9.3920	0.2466	53
8	9.2605	0.1822	2841	1924	9.3070	2028	3292	2134	9.3509	2243	3719	2355	9.3924	0.2468	52
9	9.2609	0.1824	2845	1925	9.3074	2029	3296	2136	9.3512	2245	3723	2356	9.3927	0.2470	51
10	9.2613	0.1825	2849	1927	9.3077	2031	3300	2138	9.3516	2247	3726	2358	9.3931	0.2472	50
11	9.2617	0.1827	2853	1929	9.3081	2033	3303	2140	9.3519	2249	3729	2360	9.3934	0.2474	49
12	9.2621	0.1829	2856	1930	9.3085	2035	3307	2141	9.3523	2251	3733	2362	9.3937	0.2476	48
13	9.2625	0.1830	2860	1932	9.3089	2036	3311	2143	9.3526	2252	3736	2364	9.3941	0.2478	47
14	9.2629	0.1832	2864	1934	9.3093	2038	3314	2145	9.3530	2254	3740	2366	9.3944	0.2480	46
15	9.2633	0.1834	2868	1936	9.3096	2040	3318	2147	9.3534	2256	3743	2368	9.3947	0.2482	45
16	9.2637	0.1835	2872	1937	9.3100	2042	3322	2149	9.3537	2258	3747	2370	9.3951	0.2484	44
17	9.2641	0.1837	2876	1939	9.3104	2044	3325	2150	9.3541	2260	3750	2371	9.3954	0.2485	43
18	9.2645	0.1839	2880	1941	9.3107	2045	3329	2152	9.3544	2262	3754	2373	9.3957	0.2487	42
19	9.2649	0.1840	2883	1942	9.3111	2047	3333	2154	9.3548	2263	3757	2375	9.3961	0.2489	41
20	9.2653	0.1842	2887	1944	9.3115	2049	3336	2156	9.3551	2265	3760	2377	9.3964	0.2491	40
21	9.2657	0.1844	2891	1946	9.3119	2051	3340	2158	9.3555	2267	3764	2379	9.3967	0.2493	39
22	9.2661	0.1845	2895	1948	9.3122	2052	3343	2159	9.3558	2269	3767	2381	9.3971	0.2495	38
23	9.2665	0.1847	2899	1949	9.3126	2054	3347	2161	9.3562	2271	3771	2383	9.3974	0.2497	37
24	9.2669	0.1849	2903	1951	9.3130	2056	3351	2163	9.3565	2273	3774	2385	9.3977	0.2499	36
25	9.2673	0.1850	2907	1953	9.3134	2058	3354	2165	9.3569	2275	3778	2387	9.3981	0.2501	35
26	9.2677	0.1852	2910	1955	9.3137	2059	3358	2167	9.3572	2276	3781	2388	9.3984	0.2503	34
27	9.2681	0.1854	2914	1956	9.3141	2061	3362	2168	9.3576	2278	3784	2390	9.3987	0.2505	33
28	9.2685	0.1855	2918	1958	9.3145	2063	3365	2170	9.3579	2280	3788	2392	9.3991	0.2507	32
29	9.2688	0.1857	2922	1960	9.3149	2065	3369	2172	9.3583	2282	3791	2394	9.3994	0.2509	31
30	9.2692	0.1859	2926	1961	9.3152	2066	3372	2174	9.3586	2284	3795	2396	9.3998	0.2510	30
31	9.2696	0.1861	2930	1963	9.3156	2068	3376	2176	9.3590	2286	3798	2398	9.4001	0.2512	29
32	9.2700	0.1862	2933	1965	9.3160	2070	3380	2178	9.3594	2287	3802	2400	9.4004	0.2514	28
33	9.2704	0.1864	2937	1967	9.3163	2072	3383	2179	9.3597	2289	3805	2402	9.4008	0.2516	27
34	9.2708	0.1866	2941	1968	9.3167	2074	3387	2181	9.3601	2291	3808	2404	9.4011	0.2518	26
35	9.2712	0.1867	2945	1970	9.3171	2075	3390	2183	9.3604	2293	3812	2405	9.4014	0.2520	25
36	9.2716	0.1869	2949	1972	9.3175	2077	3394	2185	9.3608	2295	3815	2407	9.4017	0.2522	24
37	9.2720	0.1871	2953	1974	9.3178	2079	3398	2187	9.3611	2297	3819	2409	9.4021	0.2524	23
38	9.2724	0.1872	2956	1975	9.3182	2081	3401	2188	9.3615	2299	3822	2411	9.4024	0.2526	22
39	9.2728	0.1874	2960	1977	9.3186	2082	3405	2190	9.3618	2300	3826	2413	9.4027	0.2528	21
40	9.2732	0.1876	2964	1979	9.3189	2084	3409	2192	9.3622	2302	3829	2415	9.4031	0.2530	20
41	9.2736	0.1877	2968	1981	9.3193	2086	3412	2194	9.3625	2304	3832	2417	9.4034	0.2532	19
42	9.2740	0.1879	2972	1982	9.3197	2088	3416	2196	9.3629	2306	3836	2419	9.4037	0.2534	18
43	9.2744	0.1881	2975	1984	9.3201	2090	3419	2198	9.3632	2308	3839	2421	9.4041	0.2536	17
44	9.2747	0.1883	2979	1986	9.3204	2091	3423	2199	9.3636	2310	3843	2422	9.4044	0.2537	16
45	9.2751	0.1884	2983	1987	9.3208	2093	3427	2201	9.3639	2312	3846	2424	9.4047	0.2539	15
46	9.2755	0.1886	2987	1989	9.3212	2095	3430	2203	9.3643	2313	3849	2426	9.4051	0.2541	14
47	9.2759	0.1888	2991	1991	9.3215	2097	3434	2205	9.3646	2315	3853	2428	9.4054	0.2543	13
48	9.2763	0.1889	2994	1993	9.3219	2098	3437	2207	9.3650	2317	3856	2430	9.4057	0.2545	12
49	9.2767	0.1891	2998	1994	9.3223	2100	3441	2208	9.3653	2319	3860	2432	9.4061	0.2547	11
50	9.2771	0.1893	3002	1996	9.3226	2102	3444	2210	9.3657	2321	3863	2434	9.4064	0.2549	10
51	9.2775	0.1894	3006	1998	9.3230	2104	3448	2212	9.3660	2323	3866	2436	9.4067	0.2551	9
52	9.2779	0.1896	3010	2000	9.3234	2106	3452	2214	9.3664	2325	3870	2438	9.4071	0.2553	8
53	9.2783	0.1898	3013	2001	9.3237	2107	3455	2216	9.3667	2326	3873	2440	9.4074	0.2555	7
54	9.2787	0.1900	3017	2003	9.3241	2109	3459	2218	9.3670	2328	3877	2441	9.4077	0.2557	6
55	9.2790	0.1901	3021	2005	9.3245	2111	3462	2219	9.3674	2330	3880	2443	9.4080	0.2559	5
56	9.2794	0.1903	3025	2007	9.3248	2113	3466	2221	9.3677	2332	3883	2445	9.4084	0.2561	4
57	9.2798	0.1905	3028	2008	9.3252	2115	3469	2223	9.3681	2334	3887	2447	9.4087	0.2563	3
58	9.2802	0.1906	3032	2010	9.3256	2116	3473	2225	9.3684	2336	3890	2449	9.4090	0.2565	2
59	9.2806	0.1908	3036	2012	9.3259	2118	3477	2227	9.3688	2338	3893	2451	9.4094	0.2567	1
60	9.2810	0.1910	3040	2014	9.3263	2120	3480	2229	9.3691	2340	3897	2453	9.4097	0.2569	0
′	Log 324°	Nat	Log 323°	Nat	Log 322°	Nat	Log 321°	Nat	Log 320°	Nat	Log 319°	Nat	Log 318°	Nat	′

NOTE: For compactness. the leading digit is not displayed in every column. Scan other columns for necessary digit.

VERSINES

VERSINES

′	42° Log	Nat	43° Log	Nat	44° Log	Nat	45° Log	Nat	46° Log	Nat	47° Log	Nat	48° Log	Nat	′
0	9.4097	0.2569	4292	2686	9.4482	2807	4667	2929	9.4848	3053	5024	3180	9.5197	0.3309	60
1	9.4100	0.2570	4295	2688	9.4485	2809	4670	2931	9.4851	3056	5027	3182	9.5199	0.3311	59
2	9.4103	0.2572	4298	2690	9.4488	2811	4673	2933	9.4854	3058	5030	3184	9.5202	0.3313	58
3	9.4107	0.2574	4301	2692	9.4491	2813	4676	2935	9.4857	3060	5033	3186	9.5205	0.3315	57
4	9.4110	0.2576	4305	2694	9.4494	2815	4679	2937	9.4860	3062	5036	3189	9.5208	0.3317	56
5	9.4113	0.2578	4308	2696	9.4497	2817	4682	2939	9.4863	3064	5039	3191	9.5211	0.3320	55
6	9.4117	0.2580	4311	2698	9.4501	2819	4685	2941	9.4866	3066	5042	3193	9.5214	0.3322	54
7	9.4120	0.2582	4314	2700	9.4504	2821	4688	2943	9.4869	3068	5045	3195	9.5216	0.3324	53
8	9.4123	0.2584	4317	2702	9.4507	2823	4691	2945	9.4872	3070	5048	3197	9.5219	0.3326	52
9	9.4126	0.2586	4321	2704	9.4510	2825	4694	2947	9.4875	3072	5050	3199	9.5222	0.3328	51
10	9.4130	0.2588	4324	2706	9.4513	2827	4698	2950	9.4878	3074	5053	3201	9.5225	0.3330	50
11	9.4133	0.2590	4327	2708	9.4516	2829	4701	2952	9.4881	3076	5056	3203	9.5228	0.3333	49
12	9.4136	0.2592	4330	2710	9.4519	2831	4704	2954	9.4883	3079	5059	3206	9.5231	0.3335	48
13	9.4140	0.2594	4333	2712	9.4522	2833	4707	2956	9.4886	3081	5062	3208	9.5233	0.3337	47
14	9.4143	0.2596	4337	2714	9.4525	2835	4710	2958	9.4889	3083	5065	3210	9.5236	0.3339	46
15	9.4146	0.2598	4340	2716	9.4529	2837	4713	2960	9.4892	3085	5068	3212	9.5239	0.3341	45
16	9.4149	0.2600	4343	2718	9.4532	2839	4716	2962	9.4895	3087	5071	3214	9.5242	0.3343	44
17	9.4153	0.2602	4346	2720	9.4535	2841	4719	2964	9.4898	3089	5074	3216	9.5245	0.3346	43
18	9.4156	0.2604	4349	2722	9.4538	2843	4722	2966	9.4901	3091	5076	3218	9.5247	0.3348	42
19	9.4159	0.2606	4352	2724	9.4541	2845	4725	2968	9.4904	3093	5079	3221	9.5250	0.3350	41
20	9.4162	0.2608	4356	2726	9.4544	2847	4728	2970	9.4907	3095	5082	3223	9.5253	0.3352	40
21	9.4166	0.2610	4359	2728	9.4547	2849	4731	2972	9.4910	3097	5085	3225	9.5256	0.3354	39
22	9.4169	0.2612	4362	2730	9.4550	2851	4734	2974	9.4913	3100	5088	3227	9.5259	0.3356	38
23	9.4172	0.2613	4365	2732	9.4553	2853	4737	2976	9.4916	3102	5091	3229	9.5262	0.3359	37
24	9.4175	0.2615	4368	2734	9.4556	2855	4740	2978	9.4919	3104	5094	3231	9.5264	0.3361	36
25	9.4179	0.2617	4372	2736	9.4560	2857	4743	2981	9.4922	3106	5097	3233	9.5267	0.3363	35
26	9.4182	0.2619	4375	2738	9.4563	2859	4746	2983	9.4925	3108	5099	3236	9.5270	0.3365	34
27	9.4185	0.2621	4378	2740	9.4566	2861	4749	2985	9.4928	3110	5102	3238	9.5273	0.3367	33
28	9.4188	0.2623	4381	2742	9.4569	2863	4752	2987	9.4931	3112	5105	3240	9.5276	0.3369	32
29	9.4192	0.2625	4384	2744	9.4572	2865	4755	2989	9.4934	3114	5108	3242	9.5278	0.3372	31
30	9.4195	0.2627	4387	2746	9.4575	2867	4758	2991	9.4937	3116	5111	3244	9.5281	0.3374	30
31	9.4198	0.2629	4391	2748	9.4578	2870	4761	2993	9.4940	3119	5114	3246	9.5284	0.3376	29
32	9.4201	0.2631	4394	2750	9.4581	2872	4764	2995	9.4942	3121	5117	3248	9.5287	0.3378	28
33	9.4205	0.2633	4397	2752	9.4584	2874	4767	2997	9.4945	3123	5120	3251	9.5290	0.3380	27
34	9.4208	0.2635	4400	2754	9.4587	2876	4770	2999	9.4948	3125	5122	3253	9.5292	0.3383	26
35	9.4211	0.2637	4403	2756	9.4590	2878	4773	3001	9.4951	3127	5125	3255	9.5295	0.3385	25
36	9.4214	0.2639	4406	2758	9.4594	2880	4776	3003	9.4954	3129	5128	3257	9.5298	0.3387	24
37	9.4218	0.2641	4410	2760	9.4597	2882	4779	3005	9.4957	3131	5131	3259	9.5301	0.3389	23
38	9.4221	0.2643	4413	2762	9.4600	2884	4782	3008	9.4960	3133	5134	3261	9.5304	0.3391	22
39	9.4224	0.2645	4416	2764	9.4603	2886	4785	3010	9.4963	3135	5137	3263	9.5306	0.3393	21
40	9.4227	0.2647	4419	2766	9.4606	2888	4788	3012	9.4966	3138	5140	3266	9.5309	0.3396	20
41	9.4231	0.2649	4422	2768	9.4609	2890	4791	3014	9.4969	3140	5142	3268	9.5312	0.3398	19
42	9.4234	0.2651	4425	2770	9.4612	2892	4794	3016	9.4972	3142	5145	3270	9.5315	0.3400	18
43	9.4237	0.2653	4428	2772	9.4615	2894	4797	3018	9.4975	3144	5148	3272	9.5318	0.3402	17
44	9.4240	0.2655	4432	2774	9.4618	2896	4800	3020	9.4978	3146	5151	3274	9.5320	0.3404	16
45	9.4244	0.2657	4435	2776	9.4621	2898	4803	3022	9.4981	3148	5154	3276	9.5323	0.3407	15
46	9.4247	0.2659	4438	2778	9.4624	2900	4806	3024	9.4984	3150	5157	3278	9.5326	0.3409	14
47	9.4250	0.2661	4441	2780	9.4627	2902	4809	3026	9.4986	3152	5160	3281	9.5329	0.3411	13
48	9.4253	0.2663	4444	2782	9.4630	2904	4812	3028	9.4989	3155	5162	3283	9.5331	0.3413	12
49	9.4256	0.2665	4447	2784	9.4633	2906	4815	3030	9.4992	3157	5165	3285	9.5334	0.3415	11
50	9.4260	0.2667	4450	2786	9.4637	2908	4818	3033	9.4995	3159	5168	3287	9.5337	0.3417	10
51	9.4263	0.2669	4454	2788	9.4640	2910	4821	3035	9.4998	3161	5171	3289	9.5340	0.3420	9
52	9.4266	0.2671	4457	2790	9.4643	2912	4824	3037	9.5001	3163	5174	3291	9.5343	0.3422	8
53	9.4269	0.2673	4460	2792	9.4646	2915	4827	3039	9.5004	3165	5177	3294	9.5345	0.3424	7
54	9.4273	0.2675	4463	2794	9.4649	2917	4830	3041	9.5007	3167	5180	3296	9.5348	0.3426	6
55	9.4276	0.2677	4466	2797	9.4652	2919	4833	3043	9.5010	3169	5182	3298	9.5351	0.3428	5
56	9.4279	0.2679	4469	2799	9.4655	2921	4836	3045	9.5013	3172	5185	3300	9.5354	0.3431	4
57	9.4282	0.2681	4472	2801	9.4658	2923	4839	3047	9.5016	3174	5188	3302	9.5357	0.3433	3
58	9.4285	0.2682	4476	2803	9.4661	2925	4842	3049	9.5018	3176	5191	3304	9.5359	0.3435	2
59	9.4289	0.2684	4479	2805	9.4664	2927	4845	3051	9.5021	3178	5194	3307	9.5362	0.3437	1
60	9.4292	0.2686	4482	2807	9.4667	2929	4848	3053	9.5024	3180	5197	3309	9.5365	0.3439	0
′	Log 317°	Nat	Log 316°	Nat	Log 315°	Nat	Log 314°	Nat	Log 313°	Nat	Log 312°	Nat	Log 311°	Nat	′

NOTE: For compactness. the leading digit is not displayed in every column. Scan other columns for necessary digit.

VERSINES

′	49° Log	Nat	50° Log	Nat	51° Log	Nat	52° Log	Nat	53° Log	Nat	54° Log	Nat	55° Log	Nat	′
0	9.5365	0.3439	5529	3572	9.5690	3707	5847	3843	9.6001	3982	6151	4122	9.6298	0.4264	60
1	9.5368	0.3442	5532	3574	9.5693	3709	5850	3846	9.6003	3984	6154	4125	9.6301	0.4267	59
2	9.5370	0.3444	5535	3577	9.5695	3711	5852	3848	9.6006	3986	6156	4127	9.6303	0.4269	58
3	9.5373	0.3446	5537	3579	9.5698	3714	5855	3850	9.6008	3989	6159	4129	9.6306	0.4271	57
4	9.5376	0.3448	5540	3581	9.5701	3716	5857	3853	9.6011	3991	6161	4132	9.6308	0.4274	56
5	9.5379	0.3450	5543	3583	9.5703	3718	5860	3855	9.6014	3993	6164	4134	9.6311	0.4276	55
6	9.5381	0.3453	5546	3586	9.5706	3720	5863	3857	9.6016	3996	6166	4136	9.6313	0.4279	54
7	9.5384	0.3455	5548	3588	9.5709	3723	5865	3859	9.6019	3998	6169	4139	9.6315	0.4281	53
8	9.5387	0.3457	5551	3590	9.5711	3725	5868	3862	9.6021	4000	6171	4141	9.6318	0.4283	52
9	9.5390	0.3459	5554	3592	9.5714	3727	5870	3864	9.6024	4003	6174	4143	9.6320	0.4286	51
10	9.5393	0.3461	5556	3594	9.5716	3729	5873	3866	9.6026	4005	6176	4146	9.6323	0.4288	50
11	9.5395	0.3464	5559	3597	9.5719	3732	5876	3869	9.6029	4007	6178	4148	9.6325	0.4290	49
12	9.5398	0.3466	5562	3599	9.5722	3734	5878	3871	9.6031	4010	6181	4150	9.6327	0.4293	48
13	9.5401	0.3468	5564	3601	9.5724	3736	5881	3873	9.6034	4012	6183	4153	9.6330	0.4295	47
14	9.5404	0.3470	5567	3603	9.5727	3738	5883	3876	9.6036	4014	6186	4155	9.6332	0.4298	46
15	9.5406	0.3472	5570	3606	9.5730	3741	5886	3878	9.6039	4017	6188	4158	9.6335	0.4300	45
16	9.5409	0.3475	5572	3608	9.5732	3743	5888	3880	96041	4019	6191	4160	9.6337	0.4302	44
17	9.5412	0.3477	5575	3610	9.5735	3745	5891	3882	9.6044	4021	6193	4162	9.6340	0.4305	43
18	9.5415	0.3479	5578	3612	9.5738	3748	5894	3885	9.6046	4024	6196	4165	9.6342	0.4307	42
19	9.5417	0.3481	5581	3615	9.5740	3750	5896	3887	9.6049	4026	6198	4167	9.6344	0.4310	41
20	9.5420	0.3483	5583	3617	9.5743	3752	5899	3889	9.6051	4028	6201	4169	9.6347	0.4312	40
21	9.5423	0.3486	5586	3619	9.5745	3754	5901	3892	9.6054	4031	6203	4172	9.6349	0.4314	39
22	9.5426	0.3488	5589	3621	9.5748	3757	5904	3894	9.6056	4033	6206	4174	9.6352	0.4317	38
23	9.5428	0.3490	5591	3624	9.5751	3759	5906	3896	9.6059	4035	6208	4176	9.6354	0.4319	37
24	9.5431	0.3492	5594	3626	9.5753	3761	5909	3899	9.6061	4038	6210	4179	9.6356	0.4322	36
25	9.5434	0.3494	5597	3628	9.5756	3763	5912	3901	9.6064	4040	6213	4181	9.6359	0.4324	35
26	9.5437	0.3497	5599	3630	9.5759	3766	5914	3903	9.6066	4042	6215	4184	9.6361	0.4326	34
27	9.5439	0.3499	5602	3632	9.5761	3768	5917	3905	9.6069	4045	6218	4186	9.6364	0.4329	33
28	9.5442	0.3501	5605	3635	9.5764	3770	5919	3908	9.6071	4047	6220	4188	9.6366	0.4331	32
29	9.5445	0.3503	5607	3637	9.5766	3773	5922	3910	9.6074	4049	6223	4191	9.6368	0.4334	31
30	9.5448	0.3506	5610	3639	9.5769	3775	5924	3912	9.6076	4052	6225	4193	9.6371	0.4336	30
31	9.5450	0.3508	5613	3641	9.5772	3777	5927	3915	9.6079	4054	6228	4195	9.6373	0.4338	29
32	9.5453	0.3510	5615	3644	9.5774	3779	5930	3917	9.6081	4056	6230	4198	9.6376	0.4341	28
33	9.5456	0.3512	5618	3646	9.5777	3782	5932	3919	9.6084	4059	6233	4200	9.6378	0.4343	27
34	9.5458	0.3514	5621	3648	9.5779	3784	5935	3922	9.6086	4061	6235	4202	9.6380	0.4346	26
35	9.5461	0.3517	5623	3650	9.5782	3786	5937	3924	9.6089	4063	6237	4205	9.6383	0.4348	25
36	9.5464	0.3519	5626	3653	9.5785	3789	5940	3926	9.6091	4066	6240	4207	9.6385	0.4350	24
37	9.5467	0.3521	5629	3655	9.5787	3791	5942	3929	9.6094	4068	6242	4210	9.6388	0.4353	23
38	9.5469	0.3523	5631	3657	9.5790	3793	5945	3931	9.6096	4070	6245	4212	9.6390	0.4355	22
39	9.5472	0.3525	5634	3659	9.5793	3795	5947	3933	9.6099	4073	6247	4214	9.6392	0.4358	21
40	9.5475	0.3528	5637	3662	9.5795	3798	5950	3935	9.6101	4075	6250	4217	9.6395	0.4360	20
41	9.5478	0.3530	5639	3664	9.5798	3800	5953	3938	9.6104	4078	6252	4219	9.6397	0.4362	19
42	9.5480	0.3532	5642	3666	9.5800	3802	5955	3940	9.6106	4080	6255	4221	9.6400	0.4365	18
43	9.5483	0.3534	5645	3668	9.5803	3804	5958	3942	9.6109	4082	6257	4224	9.6402	0.4367	17
44	9.5486	0.3537	5647	3671	9.5806	3807	5960	3945	9.6111	4085	6259	4226	9.6404	0.4370	16
45	9.5489	0.3539	5650	3673	9.5808	3809	5963	3947	9.6114	4087	6262	4229	9.6407	0.4372	15
46	9.5491	0.3541	5653	3675	9.5811	3811	5965	3949	9.6116	4089	6264	4231	9.6409	0.4374	14
47	9.5494	0.3543	5655	3677	9.5813	3814	5968	3952	9.6119	4092	6267	4233	9.6412	0.4377	13
48	9.5497	0.3545	5658	3680	9.5816	3816	5970	3954	9.6121	4094	6269	4236	9.6414	0.4379	12
49	9.5499	0.3548	5661	3682	9.5819	3818	5973	3956	9.6124	4096	6272	4238	9.6416	0.4382	11
50	9.5502	0.3550	5663	3684	9.5821	3820	5975	3959	9.6126	4099	6274	4240	9.6419	0.4384	10
51	9.5505	0.3552	5666	3686	9.5824	3823	5978	3961	9.6129	4101	6277	4243	9.6421	0.4386	9
52	9.5508	0.3554	5669	3689	9.5826	3825	5981	3963	9.6131	4103	6279	4245	9.6423	0.4389	8
53	9.5510	0.3557	5671	3691	9.5829	3827	5983	3966	9.6134	4106	6281	4248	9.6426	0.4391	7
54	9.5513	0.3559	5674	3693	9.5832	3830	5986	3968	9.6136	4108	6284	4250	9.6428	0.4394	6
55	9.5516	0.3561	5677	3695	9.5834	3832	5988	3970	9.6139	4110	6286	4252	9.6431	0.4396	5
56	9.5518	0.3563	5679	3698	9.5837	3834	5991	3973	9.6141	4113	6289	4255	9.6433	0.4398	4
57	9.5521	0.3565	5682	3700	9.5839	3837	5993	3975	9.6144	4115	6291	4257	9.6435	0.4401	3
58	9.5524	0.3568	5685	3702	9.5842	3839	5996	3977	9.6146	4117	6294	4259	9.6438	0.4403	2
59	9.5527	0.3570	5687	3705	9.5845	3841	5998	3980	9.6149	4120	6296	4262	9.6440	0.4406	1
60	9.5529	0.3572	5690	3707	9.5847	3843	6001	3982	9.6151	4122	6298	4264	9.6442	0.4408	0
′	Log 310°	Nat	Log 309°	Nat	Log 308°	Nat	Log 307°	Nat	Log 306°	Nat	Log 305°	Nat	Log 304°	Nat	′

NOTE: For compactness. the leading digit is not displayed in every column. Scan other columns for necessary digit.

'	56° Log	56° Nat	57° Log	57° Nat	58° Log	58° Nat	59° Log	59° Nat	60° Log	60° Nat	61° Log	61° Nat	62° Log	62° Nat	'
0	9.6442	0.4408	6584	4554	9.6722	4701	6857	4850	9.6990	5000	7120	5152	9.7247	0.5305	60
1	9.6445	0.4410	6586	4556	9.6724	4703	6859	4852	9.6992	5003	7122	5154	9.7249	0.5308	59
2	9.6447	0.4413	6588	4558	9.6726	4706	6862	4855	9.6994	5005	7124	5157	9.7251	0.5310	58
3	9.6450	0.4415	6591	4561	9.6729	4708	6864	4857	9.6996	5008	7126	5160	9.7253	0.5313	57
4	9.6452	0.4418	6593	4563	9.6731	4711	6866	4860	9.6998	5010	7128	5162	9.7255	0.5316	56
5	9.6454	0.4420	6595	4566	9.6733	4713	6868	4862	9.7001	5013	7130	5165	9.7258	0.5318	55
6	9.6457	0.4423	6598	4568	9.6735	4716	6870	4865	9.7003	5015	7133	5167	9.7260	0.5321	54
7	9.6459	0.4425	6600	4571	9.6738	4718	6873	4867	9.7005	5018	7135	5170	9.7262	0.5323	53
8	9.6461	0.4427	6602	4573	9.6740	4721	6875	4870	9.7007	5020	7137	5172	9.7264	0.5326	52
9	9.6464	0.4430	6604	4576	9.6742	4723	6877	4872	9.7009	5023	7139	5175	9.7266	0.5328	51
10	9.6466	0.4432	6607	4578	9.6744	4725	6879	4875	9.7012	5025	7141	5177	9.7268	0.5331	50
11	9.6469	0.4435	6609	4580	9.6747	4728	6882	4877	9.7014	5028	7143	5180	9.7270	0.5334	49
12	9.6471	0.4437	6611	4583	9.6749	4730	6884	4880	9.7016	5030	7145	5182	9.7272	0.5336	48
13	9.6473	0.4439	6614	4585	9.6751	4733	6886	4882	9.7018	5033	7147	5185	9.7274	0.5339	47
14	9.6476	0.4442	6616	4588	9.6754	4735	6888	4885	9.7020	5035	7150	5188	9.7276	0.5341	46
15	9.6478	0.4444	6618	4590	9.6756	4738	6890	4887	9.7022	5038	7152	5190	9.7279	0.5344	45
16	9.6480	0.4447	6621	4593	9.6758	4740	6893	4890	9.7025	5040	7154	5193	9.7281	0.5346	44
17	9.6483	0.4449	6623	4595	9.6760	4743	6895	4892	9.7027	5043	7156	5195	9.7283	0.5349	43
18	9.6485	0.4452	6625	4598	9.6763	4745	6897	4895	9.7029	5045	7158	5198	9.7285	0.5352	42
19	9.6487	0.4454	6628	4600	9.6765	4748	6899	4897	9.7031	5048	7160	5200	9.7287	0.5354	41
20	9.6490	0.4456	6630	4602	9.6767	4750	6902	4900	9.7033	5050	7162	5203	9.7289	0.5357	40
21	9.6492	0.4459	6632	4605	9.6769	4753	6904	4902	9.7035	5053	7165	5205	9.7291	0.5359	39
22	9.6495	0.4461	6635	4607	9.6772	4755	6906	4905	9.7038	5056	7167	5208	9.7293	0.5362	38
23	9.6497	0.4464	6637	4610	9.6774	4758	6908	4907	9.7040	5058	7169	5211	9.7295	0.5364	37
24	9.6499	0.4466	6639	4612	9.6776	4760	6910	4910	9.7042	5061	7171	5213	9.7297	0.5367	36
25	9.6502	0.4469	6641	4615	9.6778	4763	6913	4912	9.7044	5063	7173	5216	9.7299	0.5370	35
26	9.6504	0.4471	6644	4617	9.6781	4765	6915	4915	9.7046	5066	7175	5218	9.7302	0.5372	34
27	9.6506	0.4473	6646	4620	9.6783	4768	6917	4917	9.7049	5068	7177	5221	9.7304	0.5375	33
28	9.6509	0.4476	6648	4622	9.6785	4770	6919	4920	9.7051	5071	7179	5223	9.7306	0.5377	32
29	9.6511	0.4478	6651	4625	9.6787	4773	6922	4922	9.7053	5073	7182	5226	9.7308	0.5380	31
30	9.6513	0.4481	6653	4627	9.6790	4775	6924	4925	9.7055	5076	7184	5228	9.7310	0.5383	30
31	9.6516	0.4483	6655	4629	9.6792	4777	6926	4927	9.7057	5078	7186	5231	9.7312	0.5385	29
32	9.6518	0.4485	6658	4632	9.6794	4780	6928	4930	9.7059	5081	7188	5234	9.7314	0.5388	28
33	9.6520	0.4488	6660	4634	9.6797	4782	6930	4932	9.7062	5083	7190	5236	9.7316	0.5390	27
34	9.6523	0.4490	6662	4637	9.6799	4785	6933	4935	9.7064	5086	7192	5239	9.7318	0.5393	26
35	9.6525	0.4493	6665	4639	9.6801	4787	6935	4937	9.7066	5088	7194	5241	9.7320	0.5395	25
36	9.6527	0.4495	6667	4642	9.6803	4790	6937	4940	9.7068	5091	7196	5244	9.7322	0.5398	24
37	9.6530	0.4498	6669	4644	9.6806	4792	6939	4942	9.7070	5093	7199	5246	9.7324	0.5401	23
38	9.6532	0.4500	6671	4647	9.6808	4795	6941	4945	9.7072	5096	7201	5249	9.7326	0.5403	22
39	9.6535	0.4502	6674	4649	9.6810	4797	6944	4947	9.7074	5099	7203	5251	9.7329	0.5406	21
40	9.6537	0.4505	6676	4652	9.6812	4800	6946	4950	9.7077	5101	7205	5254	9.7331	0.5408	20
41	9.6539	0.4507	6678	4654	9.6815	4802	6948	4952	9.7079	5104	7207	5257	9.7333	0.5411	19
42	9.6542	0.4510	6681	4656	9.6817	4805	6950	4955	9.7081	5106	7209	5259	9.7335	0.5414	18
43	9.6544	0.4512	6683	4659	9.6819	4807	6952	4957	9.7083	5109	7211	5262	9.7337	0.5416	17
44	9.6546	0.4515	6685	4661	9.6821	4810	6955	4960	9.7085	5111	7213	5264	9.7339	0.5419	16
45	9.6549	0.4517	6687	4664	9.6823	4812	6957	4962	9.7087	5114	7215	5267	9.7341	0.5421	15
46	9.6551	0.4520	6690	4666	9.6826	4815	6959	4965	9.7090	5116	7218	5269	9.7343	0.5424	14
47	9.6553	0.4522	6692	4669	9.6828	4817	6961	4967	9.7092	5119	7220	5272	9.7345	0.5426	13
48	9.6556	0.4524	6694	4671	9.6830	4820	6963	4970	9.7094	5121	7222	5274	9.7347	0.5429	12
49	9.6558	0.4527	6697	4674	9.6832	4822	6966	4972	9.7096	5124	7224	5277	9.7349	0.5432	11
50	9.6560	0.4529	6699	4676	9.6835	4825	6968	4975	9.7098	5126	7226	5280	9.7351	0.5434	10
51	9.6563	0.4532	6701	4679	9.6837	4827	6970	4977	9.7100	5129	7228	5282	9.7353	0.5437	9
52	9.6565	0.4534	6703	4681	9.6839	4830	6972	4980	9.7102	5132	7230	5285	9.7355	0.5439	8
53	9.6567	0.4537	6706	4684	9.6841	4832	6974	4982	9.7105	5134	7232	5287	9.7358	0.5442	7
54	9.6570	0.4539	6708	4686	9.6844	4835	6977	4985	9.7107	5137	7234	5290	9.7360	0.5445	6
55	9.6572	0.4541	6710	4688	9.6846	4837	6979	4987	9.7109	5139	7237	5292	9.7362	0.5447	5
56	9.6574	0.4544	6713	4691	9.6848	4840	6981	4990	9.7111	5142	7239	5295	9.7364	0.5450	4
57	9.6577	0.4546	6715	4693	9.6850	4842	6983	4992	9.7113	5144	7241	5298	9.7366	0.5452	3
58	9.6579	0.4549	6717	4696	9.6853	4845	6985	4995	9.7115	5147	7243	5300	9.7368	0.5455	2
59	9.6581	0.4551	6719	4698	9.6855	4847	6988	4997	9.7118	5149	7245	5303	9.7370	0.5458	1
60	9.6584	0.4554	6722	4701	9.6857	4850	6990	5000	9.7120	5152	7247	5305	9.7372	0.5460	0
'	Log 303°	Nat	Log 302°	Nat	Log 301°	Nat	Log 300°	Nat	Log 299°	Nat	Log 298°	Nat	Log 297°	Nat	'

NOTE: For compactness. the leading digit is not displayed in every column. Scan other columns for necessary digit.

'	63° Log	63° Nat	64° Log	64° Nat	65° Log	65° Nat	66° Log	66° Nat	67° Log	67° Nat	68° Log	68° Nat	69° Log	69° Nat	'
0	9.7372	0.5460	7494	5616	9.7615	5774	7732	5933	9.7848	6093	7962	6254	9.8073	0.6416	60
4	9.7380	0.5470	7503	5627	9.7623	5784	7740	5943	9.7856	6103	7969	6265	9.8080	0.6427	56
8	9.7388	0.5481	7511	5637	9.7630	5795	7748	5954	9.7863	6114	7976	6276	9.8088	0.6438	52
12	9.7397	0.5491	7519	5648	9.7638	5805	7756	5965	9.7871	6125	7984	6286	9.8095	0.6449	48
16	9.7405	0.5502	7527	5658	9.7646	5816	7764	5975	9.7879	6136	7991	6297	9.8102	0.6460	44
20	9.7413	0.5512	7535	5669	9.7654	5827	7771	5986	9.7886	6146	7999	6308	9.8110	0.6471	40
24	9.7421	0.5522	7543	5679	9.7662	5837	7779	5997	9.7894	6157	8006	6319	9.8117	0.6482	36
28	9.7429	0.5533	7551	5690	9.7670	5848	7787	6007	9.7901	6168	8014	6330	9.8124	0.6492	32
32	9.7438	0.5543	7559	5700	9.7678	5858	7794	6018	9.7909	6179	8021	6340	9.8131	0.6503	28
36	9.7446	0.5554	7567	5711	9.7686	5869	7802	6029	9.7916	6189	8029	6351	9.8139	0.6514	24
40	9.7454	0.5564	7575	5721	9.7693	5880	7810	6039	9.7924	6200	8036	6362	9.8146	0.6525	20
44	9.7462	0.5575	7583	5732	9.7701	5890	7817	6050	9.7931	6211	8043	6373	9.8153	0.6536	16
48	9.7470	0.5585	7591	5742	9.7709	5901	7825	6061	9.7939	6222	8051	6384	9.8160	0.6547	12
52	9.7478	0.5595	7599	5753	9.7717	5911	7833	6071	9.7947	6232	8058	6395	9.8168	0.6558	8
56	9.7486	0.5606	7607	5763	9.7725	5922	7840	6082	9.7954	6243	8066	6405	9.8175	0.6569	4
60	9.7494	0.5616	7615	5774	9.7732	5933	7848	6093	9.7962	6254	8073	6416	9.8182	0.6580	0
	Log 296°	Nat	Log 295°	Nat	Log 294°	Nat	Log 293°	Nat	Log 292°	Nat	Log 291°	Nat	Log 290°	Nat	
	70° Log	70° Nat	71° Log	71° Nat	72° Log	72° Nat	73° Log	73° Nat	74° Log	74° Nat	75° Log	75° Nat	76° Log	76° Nat	
0	9.8182	0.6580	8289	6744	9.8395	6910	8498	7076	9.8600	7244	8699	7412	9.8797	0.7581	60
4	9.8189	0.6591	8296	6755	9.8402	6921	8505	7087	9.8606	7255	8706	7423	9.8804	0.7592	56
8	9.8197	0.6602	8304	6766	9.8409	6932	8512	7099	9.8613	7266	8712	7434	9.8810	0.7603	52
12	9.8204	0.6613	8311	6777	9.8416	6943	8519	7110	9.8620	7277	8719	7446	9.8817	0.7615	48
16	9.8211	0.6624	8318	6788	9.8422	6954	8525	7121	9.8626	7288	8726	7457	9.8823	0.7626	44
20	9.8218	0.6635	8325	6799	9.8429	6965	8532	7132	9.8633	7300	8732	7468	9.8829	0.7637	40
24	9.8225	0.6645	8332	6810	9.8436	6976	8539	7143	9.8640	7311	8739	7479	9.8836	0.7649	36
28	9.8232	0.6656	8339	6821	9.8443	6987	8546	7154	9.8646	7322	8745	7491	9.8842	0.7660	32
32	9.8240	0.6667	8346	6832	9.8450	6998	8552	7165	9.8653	7333	8752	7502	9.8849	0.7671	28
36	9.8247	0.6678	8353	6844	9.8457	7010	8559	7177	9.8660	7344	8758	7513	9.8855	0.7683	24
40	9.8254	0.6689	8360	6855	9.8464	7021	8566	7188	9.8666	7356	8765	7524	9.8861	0.7694	20
44	9.8261	0.6700	8367	6866	9.8471	7032	8573	7199	9.8673	7367	8771	7536	9.8868	0.7705	16
48	9.8268	0.6711	8374	6877	9.8478	7043	8579	7210	9.8679	7378	8778	7547	9.8874	0.7716	12
52	9.8275	0.6722	8381	6888	9.8484	7054	8586	7221	9.8686	7389	8784	7558	9.8881	0.7728	8
56	9.8282	0.6733	8388	6899	9.8491	7065	8593	7232	9.8693	7401	8791	7569	9.8887	0.7739	4
60	9.8289	0.6744	8395	6910	9.8498	7076	8600	7244	9.8699	7412	8797	7581	9.8893	0.7750	0
	Log 289°	Nat	Log 288°	Nat	Log 287°	Nat	Log 286°	Nat	Log 285°	Nat	Log 284°	Nat	Log 283°	Nat	
	77° Log	77° Nat	78° Log	78° Nat	79° Log	79° Nat	80° Log	80° Nat	81° Log	81° Nat	82° Log	82° Nat	83° Log	83° Nat	
0	9.8893	0.7750	8988	7921	9.9081	8092	9172	8264	9.9261	8436	9349	8608	9.9436	0.8781	60
4	9.8900	0.7762	8994	7932	9.9087	8103	9178	8275	9.9267	8447	9355	8620	9.9441	0.8793	56
8	9.8906	0.7773	9000	7944	9.9093	8115	9184	8286	9.9273	8459	9361	8631	9.9447	0.8804	52
12	9.8912	0.7785	9006	7955	9.9099	8126	9190	8298	9.9279	8470	9367	8643	9.9453	0.8816	48
16	9.8919	0.7796	9013	7966	9.9105	8138	9196	8309	9.9285	8482	9372	8654	9.9458	0.8828	44
20	9.8925	0.7807	9019	7978	9.9111	8149	9202	8321	9.9291	8493	9378	8666	9.9464	0.8839	40
24	9.8931	0.7819	9025	7989	9.9117	8160	9208	8332	9.9297	8505	9384	8677	9.9470	0.8851	36
28	9.8938	0.7830	9031	8001	9.9123	8172	9214	8344	9.9302	8516	9390	8689	9.9475	0.8862	32
32	9.8944	0.7841	9037	8012	9.9129	8183	9220	8355	9.9308	8528	9395	8701	9.9481	0.8874	28
36	9.8950	0.7853	9044	8023	9.9135	8195	9226	8367	9.9314	8539	9401	8712	9.9487	0.8885	24
40	9.8956	0.7864	9050	8035	9.9141	8206	9232	8378	9.9320	8551	9407	8724	9.9492	0.8897	20
44	9.8963	0.7875	9056	8046	9.9148	8218	9237	8390	9.9326	8562	9413	8735	9.9498	0.8908	16
48	9.8969	0.7887	9062	8058	9.9154	8229	9243	8401	9.9332	8574	9418	8747	9.9504	0.8920	12
52	9.8975	0.7898	9068	8069	9.9160	8241	9249	8413	9.9338	8585	9424	8758	9.9509	0.8932	8
56	9.8981	0.7910	9074	8080	9.9166	8252	9255	8424	9.9343	8597	9430	8770	9.9515	0.8943	4
60	9.8988	0.7921	9081	8092	9.9172	8264	9261	8436	9.9349	8608	9436	8781	9.9521	0.8955	0
'	Log 282°	Nat	Log 281°	Nat	Log 280°	Nat	Log 279°	Nat	Log 278°	Nat	Log 277°	Nat	Log 276°	Nat	'

NOTE: For compactness. the leading digit is not displayed in every column. Scan other columns for necessary digit.

59

VERSINES

′	84° Log	84° Nat	85° Log	85° Nat	86° Log	86° Nat	87° Log	87° Nat	88° Log	88° Nat	89° Log	89° Nat	90° Log	90° Nat	′
0	9.9521	0.8955	9604	9128	9.9686	9302	9767	9477	9.9846	9651	9924	9825	0.0000	1.0000	60
4	9.9526	0.8966	9609	9140	9.9691	9314	9772	9488	9.9851	9663	9929	9837	0.0005	1.0012	56
8	9.9532	0.8978	9615	9152	9.9697	9326	9777	9500	9.9856	9674	9934	9849	0.0010	1.0023	52
12	9.9537	0.8989	9620	9163	9.9702	9337	9782	9512	9.9861	9686	9939	9860	0.0015	1.0035	48
16	9.9543	0.9001	9626	9175	9.9708	9349	9788	9523	9.9867	9698	9944	9872	0.0020	1.0047	44
20	9.9548	0.9013	9631	9186	9.9713	9360	9793	9535	9.9872	9709	9949	9884	0.0025	1.0058	40
24	9.9554	0.9024	9637	9198	9.9718	9372	9798	9546	9.9877	9721	9954	9895	0.0030	1.0070	36
28	9.9560	0.9036	9642	9210	9.9724	9384	9804	9558	9.9882	9732	9959	9907	0.0035	1.0081	32
32	9.9565	0.9047	9648	9221	9.9729	9395	9809	9570	9.9887	9744	9964	9919	0.0040	1.0093	28
36	9.9571	0.9059	9653	9233	9.9734	9407	9814	9581	9.9893	9756	9970	9930	0.0045	1.0105	24
40	9.9576	0.9071	9659	9244	9.9740	9419	9819	9593	9.9898	9767	9975	9942	0.0050	1.0116	20
44	9.9582	0.9082	9664	9256	9.9745	9430	9825	9604	9.9903	9779	9980	9953	0.0055	1.0128	16
48	9.9587	0.9094	9670	9268	9.9751	9442	9830	9616	9.9908	9791	9985	9965	0.0060	1.0140	12
52	9.9593	0.9105	9675	9279	9.9756	9453	9835	9628	9.9913	9802	9990	9977	0.0065	1.0151	8
56	9.9598	0.9117	9681	9291	9.9761	9465	9840	9639	9.9918	9814	9995	9988	0.0070	1.0163	4
60	9.9604	0.9128	9686	9302	9.9767	9477	9846	9651	9.9924	9825	0000	0000	0.0075	1.0175	0
	Log	Nat	Log	Nat	Log	Nat	Log	Nat	Log	Nat	Log	Nat	Log	Nat	
	275°		274°		273°		272°		271°		270°		269°		

′	91° Log	91° Nat	92° Log	92° Nat	93° Log	93° Nat	94° Log	94° Nat	95° Log	95° Nat	96° Log	96° Nat	97° Log	97° Nat	′
0	0.0075	1.0175	0149	0349	0.0222	0523	0293	0698	0.0363	0872	0432	1045	0.0499	1.1219	60
4	0.0080	1.0186	0154	0361	0.0226	0535	0298	0709	0.0368	0883	0436	1057	0.0504	1.1230	56
8	0.0085	1.0198	0159	0372	0.0231	0547	0302	0721	0.0372	0895	0441	1068	0.0508	1.1242	52
12	0.0090	1.0209	0164	0384	0.0236	0558	0307	0732	0.0377	0906	0445	1080	0.0513	1.1253	48
16	0.0095	1.0221	0168	0396	0.0241	0570	0312	0744	0.0381	0918	0450	1092	0.0517	1.1265	44
20	0.0100	1.0233	0173	0407	0.0245	0581	0316	0756	0.0386	0929	0454	1103	0.0522	1.1276	40
24	0.0105	1.0244	0178	0419	0.0250	0593	0321	0767	0.0391	0941	0459	1115	0.0526	1.1288	36
28	0.0110	1.0256	0183	0430	0.0255	0605	0326	0779	0.0395	0953	0463	1126	0.0531	1.1299	32
32	0.0115	1.0268	0188	0442	0.0260	0616	0330	0790	0.0400	0964	0468	1138	0.0535	1.1311	28
36	0.0120	1.0279	0193	0454	0.0264	0628	0335	0802	0.0404	0976	0473	1149	0.0539	1.1323	24
40	0.0125	1.0291	0197	0465	0.0269	0640	0340	0814	0.0409	0987	0477	1161	0.0544	1.1334	20
44	0.0129	1.0302	0202	0477	0.0274	0651	0344	0825	0.0414	0999	0481	1172	0.0548	1.1346	16
48	0.0134	1.0314	0207	0488	0.0279	0663	0349	0837	0.0418	1011	0486	1184	0.0553	1.1357	12
52	0.0139	1.0326	0212	0500	0.0283	0674	0354	0848	0.0423	1022	0490	1196	0.0557	1.1369	8
56	0.0144	1.0337	0217	0512	0.0288	0686	0358	0860	0.0427	1034	0495	1207	0.0562	1.1380	4
60	0.0149	1.0349	0222	0523	0.0293	0698	0363	0872	0.0432	1045	0499	1219	0.0566	1.1392	0
	Log	Nat	Log	Nat	Log	Nat	Log	Nat	Log	Nat	Log	Nat	Log	Nat	
	268°		267°		266°		265°		264°		263°		262°		

′	98° Log	98° Nat	99° Log	99° Nat	100° Log	100° Nat	101° Log	101° Nat	102° Log	102° Nat	103° Log	103° Nat	104° Log	104° Nat	′
0	0.0566	1.1392	0631	1564	0.0695	1736	0758	1908	0.0820	2079	0881	2250	0.0941	1.2419	60
4	0.0570	1.1403	0636	1576	0.0700	1748	0763	1920	0.0824	2090	0885	2261	0.0945	1.2431	56
8	0.0575	1.1415	0640	1587	0.0704	1759	0767	1931	0.0829	2102	0889	2272	0.0949	1.2442	52
12	0.0579	1.1426	0644	1599	0.0708	1771	0771	1942	0.0833	2113	0893	2284	0.0953	1.2453	48
16	0.0583	1.1438	0648	1610	0.0712	1782	0775	1954	0.0837	2125	0897	2295	0.0957	1.2464	44
20	0.0588	1.1449	0653	1622	0.0717	1794	0779	1965	0.0841	2136	0901	2306	0.0961	1.2476	40
24	0.0592	1.1461	0657	1633	0.0721	1805	0783	1977	0.0845	2147	0905	2317	0.0965	1.2487	36
28	0.0597	1.1472	0661	1645	0.0725	1817	0787	1988	0.0849	2159	0909	2329	0.0968	1.2498	32
32	0.0601	1.1484	0666	1656	0.0729	1828	0792	1999	0.0853	2170	0913	2340	0.0972	1.2509	28
36	0.0605	1.1495	0670	1668	0.0733	1840	0796	2011	0.0857	2181	0917	2351	0.0976	1.2521	24
40	0.0610	1.1507	0674	1679	0.0738	1851	0800	2022	0.0861	2193	0921	2363	0.0980	1.2532	20
44	0.0614	1.1518	0678	1691	0.0742	1862	0804	2034	0.0865	2204	0925	2374	0.0984	1.2543	16
48	0.0618	1.1530	0683	1702	0.0746	1874	0808	2045	0.0869	2215	0929	2385	0.0988	1.2554	12
52	0.0623	1.1541	0687	1714	0.0750	1885	0812	2056	0.0873	2227	0933	2397	0.0992	1.2566	8
56	0.0627	1.1553	0691	1725	0.0754	1897	0816	2068	0.0877	2238	0937	2408	0.0996	1.2577	4
60	0.0631	1.1564	0695	1736	0.0758	1908	0820	2079	0.0881	2250	0941	2419	0.1000	1.2588	0
′	Log	Nat	Log	Nat	Log	Nat	Log	Nat	Log	Nat	Log	Nat	Log	Nat	′
	261°		260°		259°		258°		257°		256°		255°		

NOTE: For compactness. the leading digit is not displayed in every column. Scan other columns for necessary digit.

VERSINES

′	105° Log	105° Nat	106° Log	106° Nat	107° Log	107° Nat	108° Log	108° Nat	109° Log	109° Nat	110° Log	110° Nat	111° Log	111° Nat	′
0	0.1000	1.2588	1057	2756	0.1114	2924	1169	3090	0.1224	3256	1278	3420	0.1330	1.3584	60
4	0.1004	1.2599	1061	2768	0.1118	2935	1173	3101	0.1228	3267	1281	3431	0.1334	1.3595	56
8	0.1007	1.2611	1065	2779	0.1121	2946	1177	3112	0.1231	3278	1285	3442	0.1337	1.3605	52
12	0.1011	1.2622	1069	2790	0.1125	2957	1180	3123	0.1235	3289	1288	3453	0.1341	1.3616	48
16	0.1015	1.2633	1072	2801	0.1129	2968	1184	3134	0.1238	3300	1292	3464	0.1344	1.3627	44
20	0.1019	1.2644	1076	2812	0.1133	2979	1188	3145	0.1242	3311	1295	3475	0.1347	1.3638	40
24	0.1023	1.2656	1080	2823	0.1136	2990	1191	3156	0.1246	3322	1299	3486	0.1351	1.3649	36
28	0.1027	1.2667	1084	2835	0.1140	3002	1195	3168	0.1249	3333	1302	3497	0.1354	1.3660	32
32	0.1031	1.2678	1088	2846	0.1144	3013	1199	3179	0.1253	3344	1306	3508	0.1358	1.3670	28
36	0.1034	1.2689	1091	2857	0.1147	3024	1202	3190	0.1256	3355	1309	3518	0.1361	1.3681	24
40	0.1038	1.2700	1095	2868	0.1151	3035	1206	3201	0.1260	3365	1313	3529	0.1365	1.3692	20
44	0.1042	1.2712	1099	2879	0.1155	3046	1210	3212	0.1263	3376	1316	3540	0.1368	1.3703	16
48	0.1046	1.2723	1103	2890	0.1158	3057	1213	3223	0.1267	3387	1320	3551	0.1372	1.3714	12
52	0.1050	1.2734	1106	2901	0.1162	3068	1217	3234	0.1271	3398	1323	3562	0.1375	1.3724	8
56	0.1053	1.2745	1110	2913	0.1166	3079	1220	3245	0.1274	3409	1327	3573	0.1378	1.3735	4
60	0.1057	1.2756	1114	2924	0.1169	3090	1224	3256	0.1278	3420	1330	3584	0.1382	1.3746	0
	Log	Nat	Log	Nat	Log	Nat	Log	Nat	Log	Nat	Log	Nat	Log	Nat	
	254°		253°		252°		251°		250°		249°		248°		

′	112° Log	112° Nat	113° Log	113° Nat	114° Log	114° Nat	115° Log	115° Nat	116° Log	116° Nat	117° Log	117° Nat	118° Log	118° Nat	′
0	0.1382	1.3746	1432	3907	0.1482	4067	1531	4226	0.1579	4384	1626	4540	0.1672	1.4695	60
4	0.1385	1.3757	1436	3918	0.1485	4078	1534	4237	0.1582	4394	1629	4550	0.1675	1.4705	56
8	0.1389	1.3768	1439	3929	0.1489	4089	1537	4247	0.1585	4405	1632	4561	0.1678	1.4715	52
12	0.1392	1.3778	1442	3939	0.1492	4099	1541	4258	0.1588	4415	1635	4571	0.1681	1.4726	48
16	0.1395	1.3789	1446	3950	0.1495	4110	1544	4268	0.1591	4425	1638	4581	0.1684	1.4736	44
20	0.1399	1.3800	1449	3961	0.1498	4120	1547	4279	0.1594	4436	1641	4592	0.1687	1.4746	40
24	0.1402	1.3811	1452	3971	0.1502	4131	1550	4289	0.1598	4446	1644	4602	0.1690	1.4756	36
28	0.1406	1.3821	1456	3982	0.1505	4142	1553	4300	0.1601	4457	1647	4612	0.1693	1.4766	32
32	0.1409	1.3832	1459	3993	0.1508	4152	1557	4310	0.1604	4467	1650	4623	0.1696	1.4777	28
36	0.1412	1.3843	1462	4003	0.1511	4163	1560	4321	0.1607	4478	1653	4633	0.1699	1.4787	24
40	0.1416	1.3854	1466	4014	0.1515	4173	1563	4331	0.1610	4488	1656	4643	0.1702	1.4797	20
44	0.1419	1.3864	1469	4025	0.1518	4184	1566	4342	0.1613	4498	1659	4654	0.1705	1.4807	16
48	0.1422	1.3875	1472	4035	0.1521	4195	1569	4352	0.1616	4509	1662	4664	0.1708	1.4818	12
52	0.1426	1.3886	1476	4046	0.1524	4205	1572	4363	0.1619	4519	1666	4674	0.1711	1.4828	8
56	0.1429	1.3897	1479	4057	0.1528	4216	1576	4373	0.1623	4530	1669	4684	0.1714	1.4838	4
60	0.1432	1.3907	1482	4067	0.1531	4226	1579	4384	0.1626	4540	1672	4695	0.1717	1.4848	0
	Log	Nat	Log	Nat	Log	Nat	Log	Nat	Log	Nat	Log	Nat	Log	Nat	
	247°		246°		245°		244°		243°		242°		241°		

′	119° Log	119° Nat	120° Log	120° Nat	121° Log	121° Nat	122° Log	122° Nat	123° Log	123° Nat	124° Log	124° Nat	125° Log	125° Nat	′
0	0.1717	1.4848	1761	5000	0.1804	5150	1847	5299	0.1888	5446	1929	5592	0.1969	1.5736	60
4	0.1720	1.4858	1764	5010	0.1807	5160	1849	5309	0.1891	5456	1932	5602	0.1972	1.5745	56
8	0.1723	1.4868	1767	5020	0.1810	5170	1852	5319	0.1894	5466	1934	5611	0.1974	1.5755	52
12	0.1726	1.4879	1770	5030	0.1813	5180	1855	5329	0.1896	5476	1937	5621	0.1977	1.5764	48
16	0.1729	1.4889	1773	5040	0.1816	5190	1858	5339	0.1899	5485	1940	5630	0.1979	1.5774	44
20	0.1732	1.4899	1775	5050	0.1818	5200	1861	5348	0.1902	5495	1942	5640	0.1982	1.5783	40
24	0.1734	1.4909	1778	5060	0.1821	5210	1863	5358	0.1905	5505	1945	5650	0.1985	1.5793	36
28	0.1737	1.4919	1781	5070	0.1824	5220	1866	5368	0.1907	5515	1948	5659	0.1987	1.5802	32
32	0.1740	1.4929	1784	5080	0.1827	5230	1869	5378	0.1910	5524	1950	5669	0.1990	1.5812	28
36	0.1743	1.4939	1787	5090	0.1830	5240	1872	5388	0.1913	5534	1953	5678	0.1992	1.5821	24
40	0.1746	1.4950	1790	5100	0.1833	5250	1875	5398	0.1916	5544	1956	5688	0.1995	1.5831	20
44	0.1749	1.4960	1793	5110	0.1835	5260	1877	5407	0.1918	5553	1958	5698	0.1998	1.5840	16
48	0.1752	1.4970	1796	5120	0.1838	5270	1880	5417	0.1921	5563	1961	5707	0.2000	1.5850	12
52	0.1755	1.4980	1799	5130	0.1841	5279	1883	5427	0.1924	5573	1964	5717	0.2003	1.5859	8
56	0.1758	1.4990	1801	5140	0.1844	5289	1886	5437	0.1926	5582	1966	5726	0.2005	1.5868	4
60	0.1761	1.5000	1804	5150	0.1847	5299	1888	5446	0.1929	5592	1969	5736	0.2008	1.5878	0
′	Log	Nat	Log	Nat	Log	Nat	Log	Nat	Log	Nat	Log	Nat	Log	Nat	′
	240°		239°		238°		237°		236°		235°		234°		

NOTE: For compactness. the leading digit is not displayed in every column. Scan other columns for necessary digit.

'	126° Log	126° Nat	127° Log	127° Nat	128° Log	128° Nat	129° Log	129° Nat	130° Log	130° Nat	131° Log	131° Nat	132° Log	132° Nat	'
0	0.2008	1.5878	2046	6018	0.2084	6157	2120	6293	0.2156	6428	2191	6561	0.2225	1.6691	60
6	0.2012	1.5892	2050	6032	0.2087	6170	2124	6307	0.2159	6441	2194	6574	0.2228	1.6704	54
12	0.2016	1.5906	2054	6046	0.2091	6184	2127	6320	0.2163	6455	2198	6587	0.2232	1.6717	48
18	0.2019	1.5920	2057	6060	0.2095	6198	2131	6334	0.2166	6468	2201	6600	0.2235	1.6730	42
24	0.2023	1.5934	2061	6074	0.2098	6211	2134	6347	0.2170	6481	2205	6613	0.2238	1.6743	36
30	0.2027	1.5948	2065	6088	0.2102	6225	2138	6361	0.2173	6494	2208	6626	0.2242	1.6756	30
36	0.2031	1.5962	2069	6101	0.2106	6239	2142	6374	0.2177	6508	2211	6639	0.2245	1.6769	24
42	0.2035	1.5976	2072	6115	0.2109	6252	2145	6388	0.2180	6521	2215	6652	0.2248	1.6782	18
48	0.2039	1.5990	2076	6129	0.2113	6266	2149	6401	0.2184	6534	2218	6665	0.2252	1.6794	12
54	0.2042	1.6004	2080	6143	0.2116	6280	2152	6414	0.2187	6547	2222	6678	0.2255	1.6807	6
60	0.2046	1.6018	2084	6157	0.2120	6293	2156	6428	0.2191	6561	2225	6691	0.2258	1.6820	0
	Log 233°	Nat 233°	Log 232°	Nat 232°	Log 231°	Nat 231°	Log 230°	Nat 230°	Log 229°	Nat 229°	Log 228°	Nat 228°	Log 227°	Nat 227°	

'	133° Log	133° Nat	134° Log	134° Nat	134° Log	134° Nat	136° Log	136° Nat	137° Log	137° Nat	138° Log	138° Nat	139° Log	139° Nat	'
0	0.2258	1.6820	2291	6947	0.2323	7071	2354	7193	0.2384	7314	2413	7431	0.2442	1.7547	60
6	0.2262	1.6833	2294	6959	0.2326	7083	2357	7206	0.2387	7325	2416	7443	0.2445	1.7559	54
12	0.2265	1.6845	2297	6972	0.2329	7096	2360	7218	0.2390	7337	2419	7455	0.2448	1.7570	48
18	0.2268	1.6858	2300	6984	0.2332	7108	2363	7230	0.2393	7349	2422	7466	0.2451	1.7581	42
24	0.2271	1.6871	2304	6997	0.2335	7120	2366	7242	0.2396	7361	2425	7478	0.2453	1.7593	36
30	0.2275	1.6864	2307	7009	0.2338	7133	2369	7254	0.2399	7373	2428	7490	0.2456	1.7604	30
36	0.2278	1.6896	2310	7022	0.2341	7145	2372	7266	0.2402	7385	2431	7501	0.2459	1.7615	24
42	0.2281	1.6909	2313	7034	0.2344	7157	2375	7278	0.2405	7396	2434	7513	0.2462	1.7627	18
48	0.2284	1.6921	2316	7046	0.2347	7169	2378	7290	0.2408	7408	2436	7524	0.2464	1.7638	12
54	0.2288	1.6934	2319	7059	0.2351	7181	2381	7302	0.2410	7420	2439	7536	0.2467	1.7649	6
60	0.2291	1.6947	2323	7071	0.2354	7193	2384	7314	0.2413	7431	2442	7547	0.2470	1.7660	0
	Log 226°	Nat 226°	Log 225°	Nat 225°	Log 224°	Nat 224°	Log 223°	Nat 223°	Log 222°	Nat 222°	Log 221°	Nat 221°	Log 220°	Nat 220°	

'	140° Log	140° Nat	141° Log	141° Nat	142° Log	142° Nat	143° Log	143° Nat	144° Log	144° Nat	145° Log	145° Nat	146° Log	146° Nat	'
0	0.2470	1.7660	2497	7771	0.2524	7880	2549	7986	0.2574	8090	2599	8192	0.2622	1.8290	60
6	0.2473	1.7672	2500	7782	0.2526	7891	2552	7997	0.2577	8100	2601	8202	0.2625	1.8300	54
12	0.2476	1.7683	2503	7793	0.2529	7902	2554	8007	0.2579	8111	2603	8211	0.2627	1.8310	48
18	0.2478	1.7694	2505	7804	0.2531	7912	2557	8018	0.2582	8121	2606	8221	0.2629	1.8320	42
24	0.2481	1.7705	2508	7815	0.2534	7923	2560	8028	0.2584	8131	2608	8231	0.2631	1.8329	36
30	0.2484	1.7716	2511	7826	0.2537	7934	2562	8039	0.2587	8141	2611	8241	0.2634	1.8339	30
36	0.2486	1.7727	2513	7837	0.2539	7944	2565	8049	0.2589	8151	2613	8251	0.2636	1.8348	24
42	0.2489	1.7738	2516	7848	0.2542	7955	2567	8059	0.2591	8161	2615	8261	0.2638	1.8358	18
48	0.2492	1.7749	2518	7859	0.2544	7965	2569	8070	0.2594	8171	2618	8271	0.2641	1.8368	12
54	0.2495	1.7760	2521	7869	0.2547	7976	2572	8080	0.2596	8181	2620	8281	0.2643	1.8377	6
60	0.2497	1.7771	2524	7880	0.2549	7986	2574	8090	0.2599	8192	2622	8290	0.2645	1.8387	0
	Log 219°	Nat 219°	Log 218°	Nat 218°	Log 217°	Nat 217°	Log 216°	Nat 216°	Log 215°	Nat 215°	Log 214°	Nat 214°	Log 213°	Nat 213°	

'	147° Log	147° Nat	148° Log	148° Nat	149° Log	149° Nat	150° Log	150° Nat	151° Log	151° Nat	152° Log	152° Nat	153° Log	153° Nat	'
0	0.2645	1.8387	2667	8480	0.2689	8572	2709	8660	0.2729	8746	2748	8829	0.2767	1.8910	60
6	0.2647	1.8396	2669	8490	0.2691	8581	2711	8669	0.2731	8755	2750	8838	0.2769	1.8918	54
12	0.2650	1.8406	2671	8499	0.2693	8590	2713	8678	0.2733	8763	2752	8846	0.2771	1.8926	48
18	0.2652	1.8415	2674	8508	0.2695	8599	2715	8686	0.2735	8771	2754	8854	0.2772	1.8934	42
24	0.2654	1.8425	2676	8517	0.2697	8607	2717	8695	0.2737	8780	2756	8862	0.2774	1.8942	36
30	0.2656	1.8434	2678	8526	0.2699	8616	2719	8704	0.2739	8788	2758	8870	0.2776	1.8949	30
36	0.2658	1.8443	2680	8536	0.2701	8625	2721	8712	0.2741	8796	2760	8878	0.2778	1.8957	24
42	0.2661	1.8453	2682	8545	0.2703	8634	2723	8721	0.2743	8805	2761	8886	0.2779	1.8965	18
48	0.2663	1.8462	2684	8554	0.2705	8643	2725	8729	0.2745	8813	2763	8894	0.2781	1.8973	12
54	0.2665	1.8471	2686	8563	0.2707	8652	2727	8738	0.2746	8821	2765	8902	0.2783	1.8980	6
60	0.2667	1.8480	2689	8572	0.2709	8660	2729	8746	0.2748	8829	2767	8910	0.2785	1.8988	0
'	Log 212°	Nat 212°	Log 211°	Nat 211°	Log 210°	Nat 210°	Log 209°	Nat 209°	Log 208°	Nat 208°	Log 207°	Nat 207°	Log 206°	Nat 206°	'

NOTE: For compactness. the leading digit is not displayed in every column. Scan other columns for necessary digit.

VERSINES

'	154° Log	154° Nat	155° Log	155° Nat	156° Log	156° Nat	157° Log	157° Nat	158° Log	158° Nat	159° Log	159° Nat	160° Log	160° Nat	'
0	0.2785	1.8988	2802	9063	0.2818	9135	2834	9205	0.2849	9272	2864	9336	0.2877	1.9397	60
6	0.2787	1.8996	2804	9070	0.2820	9143	2836	9212	0.2851	9278	2865	9342	0.2879	1.9403	54
12	0.2788	1.9003	2805	9078	0.2822	9150	2837	9219	0.2852	9285	2866	9348	0.2880	1.9409	48
18	0.2790	1.9011	2807	9085	0.2823	9157	2839	9225	0.2854	9291	2868	9354	0.2881	1.9415	42
24	0.2792	1.9018	2809	9092	0.2825	9164	2840	9232	0.2855	9298	2869	9361	0.2883	1.9421	36
30	0.2793	1.9026	2810	9100	0.2826	9171	2842	9239	0.2857	9304	2871	9367	0.2884	1.9426	30
36	0.2795	1.9033	2812	9107	0.2828	9178	2843	9245	0.2858	9311	2872	9373	0.2885	1.9432	24
42	0.2797	1.9041	2814	9114	0.2829	9184	2845	9252	0.2859	9317	2873	9379	0.2887	1.9438	18
48	0.2799	1.9048	2815	9121	0.2831	9191	2846	9259	0.2861	9323	2875	9385	0.2888	1.9444	12
54	0.2800	1.9056	2817	9128	0.2833	9198	2848	9265	0.2862	9330	2876	9391	0.2889	1.9449	6
60	0.2802	1.9063	2818	9135	0.2834	9205	2849	9272	0.2864	9336	2877	9397	0.2890	1.9455	0
	Log 205°	Nat 205°	Log 204°	Nat 204°	Log 203°	Nat 203°	Log 202°	Nat 202°	Log 201°	Nat 201°	Log 200°	Nat 200°	Log 199°	Nat 199°	

'	161° Log	161° Nat	162° Log	162° Nat	163° Log	163° Nat	164° Log	164° Nat	165° Log	165° Nat	166° Log	166° Nat	167° Log	167° Nat	'
0	0.2890	1.9455	2903	9511	0.2914	9563	2925	9613	0.2936	9659	2945	9703	0.2954	1.9744	60
6	0.2892	1.9461	2904	9516	0.2915	9568	2926	9617	0.2937	9664	2946	9707	0.2955	1.9748	54
12	0.2893	1.9466	2905	9521	0.2917	9573	2927	9622	0.2938	9668	2947	9711	0.2956	1.9751	48
18	0.2894	1.9472	2906	9527	0.2918	9578	2929	9627	0.2939	9673	2948	9715	0.2957	1.9755	42
24	0.2895	1.9478	2907	9532	0.2919	9583	2930	9632	0.2940	9677	2949	9720	0.2958	1.9759	36
30	0.2897	1.9483	2909	9537	0.2920	9588	2931	9636	0.2941	9681	2950	9724	0.2959	1.9763	30
36	0.2898	1.9489	2910	9542	0.2921	9593	2932	9641	0.2942	9686	2951	9728	0.2959	1.9767	24
42	0.2899	1.9494	2911	9548	0.2922	9598	2933	9646	0.2942	9690	2952	9732	0.2960	1.9770	18
48	0.2900	1.9500	2912	9553	0.2923	9603	2934	9650	0.2943	9694	2953	9736	0.2961	1.9774	12
54	0.2901	1.9505	2913	9558	0.2924	9608	2935	9655	0.2944	9699	2953	9740	0.2962	1.9778	6
60	0.2903	1.9511	2914	9563	0.2925	9613	2936	9659	0.2945	9703	2954	9744	0.2963	1.9781	0
	Log 198°	Nat 198°	Log 197°	Nat 197°	Log 196°	Nat 196°	Log 195°	Nat 195°	Log 194°	Nat 194°	Log 193°	Nat 193°	Log 192°	Nat 192°	

'	168° Log	168° Nat	169° Log	169° Nat	170° Log	170° Nat	171° Log	171° Nat	172° Log	172° Nat	173° Log	173° Nat	174° Log	174° Nat	'
0	0.2963	1.9781	2970	9816	0.2977	9848	2983	9877	0.2989	9903	2994	9925	0.2998	1.9945	60
6	0.2963	1.9785	2971	9820	0.2978	9851	2984	9880	0.2990	9905	2995	9928	0.2999	1.9947	54
12	0.2964	1.9789	2972	9823	0.2978	9854	2985	9882	0.2990	9907	2995	9930	0.2999	1.9949	48
18	0.2965	1.9792	2972	9826	0.2979	9857	2985	9885	0.2991	9910	2995	9932	0.3000	1.9951	42
24	0.2966	1.9796	2973	9829	0.2980	9860	2986	9888	0.2991	9912	2996	9934	0.3000	1.9952	36
30	0.2966	1.9799	2974	9833	0.2980	9863	2986	9890	0.2992	9914	2996	9936	0.3000	1.9954	30
36	0.2967	1.9803	2974	9836	0.2981	9866	2987	9893	0.2992	9917	2997	9938	0.3001	1.9956	24
42	0.2968	1.9806	2975	9839	0.2982	9869	2987	9895	0.2993	9919	2997	9940	0.3001	1.9957	18
48	0.2969	1.9810	2976	9842	0.2982	9871	2988	9898	0.2993	9921	2998	9942	0.3001	1.9959	12
54	0.2969	1.9813	2977	9845	0.2983	9874	2989	9900	0.2994	9923	2998	9943	0.3002	1.9960	6
60	0.2970	1.9816	2977	9848	0.2983	9877	2989	9903	0.2994	9925	2998	9945	0.3002	1.9962	0
	Log 191°	Nat 191°	Log 190°	Nat 190°	Log 189°	Nat 189°	Log 188°	Nat 188°	Log 187°	Nat 187°	Log 186°	Nat 186°	Log 185°	Nat 185°	

'	175° Log	175° Nat	176° Log	176° Nat	177° Log	177° Nat	178° Log	178° Nat	179° Log	179° Nat	'
0	0.3002	1.9962	3005	9976	0.3007	9986	3009	9994	0.3010	9998	60
6	0.3002	1.9963	3005	9977	0.3008	9987	3009	9995	0.3010	9999	54
12	0.3003	1.9965	3006	9978	0.3008	9988	3009	9995	0.3010	9999	48
18	0.3003	1.9966	3006	9979	0.3008	9989	3009	9996	0.3010	9999	42
24	0.3003	1.9968	3006	9980	0.3008	9990	3009	9996	0.3010	9999	36
30	0.3004	1.9969	3006	9981	0.3008	9990	3010	9997	0.3010	0000	30
36	0.3004	1.9971	3006	9982	0.3008	9991	3010	9997	0.3010	0000	24
42	0.3004	1.9972	3007	9983	0.3009	9992	3010	9997	0.3010	0000	18
48	0.3004	1.9973	3007	9984	0.3009	9993	3010	9998	0.3010	0000	12
54	0.3005	1.9974	3007	9985	0.3009	9993	3010	9998	0.3010	0000	6
60	0.3005	1.9976	3007	9986	0.3009	9994	3010	9998	0.3010	0000	0
'	Log 184°	Nat 184°	Log 183°	Nat 183°	Log 182°	Nat 182°	Log 181°	Nat 181°	Log 180°	Nat 180°	'

NOTE: For compactness. the leading digit is not displayed in every column. Scan other columns for necessary digit.

LOG COSINES

	0°	1°	2°	3°	4°	5°	6°	7°	8°	9°	10°	11°	12°	13°	14°	
0	0.0000	9999	9997	9994	9989	9.9983	9976	9968	9958	9.9946	9934	9919	9904	9887	9.9869	60
1	0.0000	9999	9997	9994	9989	9.9983	9976	9967	9957	9.9946	9933	9919	9904	9887	9.9869	59
2	0.0000	9999	9997	9994	9989	9.9983	9976	9967	9957	9.9946	9933	9919	9904	9887	9.9868	58
3	0.0000	9999	9997	9994	9989	9.9983	9976	9967	9957	9.9946	9933	9919	9903	9886	9.9868	57
4	0.0000	9999	9997	9994	9989	9.9983	9976	9967	9957	9.9945	9933	9918	9903	9886	9.9868	56
5	0.0000	9999	9997	9994	9989	9.9983	9975	9967	9957	9.9945	9932	9918	9903	9886	9.9867	55
6	0.0000	9999	9997	9994	9989	9.9983	9975	9967	9956	9.9945	9932	9918	9902	9885	9.9867	54
7	0.0000	9999	9997	9994	9989	9.9983	9975	9966	9956	9.9945	9932	9918	9902	9885	9.9867	53
8	0.0000	9999	9997	9994	9989	9.9983	9975	9966	9956	9.9945	9932	9917	9902	9885	9.9867	52
9	0.0000	9999	9997	9993	9989	9.9982	9975	9966	9956	9.9944	9931	9917	9902	9885	9.9866	51
10	0.0000	9999	9997	9993	9989	9.9982	9975	9966	9956	9.9944	9931	9917	9901	9884	9.9866	50
11	0.0000	9999	9997	9993	9988	9.9982	9975	9966	9956	9.9944	9931	9917	9901	9884	9.9866	49
12	0.0000	9999	9997	9993	9988	9.9982	9975	9966	9955	9.9944	9931	9916	9901	9884	9.9865	48
13	0.0000	9999	9997	9993	9988	9.9982	9974	9965	9955	9.9944	9931	9916	9901	9883	9.9865	47
14	0.0000	9999	9997	9993	9988	9.9982	9974	9965	9955	9.9943	9930	9916	9900	9883	9.9865	46
15	0.0000	9999	9997	9993	9988	9.9982	9974	9965	9955	9.9943	9930	9916	9900	9883	9.9864	45
16	0.0000	9999	9997	9993	9988	9.9982	9974	9965	9955	9.9943	9930	9915	9900	9883	9.9864	44
17	0.0000	9999	9997	9993	9988	9.9982	9974	9965	9954	9.9943	9930	9915	9899	9882	9.9864	43
18	0.0000	9999	9996	9993	9988	9.9981	9974	9965	9954	9.9943	9929	9915	9899	9882	9.9863	42
19	0.0000	9999	9996	9993	9988	9.9981	9974	9964	9954	9.9942	9929	9915	9899	9882	9.9863	41
20	0.0000	9999	9996	9993	9988	9.9981	9973	9964	9954	9.9942	9929	9914	9899	9881	9.9863	40
21	0.0000	9999	9996	9993	9987	9.9981	9973	9964	9954	9.9942	9929	9914	9898	9881	9.9862	39
22	0.0000	9999	9996	9992	9987	9.9981	9973	9964	9954	9.9942	9929	9914	9898	9881	9.9862	38
23	0.0000	9999	9996	9992	9987	9.9981	9973	9964	9953	9.9941	9928	9914	9898	9880	9.9862	37
24	0.0000	9999	9996	9992	9987	9.9981	9973	9964	9953	9.9941	9928	9913	9897	9880	9.9861	36
25	0.0000	9999	9996	9992	9987	9.9981	9973	9964	9953	9.9941	9928	9913	9897	9880	9.9861	35
26	0.0000	9999	9996	9992	9987	9.9980	9973	9963	9953	9.9941	9928	9913	9897	9880	9.9861	34
27	0.0000	9999	9996	9992	9987	9.9980	9972	9963	9953	9.9941	9927	9913	9897	9879	9.9860	33
28	0.0000	9999	9996	9992	9987	9.9980	9972	9963	9952	9.9940	9927	9912	9896	9879	9.9860	32
29	0.0000	9999	9996	9992	9987	9.9980	9972	9963	9952	9.9940	9927	9912	9896	9879	9.9860	31
30	0.0000	9999	9996	9992	9987	9.9980	9972	9963	9952	9.9940	9927	9912	9896	9878	9.9859	30
31	0.0000	9998	9996	9992	9986	9.9980	9972	9963	9952	9.9940	9926	9912	9896	9878	9.9859	29
32	0.0000	9998	9996	9992	9986	9.9980	9972	9962	9952	9.9940	9926	9911	9895	9878	9.9859	28
33	0.0000	9998	9996	9992	9986	9.9980	9972	9962	9951	9.9939	9926	9911	9895	9877	9.9858	27
34	0.0000	9998	9996	9992	9986	9.9979	9971	9962	9951	9.9939	9926	9911	9895	9877	9.9858	26
35	0.0000	9998	9996	9992	9986	9.9979	9971	9962	9951	9.9939	9925	9911	9894	9877	9.9858	25
36	0.0000	9998	9996	9991	9986	9.9979	9971	9962	9951	9.9939	9925	9910	9894	9876	9.9857	24
37	0.0000	9998	9995	9991	9986	9.9979	9971	9962	9951	9.9939	9925	9910	9894	9876	9.9857	23
38	0.0000	9998	9995	9991	9986	9.9979	9971	9961	9951	9.9938	9925	9910	9894	9876	9.9857	22
39	0.0000	9998	9995	9991	9986	9.9979	9971	9961	9950	9.9938	9925	9910	9893	9876	9.9856	21
40	0.0000	9998	9995	9991	9986	9.9979	9971	9961	9950	9.9938	9924	9909	9893	9875	9.9856	20
41	0.0000	9998	9995	9991	9985	9.9979	9970	9961	9950	9.9938	9924	9909	9893	9875	9.9856	19
42	0.0000	9998	9995	9991	9985	9.9978	9970	9961	9950	9.9937	9924	9909	9892	9875	9.9855	18
43	0.0000	9998	9995	9991	9985	9.9978	9970	9960	9950	9.9937	9924	9909	9892	9874	9.9855	17
44	0.0000	9998	9995	9991	9985	9.9978	9970	9960	9949	9.9937	9923	9908	9892	9874	9.9855	16
45	0.0000	9998	9995	9991	9985	9.9978	9970	9960	9949	9.9937	9923	9908	9892	9874	9.9854	15
46	0.0000	9998	9995	9991	9985	9.9978	9970	9960	9949	9.9937	9923	9908	9891	9873	9.9854	14
47	0.0000	9998	9995	9991	9985	9.9978	9969	9960	9949	9.9936	9923	9908	9891	9873	9.9854	13
48	0.0000	9998	9995	9990	9985	9.9978	9969	9960	9949	9.9936	9922	9907	9891	9873	9.9853	12
49	0.0000	9998	9995	9990	9985	9.9978	9969	9959	9948	9.9936	9922	9907	9890	9872	9.9853	11
50	0.0000	9998	9995	9990	9985	9.9977	9969	9959	9948	9.9936	9922	9907	9890	9872	9.9853	10
51	0.0000	9998	9995	9990	9984	9.9977	9969	9959	9948	9.9936	9922	9906	9890	9872	9.9852	9
52	0.0000	9998	9995	9990	9984	9.9977	9969	9959	9948	9.9935	9921	9906	9890	9872	9.9852	8
53	9.9999	9998	9994	9990	9984	9.9977	9969	9959	9948	9.9935	9921	9906	9889	9871	9.9852	7
54	9.9999	9998	9994	9990	9984	9.9977	9968	9959	9947	9.9935	9921	9906	9889	9871	9.9851	6
55	9.9999	9998	9994	9990	9984	9.9977	9968	9958	9947	9.9935	9921	9905	9889	9871	9.9851	5
56	9.9999	9998	9994	9990	9984	9.9977	9968	9958	9947	9.9934	9920	9905	9888	9870	9.9851	4
57	9.9999	9997	9994	9990	9984	9.9977	9968	9958	9947	9.9934	9920	9905	9888	9870	9.9850	3
58	9.9999	9997	9994	9990	9984	9.9976	9968	9958	9947	9.9934	9920	9905	9888	9870	9.9850	2
59	9.9999	9997	9994	9989	9984	9.9976	9968	9958	9946	9.9934	9920	9904	9888	9869	9.9850	1
60	9.9999	9997	9994	9989	9983	9.9976	9968	9958	9946	9.9934	9919	9904	9887	9869	9.9849	0
	89°	88°	87°	86°	85°	84°	83°	82°	81°	80°	79°	78°	77°	76°	75°	

NOTE: For compactness, the leading digit is not displayed in every column. Scan other columns for necessary digit.

LOG SINES

LOG COSINES

	15°	16°	17°	18°	19°	20°	21°	22°	23°	24°	25°	26°	27°	28°	29°	
0	9.9849	9828	9806	9782	9757	9.9730	9702	9672	9640	9.9607	9573	9537	9499	9459	9.9418	60
1	9.9849	9828	9806	9782	9756	9.9729	9701	9671	9640	9.9607	9572	9536	9498	9459	9.9417	59
2	9.9849	9828	9805	9781	9756	9.9729	9701	9671	9639	9.9606	9572	9535	9498	9458	9.9417	58
3	9.9848	9827	9805	9781	9755	9.9728	9700	9670	9639	9.9606	9571	9535	9497	9457	9.9416	57
4	9.9848	9827	9804	9780	9755	9.9728	9700	9670	9638	9.9605	9570	9534	9496	9457	9.9415	56
5	9.9848	9827	9804	9780	9755	9.9728	9699	9669	9638	9.9604	9570	9534	9496	9456	9.9415	55
6	9.9847	9826	9804	9780	9754	9.9727	9699	9669	9637	9.9604	9569	9533	9495	9455	9.9414	54
7	9.9847	9826	9803	9779	9754	9.9727	9698	9668	9636	9.9603	9569	9532	9494	9455	9.9413	53
8	9.9847	9826	9803	9779	9753	9.9726	9698	9668	9636	9.9603	9568	9532	9494	9454	9.9413	52
9	9.9846	9825	9802	9778	9753	9.9726	9697	9667	9635	9.9602	9567	9531	9493	9453	9.9412	51
10	9.9846	9825	9802	9778	9752	9.9725	9697	9667	9635	9.9602	9567	9530	9492	9453	9.9411	50
11	9.9846	9824	9802	9778	9752	9.9725	9696	9666	9634	9.9601	9566	9530	9492	9452	9.9410	49
12	9.9845	9824	9801	9777	9751	9.9724	9696	9666	9634	9.9601	9566	9529	9491	9451	9.9410	48
13	9.9845	9824	9801	9777	9751	9.9724	9695	9665	9633	9.9600	9565	9529	9490	9451	9.9409	47
14	9.9845	9823	9801	9776	9751	9.9723	9695	9664	9633	9.9599	9564	9528	9490	9450	9.9408	46
15	9.9844	9823	9800	9776	9750	9.9723	9694	9664	9632	9.9599	9564	9527	9489	9449	9.9408	45
16	9.9844	9823	9800	9775	9750	9.9722	9694	9663	9632	9.9598	9563	9527	9488	9449	9.9407	44
17	9.9844	9822	9799	9775	9749	9.9722	9693	9663	9631	9.9598	9563	9526	9488	9448	9.9406	43
18	9.9843	9822	9799	9775	9749	9.9722	9693	9662	9631	9.9597	9562	9525	9487	9447	9.9406	42
19	9.9843	9821	9799	9774	9748	9.9721	9692	9662	9630	9.9597	9561	9525	9486	9447	9.9405	41
20	9.9843	9821	9798	9774	9748	9.9721	9692	9661	9629	9.9596	9561	9524	9486	9446	9.9404	40
21	9.9842	9821	9798	9773	9747	9.9720	9691	9661	9629	9.9595	9560	9524	9485	9445	9.9403	39
22	9.9842	9820	9797	9773	9747	9.9720	9691	9660	9628	9.9595	9560	9523	9485	9444	9.9403	38
23	9.9842	9820	9797	9773	9747	9.9719	9690	9660	9628	9.9594	9559	9522	9484	9444	9.9402	37
24	9.9841	9820	9797	9772	9746	9.9719	9690	9659	9627	9.9594	9558	9522	9483	9443	9.9401	36
25	9.9841	9819	9796	9772	9746	9.9718	9689	9659	9627	9.9593	9558	9521	9483	9442	9.9401	35
26	9.9841	9819	9796	9771	9745	9.9718	9689	9658	9626	9.9593	9557	9520	9482	9442	9.9400	34
27	9.9840	9818	9795	9771	9745	9.9717	9688	9658	9626	9.9592	9557	9520	9481	9441	9.9399	33
28	9.9840	9818	9795	9770	9744	9.9717	9688	9657	9625	9.9591	9556	9519	9481	9440	9.9398	32
29	9.9839	9818	9795	9770	9744	9.9716	9687	9657	9625	9.9591	9555	9519	9480	9440	9.9398	31
30	9.9839	9817	9794	9770	9743	9.9716	9687	9656	9624	9.9590	9555	9518	9479	9439	9.9397	30
31	9.9839	9817	9794	9769	9743	9.9715	9686	9656	9623	9.9590	9554	9517	9479	9438	9.9396	29
32	9.9838	9817	9793	9769	9743	9.9715	9686	9655	9623	9.9589	9554	9517	9478	9438	9.9396	28
33	9.9838	9816	9793	9768	9742	9.9714	9685	9655	9622	9.9589	9553	9516	9477	9437	9.9395	27
34	9.9838	9816	9793	9768	9742	9.9714	9685	9654	9622	9.9588	9552	9515	9477	9436	9.9394	26
35	9.9837	9815	9792	9767	9741	9.9714	9684	9654	9621	9.9587	9552	9515	9476	9436	9.9393	25
36	9.9837	9815	9792	9767	9741	9.9713	9684	9653	9621	9.9587	9551	9514	9475	9435	9.9393	24
37	9.9837	9815	9791	9767	9740	9.9713	9683	9652	9620	9.9586	9551	9513	9475	9434	9.9392	23
38	9.9836	9814	9791	9766	9740	9.9712	9683	9652	9620	9.9586	9550	9513	9474	9433	9.9391	22
39	9.9836	9814	9791	9766	9739	9.9712	9682	9651	9619	9.9585	9549	9512	9473	9433	9.9391	21
40	9.9836	9814	9790	9765	9739	9.9711	9682	9651	9618	9.9584	9549	9512	9473	9432	9.9390	20
41	9.9835	9813	9790	9765	9739	9.9711	9681	9650	9618	9.9584	9548	9511	9472	9431	9.9389	19
42	9.9835	9813	9789	9764	9738	9.9710	9681	9650	9617	9.9583	9548	9510	9471	9431	9.9388	18
43	9.9835	9812	9789	9764	9738	9.9710	9680	9649	9617	9.9583	9547	9510	9471	9430	9.9388	17
44	9.9834	9812	9789	9764	9737	9.9709	9680	9649	9616	9.9582	9546	9509	9470	9429	9.9387	16
45	9.9834	9812	9788	9763	9737	9.9709	9679	9648	9616	9.9582	9546	9508	9469	9429	9.9386	15
46	9.9833	9811	9788	9763	9736	9.9708	9679	9648	9615	9.9581	9545	9508	9469	9428	9.9385	14
47	9.9833	9811	9787	9762	9736	9.9708	9678	9647	9615	9.9580	9545	9507	9468	9427	9.9385	13
48	9.9833	9811	9787	9762	9735	9.9707	9678	9647	9614	9.9580	9544	9506	9467	9427	9.9384	12
49	9.9832	9810	9787	9761	9735	9.9707	9677	9646	9613	9.9579	9543	9506	9467	9426	9.9383	11
50	9.9832	9810	9786	9761	9734	9.9706	9677	9646	9613	9.9579	9543	9505	9466	9425	9.9383	10
51	9.9832	9809	9786	9761	9734	9.9706	9676	9645	9612	9.9578	9542	9505	9465	9424	9.9382	9
52	9.9831	9809	9785	9760	9734	9.9705	9676	9645	9612	9.9577	9542	9504	9465	9424	9.9381	8
53	9.9831	9809	9785	9760	9733	9.9705	9675	9644	9611	9.9577	9541	9503	9464	9423	9.9380	7
54	9.9831	9808	9785	9759	9733	9.9704	9675	9643	9611	9.9576	9540	9503	9463	9422	9.9380	6
55	9.9830	9808	9784	9759	9732	9.9704	9674	9643	9610	9.9576	9540	9502	9463	9422	9.9379	5
56	9.9830	9808	9784	9758	9732	9.9703	9674	9642	9610	9.9575	9539	9501	9462	9421	9.9378	4
57	9.9830	9807	9783	9758	9731	9.9703	9673	9642	9609	9.9575	9538	9501	9461	9420	9.9377	3
58	9.9829	9807	9783	9758	9731	9.9702	9673	9641	9608	9.9574	9538	9500	9461	9420	9.9377	2
59	9.9829	9806	9782	9757	9730	9.9702	9672	9641	9608	9.9573	9537	9499	9460	9419	9.9376	1
60	9.9828	9806	9782	9757	9730	9.9702	9672	9640	9607	9.9573	9537	9499	9459	9418	9.9375	0
	74°	73°	72°	71°	70°	69°	68°	67°	66°	65°	64°	63°	62°	61°	60°	

NOTE: For compactness, the leading digit is not displayed in every column. Scan other columns for necessary digit.

LOG SINES

LOG COSINES

′	30°	31°	32°	33°	34°	35°	36°	37°	38°	39°	40°	41°	42°	43°	44°	′
0	9.9375	9331	9284	9236	9186	9.9134	9080	9023	8965	9.8905	8843	8778	8711	8641	9.8569	60
1	9.9375	9330	9283	9235	9185	9.9133	9079	9023	8964	9.8904	8841	8777	8710	8640	9.8568	59
2	9.9374	9329	9283	9234	9184	9.9132	9078	9022	8963	9.8903	8840	8776	8708	8639	9.8567	58
3	9.9373	9328	9282	9233	9183	9.9131	9077	9021	8962	9.8902	8839	8775	8707	8638	9.8566	57
4	9.9372	9328	9281	9233	9182	9.9130	9076	9020	8961	9.8901	8838	8773	8706	8637	9.8564	56
5	9.9372	9327	9280	9232	9181	9.9129	9075	9019	8960	9.8900	8837	8772	8705	8635	9.8563	55
6	9.9371	9326	9279	9231	9181	9.9128	9074	9018	8959	9.8899	8836	8771	8704	8634	9.8562	54
7	9.9370	9325	9279	9230	9180	9.9127	9073	9017	8958	9.8898	8835	8770	8703	8633	9.8561	53
8	9.9369	9325	9278	9229	9179	9.9127	9072	9016	8957	9.8897	8834	8769	8702	8632	9.8560	52
9	9.9369	9324	9277	9229	9178	9.9126	9071	9015	8956	9.8896	8833	8768	8700	8631	9.8558	51
10	9.9368	9323	9276	9228	9177	9.9125	9070	9014	8955	9.8895	8832	8767	8699	8629	9.8557	50
11	9.9367	9322	9275	9227	9176	9.9124	9069	9013	8954	9.8894	8831	8766	8698	8628	9.8556	49
12	9.9367	9322	9275	9226	9175	9.9123	9069	9012	8953	9.8893	8830	8765	8697	8627	9.8555	48
13	9.9366	9321	9274	9225	9175	9.9122	9068	9011	8952	9.8892	8829	8763	8696	8626	9.8553	47
14	9.9365	9320	9273	9224	9174	9.9121	9067	9010	8951	9.8891	8828	8762	8695	8625	9.8552	46
15.	9.9364	9319	9272	9224	9173	9.9120	9066	9009	8950	9.8890	8827	8761	8694	8624	9.8551	45
16	9.9364	9318	9272	9223	9172	9.9119	9065	9008	8949	9.8889	8825	8760	8692	8622	9.8550	44
17	9.9363	9318	9271	9222	9171	9.9119	9064	9007	8948	9.8888	8824	8759	8691	8621	9.8549	43
18	9.9362	9317	9270	9221	9170	9.9118	9063	9006	8947	9.8887	8823	8758	8690	8620	9.8547	42
19	9.9361	9316	9269	9220	9169	9.9117	9062	9005	8946	9.8885	8822	8757	8689	8619	9.8546	41
20	9.9361	9315	9268	9219	9169	9.9116	9061	9004	8945	9.8884	8821	8756	8688	8618	9.8545	40
21	9.9360	9315	9268	9219	9168	9.9115	9060	9003	8944	9.8883	8820	8755	8687	8616	9.8544	39
22	9.9359	9314	9267	9218	9167	9.9114	9059	9002	8943	9.8882	8819	8753	8686	8615	9.8542	38
23	9.9358	9313	9266	9217	9166	9.9113	9058	9001	8942	9.8881	8818	8752	8684	8614	9.8541	37
24	9.9358	9312	9265	9216	9165	9.9112	9057	9000	8941	9.8880	8817	8751	8683	8613	9.8540	36
25	9.9357	9312	9264	9215	9164	9.9111	9056	9000	8940	9.8879	8816	8750	8682	8612	9.8539	35
26	9.9356	9311	9264	9214	9163	9.9110	9056	8999	8939	9.8878	8815	8749	8681	8610	9.8537	34
27	9.9355	9310	9263	9214	9163	9.9110	9055	8998	8938	9.8877	8814	8748	8680	8609	9.8536	33
28	9.9355	9309	9262	9213	9162	9.9109	9054	8997	8937	9.8876	8813	8747	8679	8608	9.8535	32
29	9.9354	9308	9261	9212	9161	9.9108	9053	8996	8936	9.8875	8812	8746	8677	8607	9.8534	31
30	9.9353	9308	9260	9211	9160	9.9107	9052	8995	8935	9.8874	8810	8745	8676	8606	9.8532	30
31	9.9352	9307	9259	9210	9159	9.9106	9051	8994	8934	9.8873	8809	8743	8675	8604	9.8531	29
32	9.9352	9306	9259	9209	9158	9.9105	9050	8993	8933	9.8872	8808	8742	8674	8603	9.8530	28
33	9.9351	9305	9258	9209	9157	9.9104	9049	8992	8932	9.8871	8807	8741	8673	8602	9.8529	27
34	9.9350	9305	9257	9208	9156	9.9103	9048	8991	8931	9.8870	8806	8740	8672	8601	9.8527	26
35	9.9349	9304	9256	9207	9156	9.9102	9047	8990	8930	9.8869	8805	8739	8671	8600	9.8526	25
36	9.9349	9303	9255	9206	9155	9.9101	9046	8989	8929	9.8868	8804	8738	8669	8598	9.8525	24
37	9.9348	9302	9255	9205	9154	9.9101	9045	8988	8928	9.8867	8803	8737	8668	8597	9.8524	23
38	9.9347	9301	9254	9204	9153	9.9100	9044	8987	8927	9.8866	8802	8736	8667	8596	9.8522	22
39	9.9346	9301	9253	9204	9152	9.9099	9043	8986	8926	9.8865	8801	8734	8666	8595	9.8521	21
40	9.9346	9300	9252	9203	9151	9.9098	9042	8985	8925	9.8864	8800	8733	8665	8594	9.8520	20
41	9.9345	9299	9251	9202	9150	9.9097	9041	8984	8924	9.8863	8799	8732	8664	8592	9.8519	19
42	9.9344	9298	9251	9201	9149	9.9096	9041	8983	8923	9.8862	8797	8731	8662	8591	9.8517	18
43	9.9343	9298	9250	9200	9149	9.9095	9040	8982	8922	9.8860	8796	8730	8661	8590	9.8516	17
44	99343	9297	9249	9199	9148	9.9094	9039	8981	8921	9.8859	8795	8729	8660	8589	9.8515	16
45	9.9342	9296	9248	9198	9147	9.9093	9038	8980	8920	9.8858	8794	8728	8659	8588	9.8514	15
46	9.9341	9295	9247	9198	9146	9.9092	9037	8979	8919	9.8857	8793	8727	8658	8586	9.8512	14
47	9.9340	9294	9247	9197	9145	9.9091	9036	8978	8918	9.8856	8792	8725	8657	8585	9.8511	13
48	9.9340	9294	9246	9196	9144	9.9091	9035	8977	8917	9.8855	8791	8724	8655	8584	9.8510	12
49	9.9339	9293	9245	9195	9143	9.9090	9034	8976	8916	9.8854	8790	8723	8654	8583	9.8509	11
50	9.9338	9292	9244	9194	9142	9.9089	9033	8975	8915	9.8853	8789	8722	8653	8582	9.8507	10
51	9.9337	9291	9243	9193	9142	9.9088	9032	8974	8914	9.8852	8788	8721	8652	8580	9.8506	9
52	9.9337	9291	9242	9193	9141	9.9087	9031	8973	8913	9.8851	8787	8720	8651	8579	9.8505	8
53	9.9336	9290	9242	9192	9140	9.9086	9030	8972	8912	9.8850	8785	8719	8650	8578	9.8504	7
54	9.9335	9289	9241	9191	9139	9.9085	9029	8971	8911	9.8849	8784	8718	8648	8577	9.8502	6
55	9.9334	9288	9240	9190	9138	9.9084	9028	8970	8910	9.8848	8783	8716	8647	8575	9.8501	5
56	9.9334	9287	9239	9189	9137	9.9083	9027	8969	8909	9.8847	8782	8715	8646	8574	9.8500	4
57	9.9333	9287	9238	9188	9136	9.9082	9026	8968	8908	9.8846	8781	8714	8645	8573	9.8499	3
58	9.9332	9286	9238	9187	9135	9.9081	9025	8967	8907	9.8845	8780	8713	8644	8572	9.8497	2
59	9.9331	9285	9237	9187	9135	9.9080	9024	8966	8906	9.8844	8779	8712	8642	8571	9.8496	1
60	9.9331	9284	9236	9186	9134	9.9080	9023	8965	8905	9.8843	8778	8711	8641	8569	9.8495	0
	59°	58°	57°	56°	55°	54°	53°	52°	51°	50°	49°	48°	47°	46°	45°	

NOTE: For compactness, the leading digit is not displayed in every column. Scan other columns for necessary digit.

LOG SINES

LOG COSINES

′	45°	46°	47°	48°	49°	50°	51°	52°	53°	54°	55°	56°	57°	58°	59°	′
0	9.8495	8418	8338	8255	8169	9.8081	7989	7893	7795	9.7692	7586	7476	7361	7242	9.7118	60
1	9.8494	8416	8336	8254	8168	9.8079	7987	7892	7793	9.7690	7584	7474	7359	7240	9.7116	59
2	9.8492	8415	8335	8252	8167	9.8078	7986	7890	7791	9.7689	7582	7472	7357	7238	9.7114	58
3	9.8491	8414	8334	8251	8165	9.8076	7984	7889	7790	9.7687	7580	7470	7355	7236	9.7112	57
4	9.8490	8412	8332	8249	8164	9.8075	7982	7887	7788	9.7685	7579	7468	7353	7234	9.7110	56
5	9.8489	8411	8331	8248	8162	9.8073	7981	7885	7786	9.7683	7577	7466	7351	7232	9.7108	55
6	9.8487	8410	8330	8247	8161	9.8072	7979	7884	7785	9.7682	7575	7464	7349	7230	9.7106	54
7	9.8486	8409	8328	8245	8159	9.8070	7978	7882	7783	9.7680	7573	7462	7347	7228	9.7104	53
8	9.8485	8407	8327	8244	8158	9.8069	7976	7880	7781	9.7678	7571	7461	7345	7226	9.7102	52
9	9.8483	8406	8326	8242	8156	9.8067	7975	7879	7780	9.7676	7570	7459	7344	7224	9.7099	51
10	9.8482	8405	8324	8241	8155	9.8066	7973	7877	7778	9.7675	7568	7457	7342	7222	9.7097	50
11	9.8481	8403	8323	8240	8153	9.8064	7972	7876	7776	9.7673	7566	7455	7340	7220	9.7095	49
12	9.8480	8402	8322	8238	8152	9.8063	7970	7874	7774	9.7671	7564	7453	7338	7218	9.7093	48
13	9.8478	8401	8320	8237	8150	9.8061	7968	7872	7773	9.7669	7562	7451	7336	7216	9.7091	47
14	9.8477	8399	8319	8235	8149	9.8060	7967	7871	7771	9.7668	7561	7449	7334	7214	9.7089	46
15	9.8476	8398	8317	8234	8148	9.8058	7965	7869	7769	9.7666	7559	7447	7332	7212	9.7087	45
16	9.8475	8397	8316	8233	8146	9.8056	7964	7867	7768	9.7664	7557	7445	7330	7210	9.7085	44
17	9.8473	8395	8315	8231	8145	9.8055	7962	7866	7766	9.7662	7555	7444	7328	7208	9.7082	43
18	9.8472	8394	8313	8230	8143	9.8053	7960	7864	7764	9.7661	7553	7442	7326	7205	9.7080	42
19	9.8471	8393	8312	8228	8142	9.8052	7959	7863	7763	9.7659	7551	7440	7324	7203	9.7078	41
20	9.8469	8391	8311	8227	8140	9.8050	7957	7861	7761	9.7657	7550	7438	7322	7201	9.7076	40
21	9.8468	8390	8309	8225	8139	9.8049	7956	7859	7759	9.7655	7548	7436	7320	7199	9.7074	39
22	9.8467	8389	8308	8224	8137	9.8047	7954	7858	7758	9.7654	7546	7434	7318	7197	9.7072	38
23	9.8466	8387	8306	8223	8136	9.8046	7953	7856	7756	9.7652	7544	7432	7316	7195	9.7070	37
24	9.8464	8386	8305	8221	8134	9.8044	7951	7854	7754	9.7650	7542	7430	7314	7193	9.7068	36
25	9.8463	8385	8304	8220	8133	9.8043	7949	7853	7752	9.7648	7540	7428	7312	7191	9.7065	35
26	9.8462	8383	8302	8218	8131	9.8041	7948	7851	7751	9.7647	7539	7427	7310	7189	9.7063	34
27	9.8460	8382	8301	8217	8130	9.8040	7946	7849	7749	9.7645	7537	7425	7308	7187	9.7061	33
28	9.8459	8381	8300	8215	8128	9.8038	7945	7848	7747	9.7643	7535	7423	7306	7185	9.7059	32
29	9.8458	8379	8298	8214	8127	9.8037	7943	7846	7746	9.7641	7533	7421	7304	7183	9.7057	31
30	9.8457	8378	8297	8213	8125	9.8035	7941	7844	7744	9.7640	7531	7419	7302	7181	9.7055	30
31	9.8455	8377	8295	8211	8124	9.8034	7940	7843	7742	9.7638	7529	7417	7300	7179	9.7053	29
32	9.8454	8375	8294	8210	8122	9.8032	7938	7841	7740	9.7636	7528	7415	7298	7177	9.7050	28
33	9.8453	8374	8293	8208	8121	9.8031	7937	7840	7739	9.7634	7526	7413	7296	7175	9.7048	27
34	9.8451	8373	8291	8207	8120	9.8029	7935	7838	7737	9.7632	7524	7411	7294	7173	9.7046	26
35	9.8450	8371	8290	8205	8118	9.8027	7934	7836	7735	9.7631	7522	7409	7292	7171	9.7044	25
36	9.8449	8370	8289	8204	8117	9.8026	7932	7835	7734	9.7629	7520	7407	7290	7168	9.7042	24
37	9.8448	8369	8287	8203	8115	9.8024	7930	7833	7732	9.7627	7518	7406	7288	7166	9.7040	23
38	9.8446	8367	8286	8201	8114	9.8023	7929	7831	7730	9.7625	7517	7404	7286	7164	9.7037	22
39	9.8445	8366	8284	8200	8112	9.8021	7927	7830	7728	9.7624	7515	7402	7284	7162	9.7035	21
40	9.8444	8365	8283	8198	8111	9.8020	7926	7828	7727	9.7622	7513	7400	7282	7160	9.7033	20
41	9.8442	8363	8282	8197	8109	9.8018	7924	7826	7725	9.7620	7511	7398	7280	7158	9.7031	19
42	9.8441	8362	8280	8195	8108	9.8017	7922	7825	7723	9.7618	7509	7396	7278	7156	9.7029	18
43	9.8440	8361	8279	8194	8106	9.8015	7921	7823	7722	9.7616	7507	7394	7276	7154	9.7027	17
44	9.8439	8359	8277	8193	8105	9.8014	7919	7821	7720	9.7615	7505	7392	7274	7152	9.7025	16
45	9.8437	8358	8276	8191	8103	9.8012	7918	7820	7718	9.7613	7504	7390	7272	7150	9.7022	15
46	9.8436	8357	8275	8190	8102	9.8010	7916	7818	7716	9.7611	7502	7388	7270	7148	9.7020	14
47	9.8435	8355	8273	8188	8100	9.8009	7914	7816	7715	9.7609	7500	7386	7268	7146	9.7018	13
48	9.8433	8354	8272	8187	8099	9.8007	7913	7815	7713	9.7607	7498	7384	7266	7144	9.7016	12
49	9.8432	8353	8270	8185	8097	9.8006	7911	7813	7711	9.7606	7496	7382	7264	7141	9.7014	11
50	9.8431	8351	8269	8184	8096	9.8004	7910	7811	7710	9.7604	7494	7380	7262	7139	9.7012	10
51	9.8429	8350	8268	8182	8094	9.8003	7908	7810	7708	9.7602	7492	7379	7260	7137	9.7009	9
52	9.8428	8349	8266	8181	8093	9.8001	7906	7808	7706	9.7600	7491	7377	7258	7135	9.7007	8
53	9.8427	8347	8265	8180	8091	9.8000	7905	7806	7704	9.7599	7489	7375	7256	7133	9.7005	7
54	9.8426	8346	8264	8178	8090	9.7998	7903	7805	7703	9.7597	7487	7373	7254	7131	9.7003	6
55	9.8424	8345	8262	8177	8088	9.7997	7901	7803	7701	9.7595	7485	7371	7252	7129	9.7001	5
56	9.8423	8343	8261	8175	8087	97995	7900	7801	7699	9.7593	7483	7369	7250	7127	9.6998	4
57	9.8422	8342	8259	8174	8085	9.7993	7898	7800	7697	9.7591	7481	7367	7248	7125	9.6996	3
58	9.8420	8341	8258	8172	8084	9.7992	7897	7798	7696	9.7590	7479	7365	7246	7123	9.6994	2
59	9.8419	8339	8257	8171	8082	9.7990	7895	7796	7694	9.7588	7477	7363	7244	7120	9.6992	1
60	9.8418	8338	8255	8169	8081	9.7989	7893	7795	7692	9.7586	7476	7361	7242	7118	9.6990	0
	44°	43°	42°	41°	40°	39°	38°	37°	36°	35°	34°	33°	32°	31°	30°	

NOTE: For compactness, the leading digit is not displayed in every column. Scan other columns for necessary digit.

LOG SINES

LOG COSINES

′	60°	61°	62°	63°	64°	65°	66°	67°	68°	69°	70°	71°	72°	73°	74°	′
0	9.6990	6856	6716	6570	6418	9.6259	6093	5919	5736	9.5543	5341	5126	4900	4659	9.4403	60
1	9.6988	6853	6714	6568	6416	9.6257	6090	5916	5733	9.5540	5337	5123	4896	4655	9.4399	59
2	9.6985	6851	6711	6566	6413	9.6254	6087	5913	5729	9.5537	5334	5119	4892	4651	9.4395	58
3	9.6983	6849	6709	6563	6411	9.6251	6085	5910	5726	9.5533	5330	5115	4888	4647	9.4390	57
4	9.6981	6847	6707	6561	6408	9.6249	6082	5907	5723	9.5530	5327	5112	4884	4643	9.4386	56
5	9.6979	6844	6704	6558	6405	9.6246	6079	5904	5720	9.5527	5323	5108	4880	4639	9.4381	55
6	9.6977	6842	6702	6556	6403	9.6243	6076	5901	5717	9.5524	5320	5104	4876	4634	9.4377	54
7	9.6974	6840	6699	6553	6400	9.6240	6073	5898	5714	9.5520	5316	5101	4873	4630	9.4372	53
8	9.6972	6837	6697	6551	6398	9.6238	6070	5895	5711	9.5517	5313	5097	4869	4626	9.4368	52
9	9.6970	6835	6695	6548	6395	9.6235	6068	5892	5708	9.5514	5309	5093	4865	4622	9.4364	51
10	9.6968	6833	6692	6546	6392	9.6232	6065	5889	5704	9.5510	5306	5090	4861	4618	9.4359	50
11	9.6966	6831	6690	6543	6390	9.6230	6062	5886	5701	9.5507	5302	5086	4857	4614	9.4355	49
12	9.6963	6828	6687	6541	6387	9.6227	6059	5883	5698	9.5504	5299	5082	4853	4609	9.4350	48
13	9.6961	6826	6685	6538	6385	9.6224	6056	5880	5695	9.5500	5295	5078	4849	4605	9.4346	47
14	9.6959	6824	6683	6536	6382	9.6221	6053	5877	5692	9.5497	5292	5075	4845	4601	9.4341	46
15	9.6957	6821	6680	6533	6379	9.6219	6050	5874	5689	9.5494	5288	5071	4841	4597	9.4337	45
16	9.6955	6819	6678	6531	6377	9.6216	6047	5871	5685	9.5490	5285	5067	4837	4593	9.4332	44
17	9.6952	6817	6675	6528	6374	9.6213	6045	5868	5682	9.5487	5281	5064	4833	4588	9.4328	43
18	9.6950	6814	6673	6526	6371	9.6210	6042	5865	5679	9.5484	5278	5060	4829	4584	9.4323	42
19	9.6948	6812	6671	6523	6369	9.6208	6039	5862	5676	9.5480	5274	5056	4825	4580	9.4319	41
20	9.6946	6810	6668	6521	6366	9.6205	6036	5859	5673	9.5477	5270	5052	4821	4576	9.4314	40
21	9.6943	6808	6666	6518	6364	9.6202	6033	5856	5670	9.5474	5267	5049	4817	4572	9.4310	39
22	9.6941	6805	6663	6515	6361	9.6199	6030	5853	5666	9.5470	5263	5045	4813	4567	9.4305	38
23	9.6939	6803	6661	6513	6358	9.6197	6027	5850	5663	9.5467	5260	5041	4809	4563	9.4301	37
24	9.6937	6801	6659	6510	6356	9.6194	6024	5847	5660	9.5463	5256	5037	4805	4559	9.4296	36
25	9.6935	6798	6656	6508	6353	9.6191	6021	5844	5657	9.5460	5253	5034	4801	4555	9.4292	35
26	9.6932	6796	6654	6505	6350	9.6188	6019	5841	5654	9.5457	5249	5030	4797	4550	9.4287	34
27	9.6930	6794	6651	6503	6348	9.6186	6016	5838	5650	9.5453	5246	5026	4793	4546	9.4283	33
28	9.6928	6791	6649	6500	6345	9.6183	6013	5834	5647	9.5450	5242	5022	4789	4542	9.4278	32
29	9.6926	6789	6646	6498	6342	9.6180	6010	5831	5644	9.5447	5239	5019	4785	4538	9.4274	31
30	9.6923	6787	6644	6495	6340	9.6177	6007	5828	5641	9.5443	5235	5015	4781	4533	9.4269	30
31	9.6921	6784	6642	6493	6337	9.6175	6004	5825	5638	M440	5231	5011	4777	4529	9.4264	29
32	9.6919	6782	6639	6490	6335	9.6172	6001	5822	5634	9.5437	5228	5007	4773	4525	9.4260	28
33	9.6917	6780	6637	6488	6332	9.6169	5998	5819	5631	9.5433	5224	5003	4769	4521	9.4255	27
34	9.6914	6777	6634	6485	6329	9.6166	5995	5816	5628	9.5430	5221	5000	4765	4516	9.4251	26
35	9.6912	6775	6632	6483	6327	9.6163	5992	5813	5625	9.5426	5217	4996	4761	4512	9.4246	25
36	9.6910	6773	6629	6480	6324	9.6161	5990	5810	5621	9.5423	5213	4992	4757	4508	9.4242	24
37	9.6908	6770	6627	6477	6321	9.6158	5987	5807	5618	9.5420	5210	4988	4753	4503	9.4237	23
38	9.6905	6768	6625	6475	6319	9.6155	5984	5804	5615	9.5416	5206	4984	4749	4499	9.4232	22
39	9.6903	6766	6622	6472	6316	9.6152	5981	5801	5612	9.5413	5203	4981	4745	4495	9.4228	21
40	9.6901	6763	6620	6470	6313	9.6149	5978	5798	5609	9.5409	5199	4977	4741	4491	9.4223	20
41	9.6899	6761	6617	6467	6311	9.6147	5975	5795	5605	9.5406	5196	4973	4737	4486	9.4219	19
42	9.6896	6759	6615	6465	6308	9.6144	5972	5792	5602	9.5403	5192	4969	4733	4482	9.4214	18
43	9.6894	6756	6612	6462	6305	9.6141	5969	5789	5599	9.5399	5188	4965	4729	4478	9.4209	17
44	9.6892	6754	6610	6460	6303	9.6138	5966	5785	5596	9.5396	5185	4962	4725	4473	9.4205	16
45	9.6890	6752	6607	6457	6300	9.6135	5963	5782	5592	9.5392	5181	4958	4721	4469	9.4200	15
46	9.6887	6749	6605	6454	6297	9.6133	5960	5779	5589	9.5389	5177	4954	4717	4465	9.4195	14
47	9.6885	6747	6603	6452	6295	9.6130	5957	5776	5586	9.5385	5174	4950	4713	4460	9.4191	13
48	9.6883	6744	6600	6449	6292	9.6127	5954	5773	5583	9.5382	5170	4946	4709	4456	9.4186	12
49	9.6881	6742	6598	6447	6289	9.6124	5951	5770	5579	9.5379	5167	4942	4705	4452	9.4182	11
50	9.6878	6740	6595	6444	6286	9.6121	5948	5767	5576	9.5375	5163	4939	4700	4447	9.4177	10
51	9.6876	6737	6593	6442	6284	9.6119	5945	5764	5573	9.5372	5159	4935	4696	4443	9.4172	9
52	9.6874	6735	6590	6439	6281	9.6116	5943	5761	5570	9.5368	5156	4931	4692	4438	9.4168	8
53	9.6872	6733	6588	6437	6278	9.6113	5940	5758	5566	9.5365	5152	4927	4688	4434	9.4163	7
54	9.6869	6730	6585	6434	6276	9.6110	5937	5754	5563	9.5361	5148	4923	4684	4430	9.4158	6
55	9.6867	6728	6583	6431	6273	9.6107	5934	5751	5560	9.5358	5145	4919	4680	4425	9.4153	5
56	9.6865	6726	6580	6429	6270	9.6104	5931	5748	5556	9.5354	5141	4915	4676	4421	9.4149	4
57	9.6863	6723	6578	6426	6268	9.6102	5928	5745	5553	9.5351	5137	4911	4672	4417	9.4144	3
58	9.6860	6721	6575	6424	6265	9.6099	5925	5742	5550	9.5347	5134	4908	4668	4412	9.4139	2
59	9.6858	6718	6573	6421	6262	9.6096	5922	5739	5547	9.5344	5130	4904	4663	4408	9.4135	1
60	9.6856	6716	6570	6418	6259	9.6093	5919	5736	5543	9.5341	5126	4900	4659	4403	9.4130	0
	29°	28°	27°	26°	25°	24°	23°	22°	21°	20°	19°	18°	17°	16°	15°	

NOTE: For compactness, the leading digit is not displayed in every column. Scan other columns for necessary digit.

LOG SINES

LOG COSINES

′	75°	76°	77°	78°	79°	80°	81°	82°	83°	84°	85°	86°	87°	88°	89°	′
0	9.4130	3837	3521	3179	2806	9.2397	1943	1436	0859	9.0192	9403	8436	7188	5428	8.2419	60
1	9.4125	3832	3515	3173	2799	9.2390	1935	1427	0849	9.0180	9388	8418	7164	5392	8.2346	59
2	9.4121	3827	3510	3167	2793	9.2382	1927	1418	0838	9.0168	9374	8400	7140	5355	8.2271	58
3	9.4116	3822	3504	3161	2786	9.2375	1919	1409	0828	9.0156	9359	8381	7115	5318	8.2196	57
4	9.4111	3816	3499	3155	2780	9.2368	1911	1399	0818	9.0144	9345	8363	7090	5281	8.2119	56
5	9.4106	3811	3493	3149	2773	9.2361	1903	1390	0807	9.0132	9330	8345	7066	5243	8.2041	55
6	9.4102	3806	3488	3143	2767	9.2354	1895	1381	0797	9.0120	9315	8326	7041	5206	8.1961	54
7	9.4097	3801	3482	3137	2760	9.2346	1887	1372	0786	9.0107	9301	8307	7016	5167	8.1880	53
8	9.4092	3796	3477	3131	2754	9.2339	1879	1363	0776	9.0095	9286	8289	6991	5129	8.1797	52
9	9.4087	3791	3471	3125	2747	9.2332	1871	1354	0765	9.0083	9271	8270	6965	5090	8.1713	51
10	9.4083	3786	3466	3119	2740	9.2324	1863	1345	0755	9.0070	9256	8251	6940	5050	8.1627	50
11	9.4078	3781	3460	3113	2734	9.2317	1855	1336	0744	9.0058	9241	8232	6914	5011	8.1539	49
12	9.4073	3775	3455	3107	2727	9.2310	1847	1326	0734	9.0046	9226	8213	6889	4971	8.1450	48
13	9.4068	3770	3449	3101	2721	9.2303	1838	1317	0723	9.0033	9211	8194	6863	4930	8.1358	47
14	9.4063	3765	3444	3095	2714	9.2295	1830	1308	0712	9.0021	9196	8175	6837	4890	8.1265	46
15	9.4059	3760	3438	3089	2707	9.2288	1822	1299	0702	9.0008	9181	8156	6810	4848	8.1169	45
16	9.4054	3755	3432	3083	2701	9.2281	1814	1289	0691	8.9996	9166	8137	6784	4807	8.1072	44
17	9.4049	3750	3427	3077	2694	9.2273	1806	1280	0680	8.9983	9150	8117	6758	4765	8.0972	43
18	9.4044	3745	3421	3070	2687	9.2266	1797	1271	0670	8.9970	9135	8098	6731	4723	8.0870	42
19	9.4039	3739	3416	3064	2681	9.2258	1789	1261	0659	8.9958	9119	8078	6704	4680	8.0765	41
20	9.4035	3734	3410	3058	2674	9.2251	1781	1252	0648	8.9945	9104	8059	6677	4637	8.0658	40
21	9.4030	3729	3404	3052	2667	9.2244	1772	1242	0637	8.9932	9089	8039	6650	4593	8.0548	39
22	9.4025	3724	3399	3046	2661	9.2236	1764	1233	0626	8.9919	9073	8019	6622	4549	8.0435	38
23	9.4020	3719	3393	3040	2654	9.2229	1756	1224	0616	8.9907	9057	7999	6595	4504	8.0319	37
24	9.4015	3713	3387	3034	2647	9.2221	1747	1214	0605	8.9894	9042	7979	6567	4459	8.0200	36
25	9.4010	3708	3382	3027	2640	9.2214	1739	1205	0594	8.9881	9026	7959	6539	4414	8.0078	35
26	9.4006	3703	3376	3021	2634	9.2206	1731	1195	0583	8.9868	9010	7939	6511	4368	7.9952	34
27	9.4001	3698	3370	3015	2627	9.2199	1722	1186	0572	8.9855	8994	7918	6483	4322	7.9822	33
28	9.3996	3692	3365	3009	2620	9.2191	1714	1176	0561	8.9842	8978	7898	6454	4275	7.9689	32
29	9.3991	3687	3359	3003	2613	9.2184	1705	1167	0550	8.9829	8962	7877	6426	4227	7.9551	31
30	9.3986	3682	3353	2997	2606	9.2176	1697	1157	0539	8.9816	8946	7857	6397	4179	7.9408	30
31	9.3981	3677	3348	2990	2600	9.2169	1689	1147	0527	8.9803	8930	7836	6368	4131	7.9261	29
32	9.3976	3671	3342	2984	2593	9.2161	1680	1138	0516	8.9789	8914	7815	6339	4082	7.9109	28
33	9.3971	3666	3336	2978	2586	9.2153	1672	1128	0505	8.9776	8898	7794	6309	4032	7.8951	27
34	9.3966	3661	3331	2972	2579	9.2146	1663	1118	0494	8.9763	8882	7773	6279	3982	7.8787	26
35	9.3962	3655	3325	2965	2572	9.2138	1655	1109	0483	8.9750	8865	7752	6250	3931	7.8617	25
36	9.3957	3650	3319	2959	2565	9.2131	1646	1099	0472	8.9736	8849	7731	6220	3880	7.8439	24
37	9.3952	3645	3313	2953	2558	9.2123	1637	1089	0460	8.9723	8833	7710	6189	3828	7.8255	23
38	9.3947	3640	3308	2947	2551	9.2115	1629	1080	0449	8.9710	8816	7688	6159	3775	7.8061	22
39	9.3942	3634	3302	2940	2545	9.2108	1620	1070	0438	8.9696	8799	7667	6128	3722	7.7859	21
40	9.3937	3629	3296	2934	2538	9.2100	1612	1060	0426	8.9683	8783	7645	6097	3668	7.7648	20
41	9.3932	3624	3290	2928	2531	9.2092	1603	1050	0415	8.9669	8766	7623	6066	3613	7.7425	19
42	9.3927	3618	3284	2921	2524	9.2085	1594	1040	0403	8.9655	8749	7602	6035	3558	7.7190	18
43	9.3922	3613	3279	2915	2517	9.2077	1586	1030	0392	8.9642	8733	7580	6003	3502	7.6942	17
44	9.3917	3608	3273	2909	2510	9.2069	1577	1020	0380	8.9628	8716	7557	5972	3445	7.6678	16
45	9.3912	3602	3267	2902	2503	9.2061	1568	1011	0369	8.9614	8689	7535	5939	3388	7.6398	15
46	9.3907	3597	3261	2896	2496	9.2054	1560	1001	0357	8.9601	8682	7513	5907	3329	7.6099	14
47	9.3902	3591	3255	2890	2489	9.2046	1551	0991	0346	8.9587	8665	7491	5875	3270	7.5777	13
48	9.3897	3586	3250	2883	2482	9.2038	1542	0981	0334	8.9573	8647	7468	5842	3210	7.5429	12
49	9.3892	3581	3244	2877	2475	9.2030	1533	0971	0323	8.9559	8630	7445	5809	3150	7.5051	11
50	9.3887	3575	3238	2870	2468	9.2022	1525	0961	0311	8.9545	8613	7423	5776	3088	7.4637	10
51	9.3882	3570	3232	2864	2461	9.2015	1516	0951	0299	8.9531	8595	7400	5742	3025	7.4180	9
52	9.3877	3564	3226	2858	2454	9.2007	1507	0940	0287	8.9517	8578	7377	5708	2962	7.3668	8
53	9.3872	3559	3220	2851	2447	9.1999	1498	0930	0276	8.9503	8560	7354	5674	2898	7.3088	7
54	9.3867	3554	3214	2845	2439	9.1991	1489	0920	0264	8.9489	8543	7330	5640	2832	7.2419	6
55	9.3862	3548	3208	2838	2432	9.1983	1480	0910	0252	8.9475	8525	7307	5605	2766	7.1627	5
56	9.3857	3543	3202	2832	2425	9.1975	1471	0900	0240	8.9460	8508	7283	5571	2699	7.0658	4
57	9.3852	3537	3197	2825	2418	9.1967	1462	0890	0228	8.9446	8490	7260	5535	2630	6.9409	3
58	9.3847	3532	3191	2819	2411	9.1959	1453	0879	0216	8.9432	8472	7236	5500	2561	6.7648	2
59	9.3842	3526	3185	2812	2404	9.1951	1445	0869	0204	8.9417	8454	7212	5464	2490	6.4637	1
60	9.3837	3521	3179	2806	2397	9.1943	1436	0859	0192	8.9403	8436	7188	5428	2419	$-\infty$	0
	14°	13°	12°	11°	10°	9°	8°	7°	6°	5°	4°	3°	2°	1°	0°	

NOTE: For compactness, the leading digit is not displayed in every column. Scan other columns for necessary digit.

LOG SINES

A

+ if hour angle is listed at top
− if hour angle is listed at bottom

LHA	1° 359°	2° 358°	3° 357°	4° 356°	5° 355°	6° 354°	7° 353°	8° 352°	9° 351°	10° 350°	11° 349°	12° 348°	13° 347°	14° 346°	15° 345°
0	0.00	.000	.000	.000	.000	.000	.000	.000	.000	.000	.000	.000	.000	.000	.000
3	3.00	1.50	1.00	.749	.599	.499	.427	.373	.331	.297	.270	.247	.227	.210	.196
6	6.02	3.01	2.01	1.50	1.20	1.00	.856	.748	.664	.596	.541	.494	.455	.422	.392
9	9.07	4.54	3.02	2.27	1.81	1.51	1.29	1.13	1.00	.898	.815	.745	.686	.635	.591
12	12.2	6.09	4.06	3.04	2.43	2.02	1.73	1.51	1.34	1.21	1.09	1.00	.921	.853	.793
15	15.4	7.67	5.11	3.83	3.06	2.55	2.18	1.91	1.69	1.52	1.38	1.26	1.16	1.07	1.00
18	18.6	9.30	6.20	4.65	3.71	3.09	2.65	2.31	2.05	1.84	1.67	1.53	1.41	1.30	1.21
21	22.0	11.0	7.32	5.49	4.39	3.65	3.13	2.73	2.42	2.18	1.97	1.81	1.66	1.54	1.43
24	25.5	12.7	8.50	6.37	5.09	4.24	3.63	3.17	2.81	2.53	2.29	2.09	1.93	1.79	1.66
27	29.2	14.6	9.72	7.29	5.82	4.85	4.15	3.63	3.22	2.89	2.62	2.40	2.21	2.04	1.90
30	33.1	16.5	11.0	8.26	6.60	5.49	4.70	4.11	3.65	3.27	2.97	2.72	2.50	2.32	2.15
33	37.2	18.6	12.4	9.29	7.42	6.18	5.29	4.62	4.10	3.68	3.34	3.06	2.81	2.61	2.42
36	41.6	20.8	13.9	10.4	8.30	6.91	5.92	5.17	4.59	4.12	3.74	3.42	3.15	2.91	2.71
38	44.8	22.4	14.9	11.2	8.93	7.43	6.36	5.56	4.93	4.43	4.02	3.68	3.38	3.13	2.92
40	48.1	24.0	16.0	12.0	9.59	7.98	6.83	5.97	5.30	4.76	4.32	3.95	3.63	3.37	3.13
42	51.6	25.8	17.2	12.9	10.3	8.57	7.33	6.41	5.69	5.11	4.63	4.24	3.90	3.61	3.36
44	55.3	27.7	18.4	13.8	11.0	9.19	7.86	6.87	6.10	5.48	4.97	4.54	4.18	3.87	3.60
46	59.3	29.7	19.8	14.8	11.8	9.85	8.43	7.37	6.54	5.87	5.33	4.87	4.49	4.15	3.86
48	63.6	31.8	21.2	15.9	12.7	10.6	9.05	7.90	7.01	6.30	5.71	5.23	4.81	4.45	4.14
50	68.3	34.1	22.7	17.0	13.6	11.3	9.71	8.48	7.52	6.76	6.13	5.61	5.16	4.78	4.45
52	73.3	36.7	24.4	18.3	14.6	12.2	10.4	9.11	8.08	7.26	6.58	6.02	5.55	5.13	4.78
54	78.9	39.4	26.3	19.7	15.7	13.1	11.2	9.79	8.69	7.81	7.08	6.48	5.96	5.52	5.14
56	84.9	42.5	28.3	21.2	16.9	14.1	12.1	10.5	9.36	8.41	7.63	6.97	6.42	5.95	5.53
58	91.7	45.8	30.5	22.9	18.3	15.2	13.0	11.4	10.1	9.08	8.23	7.53	6.93	6.42	5.97
60	99.2	49.6	33.0	24.8	19.8	16.5	14.1	12.3	10.9	9.82	8.91	8.15	7.50	6.95	6.46
62	108	53.9	35.9	26.9	21.5	17.9	15.3	13.4	11.9	10.7	9.68	8.85	8.15	7.54	7.02
64	117	58.7	39.1	29.3	23.4	19.5	16.7	14.6	12.9	11.6	10.5	9.65	8.88	8.22	7.65
66	129	64.3	42.9	32.1	25.7	21.4	18.3	16.0	14.2	12.7	11.6	10.6	9.72	9.01	8.38
LHA	179° 181°	178° 183°	177° 183°	176° 184°	175° 185°	174° 186°	173° 187°	172° 188°	171° 189°	170° 190°	169° 191°	168° 192°	167° 193°	166° 194°	165° 195°

B

− if latitude and declination have same name
+ if latitude and declination have different names

LHA	1° 359°	2° 358°	3° 357°	4° 356°	5° 355°	6° 354°	7° 353°	8° 352°	9° 351°	10° 350°	11° 349°	12° 348°	13° 347°	14° 346°	15° 345°
0	0.00	.000	.000	.000	.000	.000	.000	.000	.000	.000	.000	.000	.000	.000	.000
3	3.00	1.50	1.00	.751	.601	.501	.430	.377	.335	.302	.275	.252	.233	.217	.202
6	6.02	3.01	2.01	1.51	1.21	1.01	.862	.755	.672	.605	.551	.506	.467	.434	.406
9	9.08	4.54	3.03	2.27	1.82	1.52	1.30	1.14	1.01	.912	.830	.762	.704	.655	.612
12	12.2	6.09	4.06	3.05	2.44	2.03	1.74	1.53	1.36	1.22	1.11	1.02	.945	.879	.821
15	15.4	7.68	5.12	3.84	3.07	2.56	2.20	1.93	1.71	1.54	1.40	1.29	1.19	1.11	1.04
18	18.6	9.31	6.21	4.66	3.73	3.11	2.67	2.33	2.08	1.87	1.70	1.56	1.44	1.34	1.26
21	22.0	11.0	7.33	5.50	4.40	3.67	3.15	2.76	2.45	2.21	2.01	1.85	1.71	1.59	1.48
24	25.5	12.8	8.51	6.38	5.11	4.26	3.65	3.20	2.85	2.56	2.33	2.14	1.98	1.84	1.72
27	29.2	14.6	9.74	7.30	5.85	4.87	4.18	3.66	3.26	2.93	2.67	2.45	2.27	2.11	1.97
30	33.1	16.5	11.0	8.28	6.62	5.52	4.74	4.15	3.69	3.32	3.03	2.78	2.57	2.39	2.23
33	37.2	18.6	12.4	9.31	7.45	6.21	5.33	4.67	4.15	3.74	3.40	3.12	2.89	2.68	2.51
36	41.6	20.8	13.9	10.4	8.34	6.95	5.96	5.22	4.64	4.18	3.81	3.49	3.23	3.00	2.81
38	44.8	22.4	14.9	11.2	8.96	7.47	6.41	5.61	4.99	4.50	4.09	3.76	3.47	3.23	3.02
40	48.1	24.0	16.0	12.0	9.63	8.03	6.89	6.03	5.36	4.83	4.40	4.04	3.73	3.47	3.24
42	51.6	25.8	17.2	12.9	10.3	8.61	7.39	6.47	5.76	5.19	4.72	4.33	4.00	3.72	3.48
44	55.3	27.7	18.5	13.8	11.1	9.24	7.92	6.94	6.17	5.56	5.06	4.64	4.29	3.99	3.73
46	59.3	29.7	19.8	14.8	11.9	9.91	8.50	7.44	6.62	5.96	5.43	4.98	4.60	4.28	4.00
48	63.6	31.8	21.2	15.9	12.7	10.6	9.11	7.98	7.10	6.40	5.82	5.34	4.94	4.59	4.29
50	68.3	34.1	22.8	17.1	13.7	11.4	9.78	8.56	7.62	6.86	6.25	5.73	5.30	4.93	4.60
52	73.3	36.7	24.5	18.3	14.7	12.2	10.5	9.20	8.18	7.37	6.71	6.16	5.69	5.29	4.95
54	78.9	39.4	26.3	19.7	15.8	13.2	11.3	9.89	8.80	7.93	7.21	6.62	6.12	5.69	5.32
56	84.9	42.5	28.3	21.3	17.0	14.2	12.2	10.7	9.48	8.54	7.77	7.13	6.59	6.13	5.73
58	91.7	45.9	30.6	22.9	18.4	15.3	13.1	11.5	10.2	9.22	8.39	7.70	7.11	6.62	6.18
60	99.2	49.6	33.1	24.8	19.9	16.6	14.2	12.5	11.1	9.97	9.08	8.33	7.70	7.16	6.69
62	108	53.9	35.9	27.0	21.6	18.0	15.4	13.5	12.0	10.8	9.86	9.05	8.36	7.77	7.27
LHA	179° 181°	178° 183°	177° 183°	176° 184°	175° 185°	174° 186°	173° 187°	172° 188°	171° 189°	170° 190°	169° 191°	168° 192°	167° 193°	166° 194°	165° 195°

ABC TABLES

A

+ if hour angle is listed at top
− if hour angle is listed at bottom

LHA	16° 344°	17° 343°	18° 342°	19° 341°	20° 340°	21° 339°	22° 338°	23° 337°	24° 336°	25° 335°	26° 334°	27° 333°	28° 332°	29° 331°	30° 330°
0	.000	.000	.000	.000	.000	.000	.000	.000	.000	.000	.000	.000	.000	.000	.000
3	.183	.171	.161	.152	.144	.137	.130	.123	.118	.112	.107	.103	.099	.095	.091
6	.367	.344	.323	.305	.289	.274	.260	.248	.236	.225	.215	.206	.198	.190	.182
9	.552	.518	.487	.460	.435	.413	.392	.373	.356	.340	.325	.311	.298	.286	.274
12	.741	.695	.654	.617	.584	.554	.526	.501	.477	.456	.436	.417	.400	.383	.368
15	.934	.876	.825	.778	.736	.698	.663	.631	.602	.575	.549	.526	.504	.483	.464
18	1.13	1.06	1.00	.944	.893	.846	.804	.765	.730	.697	.666	.638	.611	.586	.563
21	1.34	1.26	1.18	1.11	1.05	1.00	.950	.904	.862	.823	.787	.753	.722	.693	.665
24	1.55	1.46	1.37	1.29	1.22	1.16	1.10	1.05	1.00	.955	.913	.874	.837	.803	.771
27	1.78	1.67	1.57	1.48	1.40	1.33	1.26	1.20	1.14	1.09	1.04	1.00	.958	.919	.883
30	2.01	1.89	1.78	1.68	1.59	1.50	1.43	1.36	1.30	1.24	1.18	1.13	1.09	1.04	1.00
33	2.26	2.12	2.00	1.89	1.78	1.69	1.61	1.53	1.46	1.39	1.33	1.27	1.22	1.17	1.12
36	2.53	2.38	2.24	2.11	2.00	1.89	1.80	1.71	1.63	1.56	1.49	1.43	1.37	1.31	1.26
38	2.72	2.56	2.40	2.27	2.15	2.04	1.93	1.84	1.75	1.68	1.60	1.53	1.47	1.41	1.35
40	2.93	2.74	2.58	2.44	2.31	2.19	2.08	1.98	1.88	1.80	1.72	1.65	1.58	1.51	1.45
42	3.14	2.95	2.77	2.61	2.47	2.35	2.23	2.12	2.02	1.93	1.85	1.77	1.69	1.62	1.56
44	3.37	3.16	2.97	2.80	2.65	2.52	2.39	2.28	2.17	2.07	1.98	1.90	1.82	1.74	1.67
46	3.61	3.39	3.19	3.01	2.85	2.70	2.56	2.44	2.33	2.22	2.12	2.03	1.95	1.87	1.79
48	3.87	3.63	3.42	3.23	3.05	2.89	2.75	2.62	2.49	2.38	2.28	2.18	2.09	2.00	1.92
50	4.16	3.90	3.67	3.46	3.27	3.10	2.95	2.81	2.68	2.56	2.44	2.34	2.24	2.15	2.06
52	4.46	4.19	3.94	3.72	3.52	3.33	3.17	3.02	2.87	2.74	2.62	2.51	2.41	2.31	2.22
54	4.80	4.50	4.24	4.00	3.78	3.59	3.41	3.24	3.09	2.95	2.82	2.70	2.59	2.48	2.38
56	5.17	4.85	4.56	4.31	4.07	3.86	3.67	3.49	3.33	3.18	3.04	2.91	2.79	2.67	2.57
58	5.58	5.23	4.93	4.65	4.40	4.17	3.96	3.77	3.59	3.43	3.28	3.14	3.01	2.89	2.77
60	6.04	5.67	5.33	5.03	4.76	4.51	4.29	4.08	3.89	3.71	3.55	3.40	3.26	3.12	3.00
62	6.56	6.15	5.79	5.46	5.17	4.90	4.65	4.43	4.22	4.03	3.86	3.69	3.54	3.39	3.26
64	7.15	6.71	6.31	5.95	5.63	5.34	5.07	4.83	4.61	4.40	4.20	4.02	3.86	3.70	3.55
66	7.83	7.35	6.91	6.52-	6.17	5.85	5.56	5.29	5.04	4.82	4.61	4.41	4.22	4.05	3.89
LHA	164° 196°	163° 197°	162° 198°	161° 199°	160° 200°	159° 201°	158° 202°	157° 203°	156° 204°	155° 205°	154° 206°	153° 207°	152° 208°	151° 209°	150° 210°

B

− if latitude and declination have same name
+ if latitude and declination have different names

LHA	16° 344°	17° 343°	18° 342°	19° 341°	20° 340°	21° 339°	22° 338°	23° 337°	24° 336°	25° 335°	26° 334°	27° 333°	28° 332°	29° 331°	30° 330°
0	.000	.000	.000	.000	.000	.000	.000	.000	.000	.000	.000	.000	.000	.000	.000
3	.190	.179	.170	.161	.153	.146	.140	.134	.129	.124	.120	.115	.112	.108	.105
6	.381	.359	.340	.323	.307	.293	.281	.269	.258	.249	.240	.232	.224	.217	.210
9	.575	.542	.513	.486	.463	.442	.423	.405	.389	.375	.361	.349	.337	.327	.317
12	.771	.727	.688	.653	.621	.593	.567	.544	.523	.503	.485	.468	.453	.438	.425
15	.972	.916	.867	.823	.783	.748	.715	.686	.659	.634	.611	.590	.571	.553	.536
18	1.18	1.11	1.05	.998	.950	.907	.867	.832	.799	.769	.741	.716	.692	.670	.650
21	1.39	1.31	1.24	1.18	1.12	1.07	1.02	.982	.944	.908	.876	.846	.818	.792	.768
24	1.62	1.52	1.44	1.37	1.30	1.24	1.19	1.14	1.09	1.05	1.02	.981	.948	.918	.890
27	1.85	1.74	1.65	1.57	1.49	1.42	1.36	1.30	1.25	1.21	1.16	1.12	1.09	1.05	1.02
30	2.09	1.97	1.87	1.77	1.69	1.61	1.54	1.A8	1.42	1.37	1.32	1.27	1.23	1.19	1.15
33	2.36	2.22	2.10	1.99	1.90	1.81	1.73	1.66	1.60	1.54	1.48	1.43	1.38	1.34	1.30
36	2.64	2.48	2.32	2.23	2.12	2.03	1.94	1.86	1.79	1.72	1.66	1.60	1.55	1.50	1.45
38	2.83	2.67	2.53	2.40	2.28	2.18	2.09	2.00	1.92	1.85	1.78	1.72	1.66	1.61	1.56
40	3.04	2.87	2.72	2.58	2.45	2.34	2.24	2.15	2.06	1.99	1.91	1.85	1.79-	1.73	1.68
42	3.27	3.08	2.91	2.77	2.63	2.51	2.40	2.30	2.21	2.13	2.05	1.98	1.92	1.86	1.80
44	3.50	3.30	3.13	2.97	2.82	2.69	2.58	2.47	2.37	2.29	2.20	2.13	2.06	1.99	1.93
46	3.76	3.54	3.35	3.18	3.03	2.89	2.76	2.65	2.55	2.45	2.36	2.28	2.21	2.14	2.07
48	4.03	3.80	3.59	3.41	3.25	3.10	2.96	2.84	2.73	2.63	2.53	2.45	2.37	2.29	2.22
50	4.32	4.08	3.86	3.66	3.48	3.33	3.18	3.05	2.93	2.82	2.72	2.63	2.54	2.46	2.38
52	4.64	4.38	4.14	3.93	3.74	3.57	3.42	3.28	3.15	3.03	2.92	2.82	2.73	2.64	2.56
54	4.99	4.71	4.45	4.23	4.02	3.84	3.67	3.52	3.38	3.26	3.14	3.03	2.93	2.84	2.75
56	5.38	5.07	4.80	4.55	4.33	4.14	3.96	3.79	3.65	3.51	3.38	3.27	3.16	3.06	2.97
58	5.81	5.47	5.18	4.92	4.68	4.47	4.27	4.10	3.93	3.79	3.65	3.53	3.41	3.30	3.20
60	6.28	5.92	5.61	5.32	5.06	4.83	4.62	4.43	4.26	4.10	3.95	3.82	3.69	3.57	3.46
62	6.82	6.43	6.09	5.78	5.50	5.25	5.02	4.81	4.62	4.45	4.29	4.14	4.01	3.88	3.76
LHA	164° 196°	163° 197°	162° 198°	161° 199°	160° 200°	159° 201°	158° 202°	157° 203°	156° 204°	155° 205°	154° 206°	153° 207°	152° 208°	151° 209°	150° 210°

ABC TABLES

65

A

+ if hour angle is listed at top
− if hour angle is listed at bottom

LHA	32° 328°	34° 326°	36° 324°	38° 322°	40° 320°	42° 318°	44° 316°	46° 314°	48° 312°	50° 310°	52° 308°	54° 306°	56° 304°	58° 302°	60° 300°
0	.000	.000	.000	.000	.000	.000	.000	.000	.000	.000	.000	.000	.000	.000	.000
3	.084	.078	.072	.067	.062	.058	.054	.051	.047	.044	.041	.038	.035	.033	.030
6	.168	.156	.145	.135	.125	.117	.109	.101	.095	.088	.082	.076	.071	.066	.061
9	.253	.235	.218	.203	.189	.176	.164	.153	.143	.133	.124	.115	.107	.099	.091
12	.340	.315	.293	.272	.253	.236	.220	.205	.191	.178	.166	.154	.143	.133	.123
15	.429	.397	.369	.343	.319	.298	.277	.259	.241	.225	.209	.195	.181	.167	.155
18	.520	.482	.447	.416	.387	.361	.336	.314	.293	.273	.254	.236	.219	.203	.188
21	.614	.569	.528	.491	.457	.426	.398	.371	.346	.322	.300	.279	.259	.240	.222
24	.713	.660	.613	.570	.531	.494	.461	.430	.401	.374	.348	.323	.300	.278	.257
27	.815	.755	.701	.652	.607	.566	.528	.492	.459	.428	.398	.370	.344	.318	.294
30	.924	.856	.795	.739	.688	.641	.598	.558	.520	.484	.451	.419	.389	.361	.333
33	1.04	.963	.894	.831	.774	.721	.672	.627	.585	.545	.507	.472	.438	.406	.375
36	1.16	1.08	1.00	.930	.866	.807	.752	.702	.654	.610	.568	.528	.490	.454	.419
38	1.25	1.16	1.08	1.00	.931	.868	.809	.754	.703	.656	.610	.568	.527	.488	.451
40	1.34	1.24	1.15	1.07	1.00	.932	.869	.810	.756	.704	.656	.610	.566	.524	.484
42	1.44	1.33	1.24	1.15	1.07	1.00	.932	.870	.811	.756	.703	.654	.607	.563	.520
44	1.55	1.43	1.33	1.24	1.15	1.07	1.00	.933	.870	.810	.754	.702	.651	.603	.558
46	1.66	1.54	1.43	1.33	1.23	1.15	1.07	1.00	.932	.869	.809	.752	.698	.647	.598
48	1.78	1.65	1.53	1.42	1.32	1.23	1.15	1.07	1.00	.932	.868	.807	.749	.694	.641
50	1.91	1.77	1 64	1.53	1.42	1.32	1.23	1.15	1.07	1.00	.931	.866	.804	.745	.688
52	2.05	1.90	1.76	1.64	1.53	1.42	1.33	1.24	1.15	1.07	1.00	.930	.863	.800	.739
54	2.20	2.04	1.89	1.76	1.64	1.53	1.43	1.33	1.24	1.15	1.08	1.00	.928	.860	.795
56	2.37	2.20	2.04	1.90	1.77	1.65	1.54	1.43	1.33	1.24	1.16	1.08	1.00	.926	.856
58	2.56	2.37	2.20	2.05	1.91	1.78	1.66	1.55	1.44	1.34	1.25	1.16	1.08	1.00	.924
60	2.77	2.57	2.38	2.22	2.06	1.92	1.79	1.67	1.56	1.45	1.35	1.26	1.17	1.08	1.00
62	3.01	2.79	2.59	2.41	2.24	2.09	1.95	1.82	1.69	1.58	1.47	1.37	1.27	1.18	1.09
64	3.28	3.04	2.82	2.62	2.44	2.28	2.12	1.98	1.85	1.72	1.60	1.49	1.38	1.28	1.18
66	3.59	3.33	3.09	2.87	2.68	2.49	2.33	2.17	2.02	1.88	1.75	1.63	1.52	1.40	1.30
LHA	148° 212°	146° 214°	144° 216°	142° 218°	140° 220°	138° 222°	136° 224°	134° 226°	132° 228°	130° 230°	128° 232°	126° 234°	124° 236°	122° 238°	120° 240°

B

− if if latitude and declination have same name
+ if latitude and declination have different names

LHA	32° 328°	34° 326°	36° 324°	38° 322°	40° 320°	42° 318°	44° 316°	46° 314°	48° 312°	50° 310°	52° 308°	54° 306°	56° 304°	58° 302°	60° 300°
0	.000	.000	.000	.000	.000	.000	.000	.000	.000	.000	.000	.000	.000	.000	.000
3	.099	.094	.089	.085	.082	.078	.075	.073	.071	.068	.067	.065	.063	.062	.061
6	.198	.188	.179	.171	.164	.157	.151	.146	.141	.137	.133	.130	.127	.124	.121
9	.299	.283	.269	.257	.246	.237	.228	.220	.213	.207	.201	.196	.191	.187	.183
12	.401	.380	.362	.345	.331	.318	.306	.295	.286	.277	.270	.263	.256	.251	.245
15	.506	.479	.456	.435	.417	.400	.386	.372	.361	.350	.340	.331	.323	.316	.309
18	.613	.581	.553	.528	.505	.486	.468	.452	.437	.424	.412	.402	.392	.383	.375
21	.724	.686	.653	.623	.597	.574	.553	.534	.517	.501	.487	.474	.463	.453	.443
24	.840	.796	.757	.723	.693	.665	.641	.619	.599	.581	.565	.550	.537	.525	.514
27	.962	.911	.867	.828	.793	.761	.733	.708	.686	.665	.647	.630	.615	.601	.588
30	1.09	1.03	.982	.938	.898	.863	.831	.803	.777	.754	.733	.714	.696	.681	.667
33	1.23	1.16	1.11	1.05	1.01	.971	.935	.903	.874	.848	.824	.803	.783	.766	.750
36	1.37	1.30	1.24	1.18	1.13	1.09	1.05	1.01	.978	.948	.922	.898	.876	.857	.839
38	1.47	1.40	1.33	1.27	1.22	1.17	1.12	1.09	1.05	1.02	.991	.966	.942	.921	.902
40	1.58	1.50	1.43	1.36	1.31	1.25	1.21	1.17	1.13	1.10	1.06	1.04	1.01	.989	.969
42	1.70	1.61	1.53	1.46	1.40	1.35	1.30	1.25	1.21	1.18	1.14	1.11	1.09	1.06	1.04
44	1.82	1.73	1.64	1.57	1.50	1.44	1.39	1.34	1.30	1.26	1.23	1.19	1.16	1.14	1.12
46	1.95	1.85	1.76	1.68	1.61	1.55	1.49	1.44	1.39	1.35	1.31	1.28	1.25	1.22	1.20
48	2.10	1.99	1.89	1.80	1.73	1.66	1.60	1.54	1.49	1.45	1.41	1.37	1.34	1.31	1.28
50	2.25	2.13	2.03	1.94	1.85	1.78	1.72	1.66	1.60	1.56	1.51	1.47	1.44	1.41	1.38
52	2.42	2.29	2.18	2.08	1.99	1.91	1.84	1.78	1.72	1.67	1.62	1.58	1.54	1.51	1.48
54	2.60	2.46	2.34	2.24	2.14	2.06	1.98	1.91	1.85	1.80	1.75	1.70	1.66	1.62	1.59
56	2.80	2.65	2.52	2.41	2.31	2.22	2.13	2.06	2.00	1.94	1.88	1.83	1.79	1.75	1.71
58	3.02	2.86	2.72	2.60	2.49	2.39	2.30	2.22	2.15	2.09	2.03	1.98	1.93	1.89	1.85
60	3.27	3.10	2.95	2.81	2.69	2.59	2.49	2.41	2.33	2.26	2.20	2.14	2.09	2.04	2.00
62	3.55	3.36	3.20	3.05	2.93	2.81	2.71	2.61	2.53	2.46	2.39	2.32	2.27	2.22	2.17
LHA	148° 212°	146° 214°	144° 216°	142° 218°	140° 220°	138° 222°	136° 224°	134° 226°	132° 228°	130° 230°	128° 232°	126° 234°	124° 236°	122° 238°	120° 240°

A

+ if hour angle is listed at top
− if hour angle is listed at bottom

LHA	62° 298°	64° 296°	66° 294°	68° 292°	70° 290°	72° 288°	74° 286°	76° 284°	78° 282°	80° 280°	82° 278°	84° 276°	86° 274°	88° 272°	90° 270°
0	.000	.000	.000	.000	.000	.000	.000	.000	.000	.000	.000	.000	.000	.000	.000
3	.028	.026	.023	.021	.019	.017	.015	.013	.011	.009	.007	.006	.004	.002	.000
6	.056	.051	.047	.043	.038	.034	.030	.026	.022	.019	.015	.011	.007	.004	.000
9	.084	.077	.071	.064	.058	.051	.045	.039	.034	.028	.022	.017	.011	.006	.000
12	.113	.104	.065	.086	.077	.069	.061	.053	.045	.037	.030	.022	.015	.007	.000
15	.142	.131	.119	.108	.098	.087	.077	.067	.057	.047	.038	.028	.019	.009	.000
18	.173	.158	.145	.131	.118	.106	.093	.081	.069	.057	.046	.034	.023	.012	.000
21	.204	.187	.171	.155	.140	.125	.110	.096	.082	.068	.054	.040	.027	.013	.000
24	.237	.217	.198	.180	.162	.145	.128	.111	.095	.079	.063	.047	.031	.016	.000
27	.271	.249	.227	.206	.185	.166	.146	.127	.108	.090	.072	.054	.036	.018	.000
30	.307	.282	.257	.233	.210	.188	.166	.144	.123	.102	.081	.061	.040	.020	.000
33	.345	.317	.289	.262	.236	.211	.186	.162	.138	.115	.091	.068	.045	.023	.000
36	.386	.354	.323	.294	.264	.236	.208	.181	.154	.128	.102	.076	.051	.025	.000
38	.415	.381	.348	.316	.284	.254	.224	.195	.166	.138	.110	.082	.055	.027	.000
40	.446	.409	.374	.339	.305	.273	.241	.209	.178	.148	.118	.088	.059	.029	.000
42	.479	.439	.401	.364	.328	.293	.258	.224	.191	.159	.127	.095	.063	.031	.000
44	.513	.471	.430	.390	.351	.314	.277	.241	.205	.170	.136	.101	.068	.034	.000
46	.551	.505	.461	.418	.377	.336	.297	.258	.220	.183	.146	.109	.072	.036	.000
48	.591	.542	.494	.449	.404	.361	.318	.277	.236	.196	.156	.117	.078	.039	.000
50	.634	.581	.531	.481	.434	.387	.342	.297	.253	.210	.167	.125	.083	.042	.000
52	.681	.624	.570	.517	.466	.416	.367	.319	.272	.226	.180	.135	.090	.045	.000
54	.732	.671	.613	.556	.501	.447	.395	.343	.293	.243	.193	.145	.096	.048	.000
56	.788	.723	.660	.559	.540	.482	.425	.370	.315	.261	.208	.156	.104	.052	.000
58	.851	.781	.713	.647	.582	.520	.459	.399	.340	.282	.225	.168	.112	.056	.000
60	.921	.845	.771	.700	.630	.563	.497	.432	.368	.305	.243	.182	.121	.060	.000
62	1.00	.917	.837	.760	.685	.611	.539	.469	.400	.332	.264	.198	.132	.066	.000
64	1.09	1.00	.913	.828	.746	.666	.588	.511	.436	.362	.288	.215	.143	.072	.000
66	1.19	1.10	1.00	.907	.817	.730	.644	.560	.477	.396	.316	.236	.157	.078	.000
LHA	118° 242°	116° 244°	114° 246°	112° 248°	110° 250°	108° 252°	106° 254°	104° 256°	102° 258°	100° 260°	98° 262°	96° 264°	94° 266°	92° 268°	90° 270°

B

− if if latitude and declination have same name
+ if latitude and declination have different names

LHA	62° 298°	64° 296°	66° 294°	68° 292°	70° 290°	72° 288°	74° 286°	76° 284°	78° 282°	80° 280°	82° 278°	84° 276°	86° 274°	88° 272°	90° 270°
0	.000	.000	.000	.000	.000	.000	.000	.000	.000	.000	.000	.000	.000	.000	.000
3	.059	.058	.057	.057	.056	.055	.055	.054	.054	.053	.053	.053	.053	.052	.052
6	.119	.117	.115	.113	.112	.111	.109	.108	.107	.107	.106	.106	.105	.105	.105
9	.179	.176	.173	.171	.169	.167	.165	.163	.162	.161	.160	.159	.159	.158	.158
12	.241	.236	.233	.229	.226	.223	.221	.219	.217	.216	.215	.214	.213	.213	.213
15	.303	.298	.293	.289	.285	.282	.279	.276	.274	.272	.271	.269	.269	.268	.268
18	.368	.362	.356	.350	.346	.342	.338	.335	.332	.330	.328	.327	.326	.325	.325
21	.435	.427	.420	.414	.408	.404	.399	.396	.392	.390	.388	.386	.385	.384	.384
24	.504	.495	.487	.480	.474	.468	.463	.459	.455	.452	.450	.448	.446	.446	.445
27	.577	.567	.558	.550	.542	.536	.530	.525	.521	.517	.515	.512	.511	.510	.510
30	.654	.642	.632	.623	.614	.607	.601	.595	.590	.586	.583	.581	.579	.578	.577
33	.735	.723	.711	.700	.691	.683	.676	.669	.664	.659	.656	.653	.651	.650	.649
36	.823	.808	.795	.784	.773	.764	.756	.749	.743	.738	.734	.731	.728	.727	.727
38	.885	.869	.855	.843	.831	.821	.813	.805	.799	.793	.789	.786	.783	.782	.781
40	.950	.934	.919	.905	.893	.882	.873	.865	.858	.852	.847	.844	.841	.840	.839
42	1.02	1.00	.986	.971	.958	.947	.937	.928	.921	.914	.909	.905	.903	.901	.900
44	1.09	1.07	1.06	1.04	1.03	1.02	1.00	.995	.987	.981	.975	.971	.968	.966	.966
46	1.17	1.15	1.13	1.12	1.10	1.09	1.08	1.07	1.06	1.05	1.05	1.04	1.04	1.04	1.04
48	1.26	1.24	1.22	1.20	1.18	1.17	1.16	1.14	1.14	1.13	1.12	1.12	1.11	1.11	1.11
50	1.35	1.33	1.30	1.29	1.27	1.25	1.24	1.23	1.22	1.21	1.20	1.20	1.19	1.19	1.19
52	1.45	1.42	1.40	1.38	1.36	1.35	1.33	1.32	1 31	1.30	1.29	1.29	1.28	1.28	1.28
54	1.56	1.53	1.51	1.48	1.46	1.45	1.43	1.42	1 41	1.40	1.39	1.38	1.38	1.38	1.38
56	1.68	1.65	1.62	1.60	1.58	1.56	1.54	1.53	1.52	1.51	1.50	1.49	1.49	1.48	1.48
58	1.81	1.78	1.75	1.73	1.70	1.68	1.66	1.65	1.64	1.63	1.62	1.61	1.60	1.60	1.60
60	1.96	1.93	1.90	1.87	1.84	1.82	1.80	1.79	1.77	1.76	1.75	1.74	1.74	1.73	1.73
62	2.13	2.09	2.06	2.03	2.00	1.98	1.96	1.94	1.92	1.91	1.90	1.89	1.89	1.88	1.88
LHA	118° 242°	116° 244°	114° 246°	112° 248°	110° 250°	108° 252°	106° 254°	104° 256°	102° 258°	100° 260°	98° 262°	96° 264°	94° 266°	92° 268°	90° 270°

C C (correction) = A ± B

C	0.00	0.05	0.10	0.15	0.20	0.25	0.30	0.35	0.40	0.45	0.50	0.55	0.60	0.70
					Azimuth (°)									
0	90.0	87.1	84.3	81.5	78.7	76.0	73.3	70.7	68.2	65.8	63.4	61.2	59.0	55.0
10	90.0	87.2	84.4	81.6	78.9	76.2	73.5	71.0	68.5	66.1	63.8	61.6	59.4	55.4
20	90.0	87.3	84.6	82.0	79.4	76.8	74.3	71.8	69.4	67.1	64.8	62.7	60.6	56.7
24	90.0	87.4	84.8	82.2	79.6	77.1	74.7	72.3	69.9	67.7	65.5	63.3	61.3	57.4
28	90.0	87.5	85.0	82.5	80.0	77.6	75.2	72.8	70.5	68.3	66.2	64.1	62.1	58.3
30	90.0	87.5	85.1	82.6	80.2	77.8	75.4	73.1	70.9	68.7	66.6	64.5	62.5	58.8
32	90.0	87.6	85.2	82.8	80.4	78.0	75.7	73.5	71.3	69.1	67.0	65.0	63.0	59.3
34	90.0	87.6	85.3	82.9	80.6	78.3	76.0	73.8	71.7	69.5	67.5	65.5	63.6	59.9
36	90.0	87.7	85.4	83.1	80.8	78.6	76.4	74.2	72.1	70.0	68.0	66.0	64.1	60.5
38	90.0	87.7	85.5	83.3	81.0	78.9	76.7	74.6	72.5	70.5	68.5	66.6	64.7	61.1
40	90,0	87.8	85.6	83.4	81.3	79.2	77.1	75.0	73.0	71.0	69.0	67.2	65.3	61.8
42	90.0	87,9	85,7	83.6	81.5	79.5	77.4	75.4	73.4	71.5	69.6	67.8	66.0	62.5
44	90.0	87.9	85.9	83.8	81.8	79.8	77.8	75.9	73.9	72.1	70.2	68.4	66.7	63.3
46	90.0	88.0	86.0	84.1	82.1	80.1	78.2	76.3	74.5	72.6	70.8	69.1	67.4	64.1
48	90.0	88.1	86.2	84.3	82.4	80.5	78.6	76.8	75,0	73.2	71.5	69.8	68A	64.9
50	90.0	88.2	86.3	84.5	82.7	80.9	79.1	77.3	75.6	73.9	72.2	70.5	68.9	65.8
52	90.0	88.2	86.5	84.7	83.0	81.2	79.5	77.8	76.2	74.5	72.9	71.3	69.7	66.7
54	90.0	88.3	86.6	85.0	83.3	81.6	80.0	78.4	76.8	75.2	73.6	72.1	70.6	67.6
56	90.0	88.4	86.8	85.2	83.6	82.0	80.5	78.9	77.4	75.9	74.4	72.9	71.5	68.6
58	90.0	88.5	87.0	85.5	84.0	82.5	81.0	79.5	78.0	76.6	75.2	73.8	72.4	69.6
60	90.0	88.6	87.1	85.7	84.3	82.9	81.5	80.1	78.7	77.3	76.0	74.6	73.3	70.7
62	90.0	88.7	87.3	86.0	84.6	83.3	82.0	80.7	79.4	78.1	76.8	75.5	74.3	71.8
64	90.0	88.7	87.5	86.2	85.0	83.7	82.5	81.3	80.1	78.8	77.6	76.4	75.3	72.9
66	90.0	88.8	87.7	86.5	85.3	84.2	83.0	81.9	80.8	79.6	78.5	77.4	76.3	74.1
	0.00	0.05	0.10	0.15	0.20	0.25	0.30	0.35	0.40	0.45	0.50	0.55	0.60	0.70

C	0.80	0.90	1.00	1.10	1.20	1.40	1.60	1.80	2.00	2.20	2.40	2.60	2.80
					Azimuth (°)								
0	51.3	48.0	45.0	42.3	39.8	35.5	32.0	29.1	26.6	24.4	22.6	21.0	19.7
10	51.8	48.4	45.4	42.7	40.2	36.0	32.4	29.4	26.9	24.8	22.9	21.3	19.9
20	53.1	49.8	46.8	44.1	41.6	37.2	33.6	30.6	28.0	25.8	23.9	22.3	20.8
24	53.8	50.6	47.6	44.9	42.4	38.0	34.4	31.3	28.7	26.5	24.5	22.8	21.4
28	54.8	51.5	48.6	45.8	43.3	39.0	35.3	32.2	29.5	27.2	25.3	23.5	22.0
30	55.3	52.1	49.1	46.4	43.9	39.5	35.8	32.7	30.0	27.7	25.7	23.9	22.4
32	55.8	52.7	49.7	47.0	44.5	40.1	36.4	33.2	30.5	28.2	26.2	24.4	22.8
34	56.4	53.3	50.3	47.6	45.1	40.7	37.0	33.8	31.1	28.7	26.7	24.9	23.3
36	57.1	53.9	51.0	48.3	45.8	41.4	37.7	34.5	31.7	29.3	27.3	25.4	23.8
38	57.8	54.7	51.8	49.1	46.6	42.2	38.4	35.2	32.4	30.0	27.9	26.0	24.4
40	58.5	55.4	52.5	49.9	47.4	43.0	39.2	36.0	33.1	30.7	28.5	26.7	25.0
42	59.3	56.2	53.4	50.7	48.3	43.9	40.1	36.8	33.9	31.5	29.3	27.4	25.7
44	60.1	57.1	54.3	51.6	49.2	44.8	41.0	37.7	34.8	32.3	30.1	28.1	26.4
46	60.9	58.0	55.2	52.6	50.2	45.8	42.0	38.7	35.7	33.2	31.0	29.0	27.2
48	61.8	58.9	56.2	53.6	51.2	46.9	43.0	39.7	36.8	34.2	31.9	29.9	28.1
50	62.8	60.0	57.3	54.7	52.4	48.0	44.2	40.8	37.9	35.3	33.0	30.9	29.1
52	63.8	61.0	58.4	55.9	53.5	49.2	45.4	42.1	39.1	36.4	34.1	32.0	30.1
54	64.8	62.1	59.6	57.1	54.8	50.6	46.8	43.4	40.4	37.7	35.3	33.2	31.3
56	65.9	63.3	60.8	58.4	56.1	51.9	48.2	44.8	41.8	39.1	36.7	34.5	32.6
58	67.0	64.5	62.1	59.8	57.5	53.4	49.7	46.4	43.4	40.6	38.2	36.0	34.0
60	68.2	65.8	63.4	61.2	59.0	55.0	51.3	48.0	45.0	42.3	39.8	37.6	35.5
62	69.4	67.1	64.9	62.7	60.6	56.7	53.1	49.8	46.8	44.1	41.6	39.3	37.3
64	70.7	68.5	66.3	64.3	62.3	58.5	55.0	51.7	48.8	46.0	43.5	41.3	39.2
66	72.0	69.9	67.9	65.9	64.0	60.3	56.9	53.8	50.9	48.2	45.7	43.4	41.3
	0.80	0.90	1.00	1.10	1.20	1.40	1.60	1.80	2.00	2.20	2.40	2.60	2.80

Naming the azimuth: If answer is +, azimuth is **South** in north latitudes and **North** in south latitudes; if answer is –, azimuth is **North** in north latitudes and **South** in south latitudes. If hour angle is **less than 180°**, azimuth is **West; if more than 180°**, azimuth is **East.**

C C (correction) = A ± B

C	3.20	3.60	4.00	4.50	5.00	6.00	7.00	8.00	9.00	10.0	15.0	20.0	40.0
							Azimuth (°)						
0	17.4	15.5	14.0	12.5	11.3	9.5	8.1	7.1	6.3	5.7	3.8	2.9	1.4
10	17.6	15.8	14.2	12.7	11.5	9.6	8.3	7.2	6.4	5.8	3.9	2.9	1.5
20	18.4	16.5	14.9	13.3	12.0	10.1	8.6	7.6	6.7	6.1	4.1	3.0	1.5
24	18.9	16.9	15.3	13.7	12.3	10.3	8.9	7.8	6.9	6.2	4.2	3.1	1.6
28	19.5	17.5	15.8	14.1	12.8	10.7	9.2	8.1	7.2	6.5	4.3	3.2	1.6
30	19.8	17.8	16.1	14.4	13.0	10.9	9.4	8.2	7.3	6.6	4.4	3.3	1.7
32	20.2	18.1	16.4	14.7	13.3	11.1	9.6	8.4	7.5	6.7	4.5	3.4	1.7
34	20.7	18.5	16.8	15.0	13.6	11.4	9.8	8.6	7.6	6.9	4.6	3.5	1.7
36	21.1	19.0	17.2	15.4	13.9	11.6	10.0	8.8	7.8	7.0	4.7	3.5	1.8
38	21.6	19.4	17.6	15.8	14.2	11.9	10.3	9.0	8.0	7.2	4.8	3.6	1.8
40	22.2	19.9	18.1	16.2	14.6	12.3	10.6	9.3	8.3	7.4	5.0	3.7	1.9
42	22.8	20.5	18.6	16.7	15.1	12.6	10.9	9.5	8.5	7.7	5.1	3.8	1.9
44	23.5	21.1	19.2	17.2	15.5	13.0	11.2	9.9	8.8	7.9	5.3	4.0	2.0
46	24.2	21.8	19.8	17.8	16.1	13.5	11.6	10.2	9.1	8.2	5.5	4.1	2.1
48	25.0	22.5	20.5	18.4	16.6	14.0	12.1	10.6	9.4	8.5	5.7	4.3	2.1
50	25.9	23.4	21.3	19.1	17.3	14.5	12.5	11.0	9.8	8.8	5.9	4.4	2.2
52	26.9	24.3	22.1	19.9	18.0	15.1	13.1	11.5	10.2	9.2	6.2	4.6	2.3
54	28.0	25.3	23.1	20.7	18.8	15.8	13.7	12.0	10.7	9.7	6.5	4.9	2.4
56	29.2	26.4	24.1	21.7	19.7	16.6	14.3	12.6	11.2	10.1	6.8	5.1	2.6
58	30.5	27.7	25.3	22.8	20.7	17.5	15.1	13.3	11.8	10.7	7.2	5.4	2.7
60	32.0	29.1	26.6	24.0	21.8	18.4	15.9	14.0	12.5	11.3	7.6	5.7	2.9
62	33.6	30.6	28.0	25.3	23.1	19.6	16.9	14.9	13.3	12.0	8.1	6.1	3.0
64	35.5	32.4	29.7	26.9	24.5	20.8	18.1	15.9	14.2	12.9	8.6	6.5	3.3
66	37.6	34.3	31.6	28.7	26.2	22.3	19.4	17.1	15.3	13.8	9.3	7.0	3.5
	3.20	3.60	4.00	4.50	5.00	6.00	7.00	8.00	9.00	10.0	15.0	20.0	40.0

Naming the azimuth: If answer is +, azimuth is **South** in north latitudes and **North** in south latitudes; if answer is –, azimuth is **North** in north latitudes and **South** in south latitudes.
If hour angle is **less than 180°**, azimuth is **West;** if **more than 180°**, azimuth is **East.**

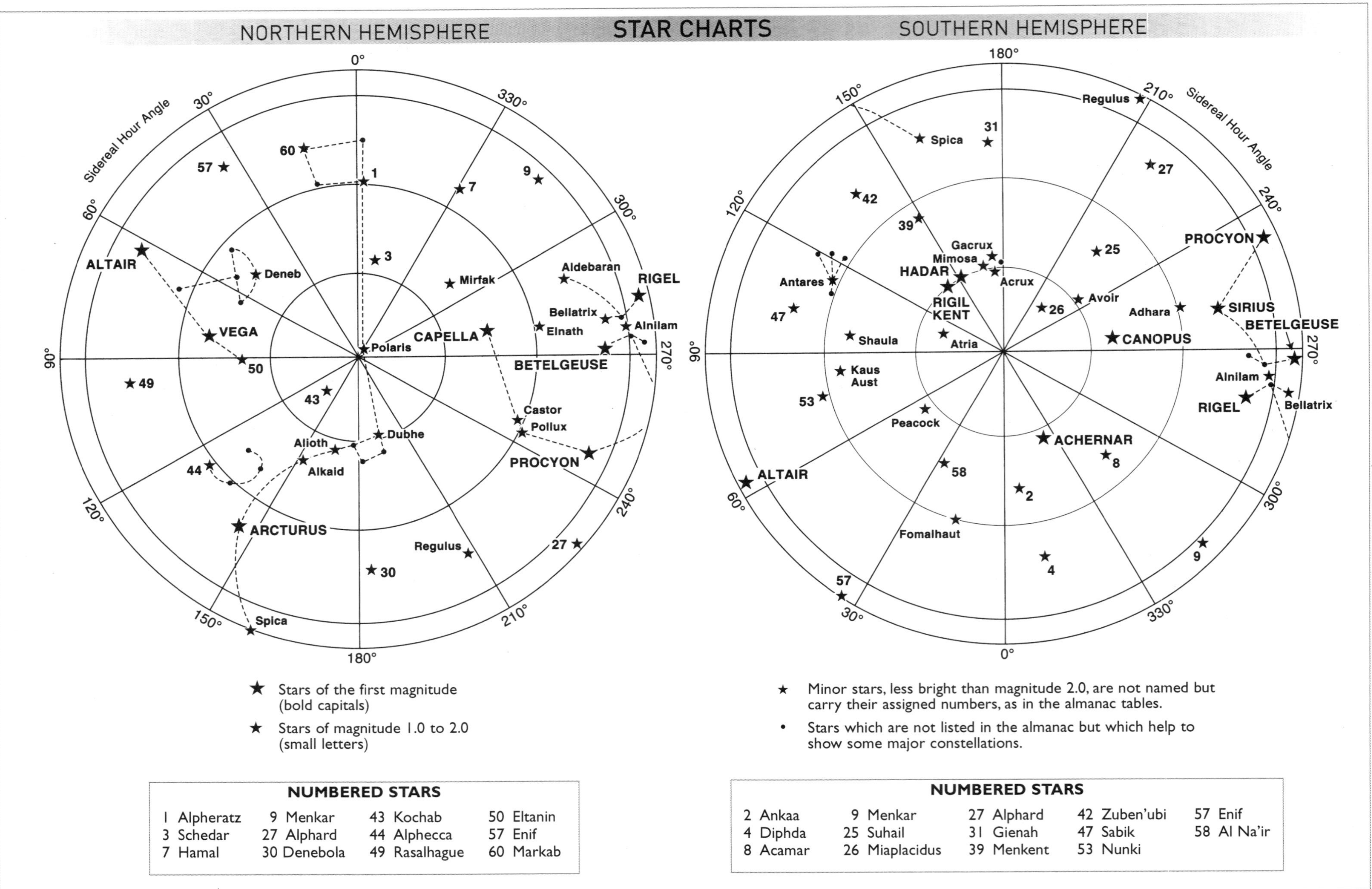

★ Stars of the first magnitude (bold capitals)

★ Stars of magnitude 1.0 to 2.0 (small letters)

NUMBERED STARS

1 Alpheratz	9 Menkar	43 Kochab	50 Eltanin
3 Schedar	27 Alphard	44 Alphecca	57 Enif
7 Hamal	30 Denebola	49 Rasalhague	60 Markab

★ Minor stars, less bright than magnitude 2.0, are not named but carry their assigned numbers, as in the almanac tables.

• Stars which are not listed in the almanac but which help to show some major constellations.

NUMBERED STARS

2 Ankaa	9 Menkar	27 Alphard	42 Zuben'ubi	57 Enif
4 Diphda	25 Suhail	31 Gienah	47 Sabik	58 Al Na'ir
8 Acamar	26 Miaplacidus	39 Menkent	53 Nunki	